MW00799721

ILLNESS, PAIN,
AND HEALTH CARE
IN EARLY CHRISTIANITY

Illness, Pain, and Health Care in Early Christianity

Helen Rhee

William B. Eerdmans Publishing Company

Grand Rapids, Michigan

Wm. B. Eerdmans Publishing Co.
4035 Park East Court SE, Grand Rapids, Michigan 49546
www.eerdmans.com

28 27 26 25 24 23 22   1 2 3 4 5 6 7

ISBN 978-0-8028-7684-3

Library of Congress Cataloging-in-Publication Data

A catalog record for this book is available from the Library of Congress.

CONTENTS

Acknowledgments

I started researching this book project in 2014, and its completion has been a long time coming. I am indebted to many along the way.

ReMeDHe, a scholarly working group on Religion, Medicine, Disability, Health, and Healing in Late Antiquity, organized and directed by Kristi Upson-Saia and Heidi Marx, provided me with opportunities for initial ideas and conversations, especially in its early years. I am also grateful to several institutions for their support of my research and writing. My home institution, Westmont College, granted me a sabbatical year: spring of 2018 and spring of 2019. I spent a significant part of the 2018 sabbatical at a beautiful cottage on the shore of Lake Michigan, enabled by generous funding from the Issachar Fund. During my stay, I benefited much from my access to the libraries of Hope College and Western Theological Seminary, and I was welcomed into the community of scholars at Western Seminary by colleagues such as Kristen Deede Johnson, Han-luen Kantzer Komline, and David Komline. I then spent the 2019 sabbatical in Biola University's Center for Christian Thoughts (CCT), which provided me with a vibrant scholarly community for interdisciplinary and collaborative research and stimulating discussions on the theme of suffering.

I want to thank Gary Ferngren, who, despite his retirement, read two of my early chapters (chaps. 1 and 2) and offered me helpful and thoughtful comments. My New Testament colleague at Westmont, Caryn Reeder, read chapters 4 and 5 and made careful and salient comments. Kya Mangrum, my English colleague at Westmont, read the entire book with meticulous attention, helping me with editing. Brenda Llewellyn Ihssen reviewed the manuscript with generous and constructive comments. Thanks to their suggestions, this book is greatly improved. I alone am responsible for any remaining errors and faults.

I researched and wrote the last three chapters through the Covid-19 pandemic and shutdowns, and it was difficult to get hold of the printed resources necessary for research and writing. Special thanks to Westmont College librarians Richard Burnweit and Jana Mayfield Mullen, in particular, for their tireless efforts in procuring me all the resources.

I want to extend my thanks to the excellent editorial team at Eerdmans: James Ernest, who initially encouraged me and pursued this project; Trevor Thompson, who made this project official and showed great patience with me as my writing became delayed because of my existing physical challenges over the years; Jenny Hoffman, who worked with me step by step through the whole editorial process with precision and care; and Tom Raabe, the copyeditor, who was scrupulous in his work.

Finally, I cannot thank my family enough—my parents, my two brothers and sisters-in-law, and even my nieces and nephews—for their unconditional love and consistent prayers and support.

ABBREVIATIONS

Primary Sources

Aelius Aristides
  Or.   *Hieroi Logoi*

Ambrose
  Ep.   *Epistulae*
  Vid.   *De viduis*

Ap. Const.   *Apostolic Constitutions*

Aristotle
  Eth. nic.   *Ethica nicomachea*

Arnobius
  Ad. nat.   *Adversus nationes*

Artemidorus
  Onir.   *Onirocritica*

Athanasius
  C. gent.   *Contra gentes*

Athenagoras
  Leg.   *Legatio pro Christianis*

Augustine
  Agon.   *De agone christiano*
  Bon. conj.   *De bono conjugali*

Catech.	*De catechizandis rudibus*
Civ.	*De civitate Dei*
Div.	*De divinatate daemonum*
Enarrat. Ps.	*Enarrationes in Psalmos*
Ep.	*Epistulae*
Gen. litt.	*De Genesi ad litteram*
Mor.	*De moribus ecclesiae catholicae et de moribus Manicheorum*
Nat. bon.	*De natura boni contra Manichaeos*
Reg. Praeceptum	*Regula: Praeceptum*
Serm.	*Sermones*
Trin.	*De Trinitate*

Aulus Gellius

Noct. att.	*Noctes atticae*

Basil

Ep.	*Epistulae*
Hom.	*Homilies*
LR	*The Longer Rules*
SR	*The Shorter Rules*

Celsus

Med.	*De Medicina*

Cicero

Div.	*De divinatione*
Fin.	*De finibus*
Tusc.	*Tusculanae disputationes*

Clement of Alexandria

Paed.	*Paedagogus*
Protr.	*Protrepticus*
Strom.	*Stromata*

Columella

Rust.	*De re rustica*

Cyprian

Eleem.	*De opera et eleemosynis*
Ep.	*Epistulae*
Laps.	*De lapsis*
Mort.	*De mortalitate*

Cyril of Alexandria

Juln.	*Contra Julianum*

Didasc.	*Didascalia Apostolorum*
Epictetus	
Diss.	*Dissertationes*
Ench.	*Enchiridion*
Epicurus	
Fr.	*Fragment*
Epiphanius	
Pan.	*Panarion (Adversus haereses)*
Eusebius	
Hist. eccl.	*Historia ecclesiastica*
Theoph.	*Theophania*
Galen	
Caus. resp.	*De causis respirationis*
Caus. symp.	*De symptomatum causis*
Hipp. Elem.	*De elementis ex Hippocrate*
Hipp. Epid.	*In Hippocratis Epidmiarum librum primum commentarius*
Ind.	*De indolentia*
Loc. affect.	*De locis affectis*
Meth. med.	*De methodo medendi*
Meth. med. Glaucon	*A Method of Medicine to Glaucon*
Opt. med.	*Quod optimus medicus sit quoque philosophus*
PHP	*De placitis Hippocratis et Platonis*
QAM	*Quod animi mores corporis temperamenta sequantur*
Simp. med.	*De simplicium medicamentorum temperamentis et facultatibus*
Symp. diff.	*De symptomatum differentiis*
Temp.	*De temperamentis*
UP	*De usu partium*
Gregory of Nazianzus	
Carm.	*Carmina*
Carm. de se ipso	*Carmen de se ipso*
Ep.	*Epistulae*
Or.	*Orationes*
Gregory of Nyssa	
Cat. magn.	*Oratio catechetica magne*
Hom. op.	*De hominis opificio*

Paup.	De pauperibus amandis
Vita	Vita Macrinae Junioris
Vita Greg. Thauma.	De vita B. Gregorii Thaumaturgi

Hippolytus

Trad. ap.	Traditio apostolica

Horsiesios

Reg.	Regulae

Ignatius

Pol.	To Polycarp
Rom.	Letter to the Romans

Irenaeus

Haer.	Adversus haereses

Jerome

Epist.	Epistulae
Vit. Hil.	Vita S. Hilarionis eremitae

John Cassian

Conf.	Conferences

John Chrysostom

De statuis hom.	Homilies on the Statutes
Hom. 1 Thess.	Homiliae in epistulam i ad Thessalonicenses
Hom. Gen.	Homilies on Genesis
Hom. John	Homilies on John
Hom. Matt.	Homiliae in Matthaeum
Hom. Tit.	Homiliae in epistulam ad Titum
Pan. Bab.	Panegyricum in Babylam martyrem
Sac.	De sacerdotio
Scand.	Ad eos qui scandalizati sunt

Julian the Apostate

Ep.	Epistulae

Justin

Dial.	Dialogus cum Tryphone
1 Apol.	First Apology

Lactantius

Mort.	De mortibus persecutorum
Opif.	De opificio Dei

Leo the Great

Serm.	Sermones

Libanius

Or.	*Orationes*
Mart. Apol.	*Martyrdom of Apollonius*
Mart. Con.	*Martyrdom of Conon*
Mart. Lyons	*Letter of the Churches of Lyons and Vienne*
Mart. Pion.	*Martyrdom of Pionius*
Mart. Pol.	*Martyrdom of Polycarp*

Minucius Felix

Oct.	*Octavius*

Nemesius

De nat. hom.	*De natura hominis*

Origen

Cels.	*Contra Celsum*
Comm. Jn.	*Commentary on the Gospel of John*
Comm. Matt.	*Commentarium in evangelium Matthaei*
Hom. Ezra	*Homilies on Ezra*
Hom. Jer.	*Homilies on Jeremiah*
Hom. Lev.	*Homilies on Leviticus*
Hom. Luke	*Homilies on Luke*
Hom. Num.	*Homilies on Numbers*
Hom. Ps.	*Homilies on Psalms*
Princ.	*De principiis*

Palladius

Dial.	*Dialogue*
HL	*Historia Lausiaca*

Pausanias

Descr.	*Graciae descriptio*

Philodemus

Lib.	*De libertate dicendi*

Plato

Gorg.	*Gorgias*
Rep.	*Republic*
Tim.	*Timaeus*

Pliny the Elder

Nat.	*Naturalis historia*

Polycarp

Phil. — To the Philippians

Pseudo-Clem. Ep. — Pseudo-Clementines Epitome

Reg. Magistri — Rule of the Master

Seneca

Clem. — De clementia

Const. — De constantia sapientis

Ep. — Epistulae morales

Prov. — De providentia

Sophocles

Phil. — Philoctetes

Sozomen

HE — Historia ecclesiastica

Strabo

Geogr. — Geographica

Tatian

Or. Graec. — Oratio ad Graecos

Tertullian

An. — De anima

Apol. — Apologeticus

Bapt. — De baptismo

Carn. Chr. — De carne Christi

Cor. — De corona militis

Idol. — De idolatria

Marc. — Adversus Marcionem

Paen. — De paenitentia

Pat. — De patientia

Scap. — Ad Scapulum

Scorp. — Scorpiace

Spect. — De spectaculis

Theodore

Instr. — Instructions

Theodoret of Cyrrhus

Curatio — Graecarum affectionum curatio

Hist. eccl. — Historia ecclesiastica

Phil. hist. — Philotheos historia

Prov. — De providentia

Theophilus

Autol.	*Ad Autolycum*
Vita	*The Life and Regimen of Blessed and Holy Teacher Syncletica*

Zosimus

Hist. Nova	*Historia Nova*

Secondary Sources

AARCC	American Academy of Religion Cultural Criticism Series
AJP	*American Journal of Philology*
ANRW	*Aufstieg und Niedergang der römischen Welt*
AugStud	*Augustinian Studies*
AUSS	*Andrews University Seminary Studies*
BA	*Biblical Archaeologist*
BAR	*Biblical Archaeology Review*
BHM	*Bulletin of the History of Medicine*
BTB	*Biblical Theology Bulletin*
CCSL	Corpus Christianorum Series Latina
CH	*Church History*
ClQ	*Classical Quarterly*
Comm	*Communio*
CRAI	Comptes rendus de l'Académie des inscriptions et belles-lettres
CSCO	Corpus Scriptorum Christianorum Orientalium
CWS	Classics of Western Spirituality
DOP	*Dumbarton Oaks Papers*
EJST	*European Journal of Science and Theology*
ExAud	*Ex Auditu*
FAT	Forschungen zum Alten Testament
FC	Fathers of the Church
GRBS	*Greek, Roman, and Byzantine Studies*
Hor	*Horizons*
HSM	Harvard Semitic Monographs
HTR	*Harvard Theological Review*
HTS	Harvard Theological Studies
Int	*Interpretation*
JAAR	*Journal of the American Academy of Religion*
JBL	*Journal of Biblical Literature*
JECS	*Journal of Early Christian Studies*

JFSR	*Journal of Feminist Studies in Religion*
JR	*Journal of Religion*
JRA	*Journal of Roman Archaeology*
JRS	*Journal of Roman Studies*
KD	*Kerygma und Dogma*
LCL	Loeb Classical Library
NTS	*New Testament Studies*
OCM	Oxford Classical Monographs
OECS	Oxford Early Christian Studies
OLA	Orientalia Lovaniensia Analecta
OSHT	Oxford Studies in Historical Theology
PG	Patrologia Graeca
PGM	*Papyri Graecae Magicae* (ed. Preisendanz)
PL	Patrologia Latina
P.Lond.	The London Papyri (in the British Museum)
P.Neph.	The Nepheros Letters
P. Oxy.	Oxyrhynchus Papyri
SC	Sources chrétiennes
SCH	Studies in Church History
SecCent	*Second Century*
SNTSMS	Society for New Testament Studies Monograph Series
ST	*Studia Theologica*
STAC	Studies and Texts in Antiquity and Christianity
TJ	*Trinity Journal*
TS	*Theological Studies*
TUGAL	Texte und Untersuchungen zur Geschichte der altchristlichen Literatur
VC	*Vigiliae Christianae*
ZAC	*Zeitschrift für Antikes Christentum*
ZPE	*Zeitschrift für Papyrologie und Epigraphik*

INTRODUCTION

This study examines the ways that early Christians understood, adopted, appropriated, and reframed Greco-Roman medicine and health care from the second century CE to the fifth century CE. In it I demonstrate a critical link between diverse and divergent forms of Greco-Roman medicine and health care and the formation of Christian identity in the early Christian church. It is my thesis that amid a massive change of its status in the early fourth century through the Constantinian revolution, and throughout the subsequent Christianization of the Roman Empire, the early Christian church's appropriation and reformulation of Greco-Roman medicine and health care were instrumental in shaping how Christians defined themselves. The concept of identity[1] is a twentieth-century notion typically associated with modern individualism, and some scholars tend to qualify its usage when speaking of the "emergence of Christian identity" in the first two centuries.[2] It is, nevertheless, still useful to speak of "Christian identity" in the early church in that the early Christian church constructed a sense of Christian continuity and common boundaries in relation to (or in terms of) otherness and differentiation in the subsequent periods; boundaries of Christian identity "involve selection out of both

---

1. This section of the paragraph is modified from Helen Rhee, *Loving the Poor, Saving the Rich: Wealth, Poverty, and Early Christian Formation* (Grand Rapids: Baker Academic, 2012), xiii–xiv.

2. For example, Judith M. Lieu, *Christian Identity in the Jewish and Graeco-Roman World* (Oxford: Oxford University Press, 2004).

similarity and difference, and promote interchange as well as distancing."[3] This (collective) identity is constructed in constant social interactions with the surrounding societies and cultures ("others") and defines and redefines those "others," such as Jews, pagans ("Greeks"), or heretics, etc. As argued by Kathryn Tanner, the distinctiveness of a Christian way of life is not just formed by the boundary but *at* the boundary. As Tanner writes, "Christian distinctiveness is something that emerges in the very cultural processes occurring *at* the boundary, processes that construct a distinctive identity for Christian social practices through the distinctive use of cultural materials *shared with others*."[4] Therefore, as is the case with other identities, Christian identity is "essentially relational"[5] and "contextualized and contingent"[6] upon time and space; and yet, Christian identity, via the selective process of self-definition, also presents and projects Christian ideals and universal claims.[7] Even in the fourth and fifth centuries, Christian leaders were concerned about what it meant to build an ideal Christian identity, or ideal Christian identities, as the empire, through imperial interventions, transitioned from pagan to Christian, a transition that early Christians navigated with a twinned sense of triumphalism and anxiety.

My pursuit of this thesis is based on fundamental assumptions that need to be made explicit at this point. First of all, early Christians were part of a larger Greco-Roman world, which means that they lived and operated within the existing social, medical, religious, and cultural framework of the Mediterranean world dominated by the Roman Empire. Therefore, understanding the varied iterations of Greco-Roman medicine (temple medicine, rational medicine, and popular medicine) and health care is not only illuminating but also critical to understanding early Christian engagement with them. Christian social and health-care practices were formed by creative processes of incorporating, engaging, and negotiating with "institutional forms from elsewhere" in dominant culture.[8] Second, there are a paucity and limitation of sources for the ancient period in general, including Greco-

3. Judith M. Lieu, "'Impregnable Ramparts and Wall of Iron': Boundary and Identity in Early 'Judaism' and 'Christianity,'" *NTS* 48 (2002): 311.

4. Kathryn Tanner, *Theories of Culture: A New Agenda for Theology* (Minneapolis: Fortress, 1997), 115 (emphasis added).

5. Tanner, *Theories of Culture*, 112.

6. Lieu, *Christian Identity*, 18.

7. Helen Rhee, *Early Christian Literature: Christ and Culture in the Second and Third Centuries* (London: Routledge, 2005), 7.

8. Tanner, *Theories of Culture*, 112; cf. Rhee, *Early Christian Literature*, 194.

Roman antiquity and late antiquity. Many of the available literary texts of the time come from the elite—the rich and powerful—and therefore inherently convey the perspectives of their authors in their writings about sickness, illness, and pain. In contrast, Christian sources such as the New Testament Apocrypha, homilies, and hagiographies are popular in nature, meaning that people from a variety of social statuses constructed and encountered these texts. Finally, regarding patient-physician relationships, most literary works—with the rare exception of the illness narratives of patients like Aelius Aristides and Marcus Aurelius—come from the Hippocratic Corpus and physicians like Galen, and therefore tend to highlight the viewpoints of the practitioners over the viewpoints of patients. Consequently, we must use caution in assessing the reliability of these literary texts and recognize the subjectivities inherent in their rhetorical construction. They need to be, when appropriate and possible, juxtaposed with and contextualized by other forms of evidence, particularly the types of evidence afforded by the study of archaeological artifacts, inscriptions, papyri, etc.

The primary sources[9] for this study include a variety of works from Greco-Roman philosophy and literature, as well as from Greco-Roman medical literature. Philosophical sources include Plato, Aristotle, Chrysippus, Plutarch, Epictetus, and Seneca for their understandings of health, disease, and pain. Literary sources include Aristides, Marcus Aurelius, Fronto, Sophocles, and Philostratus for their treatments of (their own) illness, pain, and healing. In addition, I use collections of inscriptions and papyri to supplement my readings of these sources. When I examine Greco-Roman rational medicine, I focus on medical texts found in the Hippocratic Corpus and in the writings of Galen. The former consists of more than sixty medical texts by various authors completed over two centuries, from the mid-fifth century BCE to the mid-third century BCE, under the name of Hippocrates of Cos (440–370 BCE), who is regarded as the "father of Western medicine." Galen identified Hippocrates as the main foundation and predecessor of his own medical theory and practice and wrote many commentaries on the Hippocratic Corpus. Although there is a growing appreciation for non-Hippocratic and non-Galenic medical authors in contemporary scholarship, it is Hippocratic and Galenic medicine that is the most prominent, representative, and influential out of the Greek rational

---

9. The direct quotations from the ancient writers, Hippocratic Corpus, Galen, church fathers, and other early Christian documents in the book come from the "Primary Sources" of the bibliography.

Galen = Stoic Philosophy

medical market; and it is Hippocratic and Galenic medicine that the early Christian writers engaged the most. Therefore, while acknowledging and interrogating other schools of medicine such as Methodism (Caelius Aurelianus) or other medical writers such as Aulus Cornelius Celsus when appropriate and necessary, my study will be confined to Hippocratic and Galenic medicine as representative of the church fathers' engagement with the Greco-Roman medical tradition.

The primary sources for this project also include the Hebrew and Christian Bibles and Second Temple literature, along with a variety of Christian literature ranging from the apostolic fathers in the second century to Theodoret of Cyrrhus in the mid-fifth century. For example, the respective literary categories known as the Apologies, the Martyr Acts, the New Testament Apocrypha, the gnostic writings (Nag Hammadi), and the heresiologist writings such as those of Irenaeus provide scattered and incomplete, but still substantial, records and information. They also include the literary works of the "old catholic" Greek fathers, such as Clement of Alexandria and Origen, and the Latin fathers, Tertullian and Cyprian (treatises, letters, apologies, etc.). The church histories of Eusebius and Sozomen illuminate (and sometimes complicate) a larger context and help connect the dots between individuals and events. The Nicene and post-Nicene fathers, such as Lactantius, Athanasius, the Cappadocian fathers, Jerome, Nemesius, John Chrysostom, Augustine, and Theodoret of Cyrrhus, offer crucial witnesses to the Christian adoption of and engagement with Greco-Roman medicine, on the one hand, and theological reflections on health, disease, illness, pain, and health care on the other. Furthermore, monastic literature such as *Apophthegmata patrum*, Palladius's *Lausiac History*, Theodoret's *History of the Monks of Syria*, the *Life* of Amma Syncletica, the *Bohairic Life of Pachomius*, and the short and long rules by Basil of Caesarea provides glimpses of the ascetic struggles with, endurance of, and triumphs over illness (and pain). Related to them is hagiographic literature such as the *Life of Hilarion* and the *Life of St. Martin of Tours*. The church manuals such as the *Apostolic Tradition*, the *Didascalia Apostolorum*, the *Apostolic Constitutions*, the *Canons of Hippolytus*, and the *Sacramentary of Serapion* also show how the care of the sick and ritual healing became institutionalized over the centuries. Finally, collections of Christian inscriptions and papyri complement the various literary sources.

These diverse sources testify to the vibrant and multifaceted nature and forms of Christianity over the early centuries that, while creating various communities and even competing claims among these communities, at the same time developed core doctrines, structure, and practices that defined a common Christian identity, one that stood out in relation to other identities in the Greco-Roman (and Jewish) world(s). These apparently paradoxical developments indicated significant growth and spread of Christianity over various regions of the empire (especially in urban centers) and its active interaction with and adaptation to the existing culture and milieu in those areas.

Medical Anthropology

An interdisciplinary, historical, and sociocultural work, this study places Greco-Roman medicine, literature, and philosophy (classics) in fruitful dialogue with early Christian literature and theology (patristics). This integrative approach also draws from the contemporary discipline of medical anthropology, which rejects the physicalistic reductionism of the Western biomedical model in favor of a biopsychosocial and cultural model, one that parses the distinctions among three concepts that are often misconstrued as synonymous: disease, illness, and sickness. First, according to Arthur Kleinman and other medical anthropologists, disease refers to abnormality or alteration in the biological structure and function of organs and organ systems, as diagnosed by a medical practitioner; it is a biological or psychological pathological state in terms of theories of disorder.[10] Illness, on the other hand, refers to how the sick person and the members of the family or wider social network perceive, label, experience, and respond to symptoms and disability; the illness experience includes categorizing and explaining in commonsense terms the forms of distress caused by those pathophysiological processes and is always culturally shaped and often carries multiple meanings (physiological, psychological, social, moral, cultural, religious, etc.).[11] For Kleinman, disease and illness are explanatory concepts, constructed in contexts of meaning and social relationships.[12] Disease and illness can express different interpretations of a single clinical reality, represent different aspects of a plural clinical reality, or even create different

10. Arthur Kleinman, *The Illness Narratives: Suffering, Healing, and the Human Condition* (New York: Basic Books, 1988), 5–6; Allan Young, "The Anthropologies of Illness and Sickness," *Annual Review of Anthropology* 11 (1982): 264.

11. Kleinman, *The Illness Narratives*, 3–5, 8.

12. Arthur Kleinman, *Patients and Healers in the Context of Culture: An Exploration of the Borderland between Anthropology, Medicine, and Psychiatry* (Berkeley and Los Angeles: University of California Press, 1980), 73.

clinical realities at times.[13] Lastly, "sickness" is a generic term to label events involving disease or illness, especially in relation to macrosocial—economic, political, or institutional—forces such as poverty, malnutrition, etc.[14]

Based on this distinction, Kleinman and others propose, and see value in, the construction of the illness narrative by a patient, defined as follows: "A story the patient tells, and significant others retell, to give coherence to the distinctive events and long-term course of suffering. The plot lines, core metaphors, and rhetorical devices that structure the illness narrative are drawn from cultural and personal models for arranging experiences in meaningful ways and for effectively communicating those meanings."[15] This definition highlights how a personal illness narrative is shaped by the patient's social context and informed by one's cultural models. As Linda Garro writes, "hearing narrative accounts is a principal means through which cultural understandings about illness—including possible causes, appropriate social responses, healing strategies, and characteristics of thera-peutic alternatives—are acquired, confirmed, refined, or modified."[16] For ex-ample, examining the well-developed AA (Alcoholics Anonymous) model of illness narratives, Carole Cain states, "as the AA member learns the AA story model, and learns to place the events and experiences of his own life into the model, he learns to tell and to understand his own life as an AA life, and himself as an AA alcoholic. The personal story is a cultural vehicle for identity acquisition."[17] Indeed, Ron Loewe observes that such narratives, despite a variety of people writing about diverse medical conditions, share many common elements: "mystery (disease is unexpected or difficult to diagnose), betrayal by one's own body, conflict with medical professionals or medical bureaucracies, the failure of medical science to heal, the need for self-reliance, and generally, but not always, a return to good health."[18] At the same time, while these illness narratives are socially and culturally

---

13. Kleinman, *Patients and Healers*, 73.

14. Kleinman, *The Illness Narratives*, 6; cf. Young, "The Anthropologies of Illness and Sickness," 264.

15. Kleinman, *The Illness Narratives*, 49.

16. Linda C. Garro, "Cognitive Medical Anthropology," in *Encyclopedia of Medical Anthropology: Health and Illness in the World's Cultures*, ed. C. R. Ember and M. Ember (New York: Kluwer, 2004), 19.

17. Carole Cain, "Personal Stories: Identity Acquisition and Self-Understanding in Alcoholics Anonymous," *Ethos* 19, no. 2 (1991): 215.

18. Ron Loewe, "Illness Narratives," in Ember and Ember, *Encyclopedia of Medical Anthropology*, 43.

conditioned through these common elements, they are also the means by which the patients form their distinct identities as individuals or members of social groups through their illness experiences. Furthermore, these illness narratives show how "conventional expectations about illness are altered through negotiations in different social situations and in particular webs of relationships," including "expectations about how to behave when ill."[19]

In this way the disease-illness distinction, together with the attendant illness narrative as employed in medical anthropology, helps biomedical practitioners and social scientists to be more sensitive to the multifaceted etiology of sickness and the complex nature of the illness experience.[20] Moreover, it provides the basis for examining the influence of cultural factors such as religious beliefs and social and historical contexts on the construction and experience of illness.[21]

Kleinman then presents the "core clinical functions" of health-care systems, that is, how systems of health-care knowledge and practice enable patients to do the following:[22]

1. Construct *illness* as a psychosocial experience.

2. Establish *general* criteria helpful for guiding the health-care-seeking process and evaluating the potential efficacy of different treatment approaches.

3. Manage *particular* illness episodes through communicative operations such as labeling and explaining.

4. Engage healing activities, including therapeutic interventions from drugs and surgery to psychotherapy, supportive care, and healing rituals.

5. Manage therapeutic outcomes, including cure, treatment failure, chronic illness, impairment, and death.

In this process, the meaning of healing is implicit—it is a process by which, first, disease is made into illness as a cultural construction and therefore with meaning, and second, the sufferer attains a degree of "satisfaction"

19. Kleinman, *The Illness Narratives*, 5.
20. Pieter F. Craffert, "Medical Anthropology as an Antidote for Ethnocentrism in Jesus Research? Putting the Illness-Disease Distinction into Perspective," *HTS Teologiese Studies/Theological Studies* 67, no. 1 (2011): art. #970, 7.
21. Craffert, "Medical Anthropology," 7. Craffert, however, also critiques medical anthropology and the disease-illness distinction especially when they are applied to New Testament scholarship.
22. Kleinman, *Patients and Healers*, 71–72. Also, this paragraph is indebted to Young's analysis of Kleinman: "The Anthropologies of Illness and Sickness," 265.

through reduction, or even eradication, "of the psychological, sensory, and experiential oppressiveness engendered by [one's] medical circumstances."[23] Thus, healing corresponds to illness and is more inclusive than biomedical cure of a disease per se. Here Allan Young rightly emphasizes that by implicitly distinguishing healing and cure, Kleinman is distinguishing not between mind and body but between culture and nature.[24] In Kleinman's work, healing is not a mere mental activity (as it is in the works of others), although it is bound to the feelings, perceptions, and experiences of the patient. For example, pharmacological or surgical intervention and its effects on the body are part of the healing process as they are also part of "the curing process, i.e., the process affecting pathological organic states."[25] The healing process for Kleinman "usually involves two related activities—the provision of effective control of the disease and of personal and social meaning of the experience of illness."[26] By stressing the complementarity of mind and body and also of healing and curing, Kleinman resists the rigid Cartesian dualism of the biomedical model.[27]

We need, however, in order to avoid anachronism, to be careful about how we apply modern models or approaches to ancient peoples and the ancient world;[28] and yet, we can also see how this framework of medical anthropology might help us understand the various categories of medicine in the larger Greco-Roman health-care system: rational medicine, temple medicine (religious healing), and popular medicine (folk healing). Diseases are primarily under the domain of rational medicine and, to an extent, under the domain of temple medicine and popular medicine for diagnosis; however, the notion of illness opens "a discursive space in which alternative explanations of etiology, course, and treatment of disease" are made and followed through illness narratives, especially in the context of temple medicine and popular medicine.[29] As we will see in chapter 1,

23. Young, "The Anthropologies of Illness and Sickness," 265.

24. Young, "The Anthropologies of Illness and Sickness," 265.

25. Young, "The Anthropologies of Illness and Sickness," 265.

26. Arthur Kleinman, "Concepts and a Model for the Comparison of Medical Systems as Cultural Systems," in *Concepts of Health, Illness, and Disease: A Comparative Perspective*, ed. C. Currer and M. Stacey (New York: Berg, 1986), 35.

27. Young, "The Anthropologies of Illness and Sickness," 266.

28. On the application of medical anthropology to early Christian healing and health care, see, for example, Hector Avalos, *Health Care and the Rise of Christianity* (Peabody, MA: Hendrickson, 1999), and John J. Pilch, *Healing in the New Testament: Insights from Medical and Mediterranean Anthropology* (Minneapolis: Fortress, 2000).

29. Loewe, "Illness Narratives," 44.

Greco-Roman authors such as Aelius Aristides and the emperor Marcus Aurelius constructed their own illness narratives during the Second Sophistic, when pathology belonged to the culture, and presented their distinct identities through those narratives. In chapter 2, biblical accounts of certain psalms, the book of Job, and the apostle Paul's "thorn in the flesh" can be seen as forms of illness narratives in that illness/suffering is processed within and against their socioreligious and cultural contexts. Early Christians offered supernatural and retributive etiologies and corresponding healing accounts reflecting their larger cultural context while still engaging rational medicine. In chapter 3, the pain narratives of sufferers function like the illness narratives as they put forth the sociocultural understandings of their pain experiences in distinct yet common meaning-making and identity-forming ways. In these chapters, we can also see that Greco-Roman authors and early Christian writers, by writing their own illness and pain narratives (along with pain pedagogies), already anticipated, in their respective health-care systems, Kleinman's "core clinical functions," especially functions 1 through 3. In chapters 4 and 5, both Greco-Roman and Christian health-care systems demonstrate Kleinman's "core clinical functions," particularly functions 4 and 5. In light of the complementarity of mind and body and also of healing and cure (religious and popular), healing stories typically include biomedical cures and are not independent from those cures.[30]

A Layout of Chapters

This study will proceed by outlining notions and theories of health, disease, and illness in Greco-Roman literature, medicine, and philosophy. This overview not only provides the conceptual and practical contexts and frameworks for early Christian understandings of health, disease, and illness, but also establishes the standard for the subsequent development of ideas about health, disease, and illness in the Western world. I will present

---

30. Contra John Dominic Crossan, *Jesus: A Revolutionary Biography* (San Francisco: HarperSanFrancisco, 1994), 81; John Dominic Crossan, *The Birth of Christianity: Discovering What Happened in the Years Immediately after the Execution of Jesus* (San Francisco: HarperSanFrancisco, 1998), 294; Pilch, *Healing in the New Testament*, 32, 35, 141. Employing the radical dichotomy between healing and cure (illness and disease), Crossan argues that Jesus could not cure diseases but only offered therapeutic comfort. Using the same method, Pilch also argues that Jesus's healing was only symbolic, in which the healer mediated culture and was apart from the cure of disease.

how Greco-Roman texts understand the concept of health as a mixture or balance of humors and elemental qualities within the body and as the productive and correct functioning of the body and its varied parts, a condition Greco-Roman writers deemed to be in accord with nature. Disease, by contrast, involved the isolation or imbalance of internal bodily humors and elements, and the nonproductiveness or nonfunctioning of the body and its varied parts, a condition Greco-Roman writers understood to go against nature. In this context, I will explore how somatic explanations for psychic sickness reveal the close connection between the diseases of the body and the diseases of the soul; and I will also note how the Roman public health system (or lack thereof) exposes the environmental causes and factors of disease. Finally, I will examine how Greeks and Romans created meaning out of their illness experiences in their fundamentally religious society by allowing their cultural model to inform their illness narratives.

Given these concepts of health, disease, and illness in Greco-Roman culture, chapter 2 examines ideas of health, disease, and illness as found in the Hebrew and Christian Bibles and in the theology of the early Christian church. In the Hebrew Bible, God the Creator is the sender of both health and diseases, depending on the covenantal obedience of the Israelites. In Second Temple Judaism, demons emerge as the agents of sickness; and, although a majority of Jesus's healing stories only describe medical symptoms or disabilities without mentioning any etiology, and Jesus's healing stories, in general, deny the sin-illness connection, the Synoptic Gospels continue this demon-illness tradition. As we move to the subsequent centuries, I will show how early Christians adopted the Hippocratic and Galenic construction of health, especially in their teleological functioning of human anatomy under divine providence. While they appropriated the Hippocratic and Galenic notions of physical and psychic health and diseases, they also constructed their own illness narratives around four etiologies and explanations: sin and God's punishment, God's providential pedagogy, demons, and ascetic disciplines (*askēsis*) and ambiguity toward illness. On the last point, I will analyze the changing perceptions of ascetic sickness and negotiations of the ascetic sick role as a marker of ascetic holiness.

Chapter 3 deals with the related but distinct issue of pain. Just as early Christians adopted, represented, and reformulated the ideas and experiences of health, disease, and illness in Greco-Roman and late antique society, they also assumed and reconstructed the concepts and experiences of pain in ways that were meaningful to them. I will examine pain's crucial role in Greco-Roman (rational) medicine as a diagnostic indicator and analyze pain in relation to and in convergence with suffering in literature and philosophy through the constructed pain narratives and pedagogies. In this context, I will highlight pain's innate subjectivity, (in)expressibility, and (in)shareability. The second half of the chapter is devoted to the Christian construction of pain narratives and pedagogies based upon their central pain narratives: the passion narratives of the Gospels. Early Christian perspectives on and experiences of pain and suffering developed and were shaped by these interrelated and yet divergent interpretations of Christ's passion in the changing contexts as they all tried to relate various experiences of suffering to that of Christ. Thus, whether it is the pain of violent death as found in the mimetic identification with Christ in the martyr texts, the shareable pain experienced by a community of cosufferers in ascetic texts, the ontological pain endured by sufferers in the work of Gregory of Nazianzus, or the affective pain adopted by those in solidarity with humanity in the writings of Augustine, Christians transformed the uniquely subjective unshareable pain found in the narrative of Jesus into something repeatable and shareable by his followers. This repeatable and shareable pain—and the sufferer's willingness to pursue and embrace it— was key in Christian identity formation.

In chapter 4 we will engage Greco-Roman health care as a sociocultural system and also a sociocultural phenomenon. Based on differing etiologies of illness presented in chapter 1, Greco-Roman health care sought to provide efficacious solutions and responses to illnesses. Here it is important to recognize the multifaceted nature of Greco-Roman medicine. I will show how each category of medicine—temple medicine, rational medicine, and popular medicine—operated in relation to each other and how these multiple therapeutic measures fared in the Greco-Roman open medical market. A sick person in that society would have had a wide selection of healing therapies from which to choose. In this context, I will also explore the role of Roman hospitals for slaves and soldiers. Finally, by analyzing two specific texts, Seneca's *On Anger* and Galen's *Avoiding Distress*—respective examples of the Stoic-Epicurean and the Platonic-Aristotelian approaches—I will examine how medico-philosophical therapies of the soul developed in the Greco-Roman world via medical analogies.

Chapter 5 will present Christian health care in the context of Greco-Roman health care. I will divide Christian health care into two periods: the second and third centuries, when Christians developed their understanding of Greek rational medicine, by sacrifically taking care of the sick, especially during the plagues, and attested to the reality of religious healings (cures); and the fourth and fifth centuries, when Christianity emerged as "the religion of healing" (curing) par excellence, in which the early church developed public medical charity via hospitals, established monastic infirmaries, and utilized private health care with both rational medicine and religious healing. Then I will engage with Christian therapies of the soul as the bishops and monastic superiors styled themselves as medico-philosophical physicians of the soul through the teaching of Scripture and their own writings. The Christian application of the therapies of the soul extended to heresiological and apologetic literature as believers drew up imperial Christian identities vis-à-vis "others" in late antiquity. Like the larger Greco-Roman health-care system, these forms of Christian health care were not mutually exclusive but rather synergistic, convergent, and comprehensive. In incorporating, building, and reconstructing these various elements and forms of health care, Christians continued to construct their identities around their praxis of health care, even as they reshaped the Christian Roman Empire with the building blocks from Greco-Roman *paideia*.

Lastly, I will bring this study to a close with brief thoughts about constructing Christian visions of health, constructing Christian illness narratives, constructing Christian narratives and pedagogies of pain, and constructing a Christian system of health care. Through these constructions, early Christians established their common but distinct identity even as they shaped the late antique culture and society.

1

HEALTH, DISEASE, AND ILLNESS IN GRECO-ROMAN CULTURE

In this chapter we will explore the views and understandings of health, disease, and illness found in Greco-Roman literature, medicine, and philosophy. Not only did early Christians relate to and interact with these views and understandings, but these views and understandings also set the stage for subsequent ideas about health, disease, and illness in the Western world (at least through the Renaissance). The coverage in this chapter is not meant to be exhaustive but is sufficiently representative to draw a broad yet accurate picture of these ideas as they existed in their historical and cultural contexts. I will begin with how Greco-Roman writers understood health as a balance in both bodily constituents and regimen, as well as a proper functioning of bodily organs in accordance to nature. I will then examine definitions of disease as both an imbalance and impairment of these natural functions and a phenomenon that goes against nature, especially as it relates to seasonal and environmental change. Then I will study the relationship between the diseases of the body and the diseases of the soul, as the ancients gave somatic explanations for psychic and mental diseases. After I probe some concrete phenomena of disease and environmental factors in Greco-Roman public hygiene and health, I will finally show that illness as a personal and psychosocial meaning-making experience was shaped by particular cultural, that is, Greco-Roman, contexts. It will be apparent throughout that health, disease, and illness were socioculturally shaped and conditioned notions.

Health in Greco-Roman Literature, Medicine, and Philosophy

This section treats the emerging notions of health in Greco-Roman literature, medicine, and philosophy in a broadly chronological order as health was a shared concern among these writers.

Hymns and the Hippocratic Corpus
(Fifth Century BCE–Third Century BCE)

Henry Sigerist in his *Civilization and Disease* states: "To the Greek of the 5th century B.C. and long thereafter health appeared as the highest good."[1] Health as the universal and highest good probably has resonated with most, if not all, cultures throughout history. However, for ancient Greeks, health (*hygieia*) went far beyond the concerns of the individual and was an issue of "religious, cultural and political significance."[2] As such, Health was the personified goddess of good health and naturally elicited famous hymns in her honor. Among them was the oft-quoted drinking song (*skolion*) attributed to the poet Simonides of the fifth and sixth century BCE and referred to by Socrates in Plato's *Gorgias* (451e):

> To be healthy is best for mortal man,
> second is to be of beautiful appearance,
> third is to be wealthy without trickery,
> and fourth to be young with one's friends.
>
> Simonides Fr. 651 *PGM*

The poet Ariphron of Sicyon, Hippocrates's contemporary in the late fifth and early fourth century BCE, composed another celebrated hymn to Health:

> Health, greatest of the blessed gods, may I live with you
> for the rest of my life, and may you be a willing inmate of
> my house.

---

1. Henry E. Sigerist, *Civilization and Disease* (Ithaca, NY: Cornell University Press, 1943), 68.

2. Dominic Montserrat, "'Carrying on the Work of the Earlier Firm': Doctors, Medicine and Christianity in the Thaumata of Sophronius of Jerusalem," in *Health and Antiquity*, ed. Helen King (London: Routledge, 2005), 230–42.

For if there is any joy in wealth or children,
or in a king's godlike power over men,
or in the desires which we hunt with Aphrodite's hidden nets,
or if any other delight or rest from labours
has been revealed by the gods to mortals,
it is with your help, blessed Health,
that all things flourish and shine to the Graces' murmuring.
Without you no one is happy (*eudaimōn*).

Ariphron, Hymn to Hygieia, in Athenaios 15.702 (LCL)

This hymn was inscribed on stone at the sanctuary of Asclepius at Epidaurus and elsewhere in the fourth century BCE, and Health was worshiped by healthy worshipers as part of the cult of Asclepius, the popular god of medicine and healing, and her father.[3] In the second century CE, satirist Lucian of Samosata referred to Simonides's song as a popular salutation and added this: "Then if Health is greatest of the gods, her work, the enjoyment of health, is likewise to be put before the other goods" (*De lapsu* 6).

As the ancient poets praised Health and people participated in the cult of Hygieia as the supreme good without which none of life's other goods could be enjoyed, how did contemporary physicians conceptualize health? To begin with, Hippocratic physicians shared the popular notion of health as the supreme value to humanity, affirming, "neither wealth nor anything else is of any value without health" (*Regimen IV* 69; *Affections* 1). In addition, the author of *The Oath* invoked the goddess Hygieia along with Apollo Physician, her family members, Asclepius and Panacea, and all other gods and goddesses (1). In terms of theories and definitions of health, a pre-Socratic Pythagorean physician, Alcmaeon of Croton, provided the earliest known definition of health (and disease).[4] He understood health as harmony, balance (*isonomia*), and order (*kosmos*) as part of the natural order of the universe. For Alcmaeon, balance meant *isonomia* (equality) and mixture of the body's shares of the powers (*dynameōn*); *dynameis* then denoted a "capacity"

---

3. On the cult of Hygieia in classical Greece, see Emma Stafford, "'Without You No One Is Happy': The Cult of Health in Ancient Greece," in King, *Health in Antiquity*, 120–35.

4. On Alcmaeon of Croton and his concept of health (and disease), see Stavros Kouloumentas, "The Body and Polis: Alcmaeon on Health and Disease," *British Journal for the History of Philosophy* 22, no. 5 (2014): 867–87; Jacques Jouanna, *Hippocrates*, trans. M. B. DeBevoise (Baltimore: Johns Hopkins University Press, 1999), 327–28; Vivian Nutton, *Ancient Medicine*, 2nd ed. (London: Routledge, 2013), 47–48.

or "potentiality" and referred to the body's constitutive elements, which manifested by their own properties and activities such as wet, dry, hot, cold, bitter, sweet, etc. Thus health for him was the proportionate mixture of those qualities (*tēn symmetron tōn poiōn krasin*) in "a state of permanent equilibrium" in contrast to their imbalance, that is, *monarchia* (one element dominating over the others), which would cause disease (*Fr.* 4; Aëtius 5.30.1).[5]

This constitutional definition of health laid a foundation for several Hippocratic treatises, whose concepts of health in turn became foundational and influential for the understandings of health in Greek medicine in general. According to *On Ancient Medicine*, the human body contains a mixture (*krēsis*) of a large number of fluid substances or humors (*chymou*), each with its own characteristic, for example, sweet, bitter, acidic; its own power *(dynamis)*; and its own unique effects (14.4). When the humors are mixed and blended well, none of them is manifesting more than necessary and the person is healthy; however, when one of them separates from the mixture and stands alone, disrupting the proper equilibrium, in its dominance and excess it causes the person pain (*lypē*) (14.4). In Hippocratic treatises, how many and what kinds of bodily constituents exist tend to vary.[6] The *Nature of Man* confirms health as the mixture (*krēsis*) of humors, identifying them as blood, phlegm, yellow bile, and black bile, and thus establishing the (Hippocratic) theory of the four humors (4).[7] Then health is consistently associated with *krēsis* of the humors throughout Greek medicine and philosophy, including Plato, Aristotle, and later Galen. The same *Nature of Man* also introduces another concept of health: one enjoys the "most perfect health" when those four humors "are duly proportioned (*metriōs*) to one another" with respect to blending, power, and quantity (4). Besides the mixture, health is thus the balance of or proportion among the four elements. This definition of health along with health as the absence of disease or pain would become the classic definition of health in Greek medicine as well. Hence Galen, in his *Hygiene*, states: "Almost all my predecessors defined health by the good mixture (*eukrasia*) and proportion (*symmetria*) of the elements" (1.4 [12K]).[8]

If the above Hippocratic treatises established the constitutional idea of health as the internal equilibrium, other treatises in the Hippocratic Corpus

5. Kouloumentas, "The Body and Polis," 875, 885; Jouanna, *Hippocrates*, 262, 327.

6. For example, they include: phlegm and bile in *Affections I*; hot and cold in *Fleshes* 2–3; blood, phlegm, yellow bile, and water in *Diseases IV*.

7. The four humors according to *Diseases IV* are blood, phlegm, yellow bile, and water, not black bile (4.32.1). See n. 6 above.

8. The number in brackets (12K) refers to Kühn's reference in his Greek edition.

put forth a notion of health as the equilibrium or balance between the body and external elements or factors: first between the body and its regimen (*diaita*), and then between the body and its surrounding environment.[9] First, in *Regimen III*, the due proportion (*isazein*), which brings about good health, turns out to be between food and exercise, the components of regimen (*diaita*),[10] while the "overpowering" (*krasteisthai*) of one or the other causes diseases (*ponous*) (69). Since predominance of the humors varies according to age and the seasons,[11] and since exercise (which uses up what exists) and food and drink (which replenishes what is lost) work together to produce health (*Regimen I* 2), this balanced regimen is a practical and preventative means of achieving and maintaining good health by rectifying any imbalance of the humors through adjustment of food and exercise. The author of *Regimen* exhorts his fellow physicians to "discern the power (*dynamin*) of the various exercises, both natural (*kata physin*) exercises and artificial, to know which of them tends to increase flesh and which to lessen it" (*Regimen I* 2). Furthermore, he exhorts them "to proportion (*symmetrias*) exercise to bulk of food, to the constitution (*physin*) of the patient, to the age of the individual, to the season of the year, to the changes (*metabolas*) of the winds, to the situation of the region in which the patient resides, and to the constitution of the year" (*Regimen I* 2). As one can see, overall health here depends on the nature of each individual, "who needs a different blending of bodily constituents in a different season, age, and place, and in accordance with his/her diet, [exercise], habits, sex, and disposition to certain disease."[12] Eventually, the ideal is "to discover for the constitution (*physin*) of each individual a due proportion (*symmetros*) of food to exercise, with no inaccuracy either of excess or of defect" (*Regimen I* 2). The importance of a balanced diet and exercise in preserving health and preventing and treating diseases in Greek medicine is apparent in the number of treatises devoted to those topics, including *Regimens I–IV*,[13] *Humors and Nutrition*, *Regimen in Good Health*, and *Regimen in Acute Diseases* in the Hippocratic Corpus; Diphilos of Siphnos's *Suitable Foods for the Sick and the Healthy*; Diocles of Carystus's *Healthy Regimen*; and Galen's *Hygiene*, *On the Properties of Food-*

---

9. Cf. Jouanna, *Hippocrates*, 328.

10. On regimen in Hippocratic Corpus, see Jacques Jouanna, "Dietetics in Hippocratic Medicine: Definition, Main Problems, Discussion," in *Greek Medicine from Hippocrates to Galen*, ed. Philip J. van der Eijk (Leiden: Brill, 2012), 137–53.

11. See the second point below.

12. Kouloumentas, "The Body and Polis," 874; cf. *Epidemics I* 23; *Nature of Man* 9.

13. *Regimen II* 39ff. provides systematic catalogues of food with their various properties. *Affections* 47–60 also provides the catalogue of foods.

stuffs, On the Thinning Diet, and *On the Exercise with a Small Ball* for both the healthy and the sick.

Second, health as the balance between the body and its environment is based upon the observation that "the health of human beings depends . . . not only on the manner in which they live [e.g., regimen] but on a whole series of natural factors that impose themselves upon every individual, no matter what a person's particular way of life may be."[14] Here the natural factors include seasons and people's local environment such as winds, waters (marshy, soft, hard, etc.), the four directions, and soil (dry, wooded, watered, etc.). According to *Nature of Man,* the four humors that permanently constitute the human body go through changes—they increase or decrease—inside the body with the changing rhythm of the four seasons (7). Thus, phlegm, cold and wet, increases in winter; blood, warm and wet, increases in spring; yellow bile, warm and dry, dominates in summer; and black bile, cold and dry, dominates in fall (7). Physicians are to know these natural factors and both internal and external changes to discern any epidemic diseases in any city they need to go to, so that they "succeed best in securing health" and achieve the triumphs of their medical art (*Airs, Waters, Places* 1–2). The healthiest city is oriented toward the east with moderate heat and cold, and clear water in the moderate season of spring (5). And the year is likely to be very healthy "if the signs prove normal when the stars set and rise; if there be rains in autumn, if the winter be moderate, neither too mild nor unseasonably cold, and if the rains be seasonable in spring and in summer" (10). As Jacques Jouanna rightly points out, the common factor of the healthy city and the healthy year is "a sense of proportion, or moderation (*metriotēs*)"—for the healthy city, a balance between hot and cold, and for the healthy year, a sense of proportion both within and between seasons.[15] With this knowledge of how important the moderation or balance between the human being and her environment and within the environment is, one should "be especially on one's guard against the most violent changes (*metabolas*) of the seasons" (11) as well as "the most violent changes (*metabolai*) in what concerns our constitutions and habits"; for they are the chief causes of diseases (*Regimen in Acute Diseases* 35).

Hippocratic authors defined health fundamentally as balance or equilibrium; however, this internal and external equilibrium was always fragile because of the inevitability of the aforementioned changes occurring within

---

14. Jouanna, *Hippocrates,* 212; *Airs, Waters, Places* 1–2.
15. Jouanna, *Hippocrates,* 214.

and outside of the body. In light of this, Jouanna sees three levels or degrees of health in the continuum of health and disease in Hippocrates: first, the fragility of excessively good health; second, the appearance of good health; finally, relative health.[16] Athletes are the quintessential example of human beings in extremely good health, and yet, according to *Aphorism I*, "in athletes a perfect condition that is at its highest pitch is dangerous. Such conditions cannot remain the same or be at rest, and change for the better being impossible, the only possible change is for the worse" (3). Therefore, "it is an advantage to reduce the fine condition quickly, in order that the body may make a fresh beginning of growth" (3). An athlete's perfect health is good health pushed to excess, a health that because it has reached its peak has nowhere to go but down. Thus, the top athlete's perfect health risks tipping over into sickness.

With regard to Jouanna's second degree of health, we are sometimes deceived about what we consider good health. We have encountered *Regimen*'s emphasis on a due balance between food and exercise for health and a warning against their imbalance producing disease. Elsewhere its author mentions the reality that people in a normal state (i.e., healthy state) with more or less regular regimen appear to be in good health but are not necessarily so. "For diseases do not arise among men all at once; they gather themselves together gradually before appearing with a sudden spring" to the nonspecialists, that is, regular people (*Regimen I* 2). Only a trained physician like the author can detect the first signs of disequilibrium and forecast an illness before the patient falls sick (2).

Lastly, as health is on one end of a continuum and disease is on the opposite end, a healthy state is relative, and measured in degrees. The author of *On Ancient Medicine* compares how the relative health of healthy and unhealthy people might be measured in degrees with the following illustration: "For the great majority of men can follow indifferently either the one habit or the other, and can take lunch or only one daily meal" (10). Others, "if they were to do anything outside what is beneficial, would not get off easily, but if they change their respective ways for a single day, nay, for a part of a single day, they suffer excessive discomfort" (11). The author then contends that "such constitutions (*physias*), . . . that rapidly and severely feel the effects of errors, are weaker than the others. A weak man is but one step removed from a sickly man, but a sickly man is weaker still, and is more apt to suffer distress whenever he misses the due season" (12). There is a strong

---

16. In this and the next two paragraphs, I broadly follow Jouanna, *Hippocrates*, 331–34.

sense of gradation and continuity here, and what separates the "normal" (healthy) and the pathological is only a "difference of degree" "to the extent that the body, whether healthy or ill may be defined by its greater or lesser capacity to overcome external influences and to resist change."[17]

In this way, Hippocratic treatises conceptualize health as a natural and normal (normative) state opposite disease and pain through the good mixture and balance of the body's constitutive elements (humors) and regimen and also through the balance between the body and its external environment. Here health as a natural state in the Hippocratic Corpus meant "according to nature (*kata physin*)," and nature was the natural order of the physical world, including the human body and its nature (i.e., constitution). Human nature then indicated "*normal patterns* of bodily organization, whether at the elementary, anatomical, or physiological level, and a *normal capacity* for reacting to external influences, whether regimental or environmental. The natural state, like the state of good health, was the *normal state*."[18] Thus one of the Hippocratic treatises could draw the following conclusion: "The body's nature is the physician in disease (*nousōn physies iētroi*). Nature (*physis*) finds the way for herself, not from thought. For example, blinking, and the tongue offers its assistance, and all similar things. Well trained, readily and without instruction, nature does what is needed (*hē physis hekousa ou mathousa ta deonta poiei*)" (*Epidemics VI* 5.1). The unaided human nature in a healthy state protects itself against disease, and hence "all the healing processes are natural [i.e., not supernatural] processes."[19] Nonetheless, the Hippocratic authors also acknowledged that this normative idea of health was still a fragile and relative reality in a continuum with sickness to the extent of the body's capacity to deal with and resist the internal and external change, the fundamental principle of Hippocratic pathology.

Classical Philosophers Plato, Aristotle, and Chrysippus (Fifth Century BCE–Third Century BCE)

The subject of health (and disease) was the domain of not only physicians but also natural philosophers. Hippocrates's contemporary Plato under-

---

17. Jouanna, *Hippocrates*, 334–35.

18. Jouanna, *Hippocrates*, 345 (emphasis added).

19. Owsei Temkin, *Hippocrates in a World of Pagans and Christians* (Baltimore: Johns Hopkins University Press, 1991), 193n74.

stood health, as did Alcmaeon and the Hippocratic doctors, as the established state of four bodily elements of earth, fire, water, and air in a "natural relation" (*kata physin*) of harmony, order, and balance ("dominating and being dominated by one another"), as opposed to disease, a state contrary to nature (*para physin*), in which elements exist in excess, at deficient levels, or as a monarchy (in which "one [element] rules or is ruled by the other").[20] In *Timaeus*, Plato similarly stresses health as stability in proportion: "Only when that which arrives at or leaves a particular bodily part is the same as that part (*tauton tautōi*), consistent (*kata tauton*), uniform (*hōsautōs*), and in proper proportion with it (*ana logon*), will the body be allowed to remain stable, sound, and healthy (*tauton on hautōi sōn kai hygies menein*)" (82b2–5). He further reiterates this notion of health as the order (*taxis*) of the "natural movements" (*kata physin*) of the bodily substances and "the established regular, stable, constitution of the body (*tōi synestōti de tou sōmatos kai menonti*)" (83a1–5). This notion has a cosmic significance for Plato, since the universe (*kosmos*) created by the Demiurge is in a state of health;[21] in an earlier part of *Timaeus*, the Demiurge used up all the existing four elements (earth, fire, water, and air) in fashioning the cosmos because if anything had been left outside, the cosmos would have been vulnerable to attack by that thing and thus would have suffered from disease and old age (33a). Timaeus then recommends that the best way to maintain health is "in imitation of the form of the universe (*to tou pantos apomimoumenon eidos*)" (88c7–88d1).

Plato's pupil Aristotle also wrote on health and disease, and through animal dissections developed the fields of biology and zoology. His understanding of health echoed the established medico-philosophical thought of the time: health is a balance or proportion (*symmetria*) between opposites like hot and cold, according to *Topics*.[22] In addition, Aristotle conceptualized health in terms of smooth functions of the body and its parts according to nature. In *Movement of Animals*, Aristotle likened the constitution of an animal to "a well-governed city-state. For once order is established in the city, there is no need of a monarch to be present at every activity, but individuals each play their own part as they are ordered, and one thing follows

---

20. *Rep.* 4.444d4–6; *Tim.* 81e6–82a26. Cf. G. E. R. Lloyd, *In the Grip of Disease: Studies in the Greek Imagination* (Oxford: Oxford University Press, 2003), 155.

21. Brian D. Prince, "The Metaphysics of Bodily Health and Disease in Plato's *Timaeus*," *British Journal for the History of Philosophy* 22, no. 5 (2014): 915.

22. *Topics* 139b20f.; 145b7ff.; also *Physics* 246b4f.

another because of habituation" (703a). So he saw in animals, "because of nature (*dia tēn physin*), each part naturally doing its own work as constituted by nature. So there is no need for soul in each part, but it resides in a kind of source of authority (*archēi*) in the body, and the other parts live by being naturally connected to it and they perform their own work because of nature (*poiein de to ergon to autōn dia tēn physin*)" (703a–b). This teleological understanding of an animal's anatomy and the function of that anatomy would have a far-reaching impact, particularly on Galen's notion of health as optimal function designed by nature.[23] The Stoic Chrysippus also confirmed the classic definition of health in Greek medicine as the good mixture (*eukrasia*) and proportion (*symmetria*) of the hot, cold, dry, and wet in the body.[24]

Back to Medicine with Galen (129–c. 216 CE)

We now move to Galen, the "Prince of Medicine,"[25] in the late second century CE, for his concept of health. As briefly mentioned earlier, Galen basically affirmed the traditional medico-philosophical notions of health and developed and nuanced them further. Influenced by Plato's tripartite human nature and Aristotle's biology and teleology of nature, Galen agreed that, first, in terms of constitutional health, the ancient doctors and philosophers were right in their beliefs that the best constitution of our bodies was the same as the best mixture (*The Best Constitution of Our Bodies* 4.737K)—but with a major departure and addition. Galen followed Aristotle's view of the different levels of human (animal) composition.[26] In the primary microscopic level, following his notable predecessors Hippocrates and Plato (in addition to Aristotle), Galen thought of the most basic substances in the bodies as a mixture (*krasis*) of the four elemental substances of earth, fire, water, and air, and the four elemental qualities associated with them—hot, cold, wet, and dry.[27] Thus, an observable macroscopic level

23. However, Galen rejected Aristotle's view of a single controlling organ following Plato's tripartite physiological system.

24. Jouanna, *Hippocrates*, 328.

25. This title comes from Susan P. Mattern, *The Prince of Medicine: Galen in the Roman Empire* (Oxford: Oxford University Press, 2013).

26. Aristotle discusses three levels of animal composition in book 2 of *Parts of Animals*: (1) primary substances; (2) uniform *homoiomerous* parts; (3) complex organic parts.

27. See *Hipp. Elem.* 1.480.126, 7–12; 483–86; 492. On the mixture of four qualities,

consists of the following: first, the homogeneous or uniform (*homoiomerous*)[28] parts as the lower level, simpler bodies, such as blood, bone, muscle, and cartilage; second, the complex organic parts with particular functions such as heart, brain, or liver;[29] and finally the whole body. According to Galen, "while the best constitution is necessarily one with the best mixture, a constitution with the best mixture is not necessarily the best. For the health of our homogeneous parts consists in a well-proportioned mixture (*eukrasia*) of hot, cold, dry, and wet; but the construction of the animal on the basis of all these parts [i.e., the organic parts] consists in the position, size, shape, and number of the component elements" (4.737K). Then the construction of the complex organic parts (or the body as a whole)[30] "composed of a very large number of well-balanced parts may nevertheless have some defect in respect of their size, number, construction, or relationship with each other" (4.738K). In other words, the correct composition of the complex organic parts (and the whole body) involves more than a well-proportioned mixture and the shaping and (causal) relationships with other constituents. The health of the whole body then should consist of the well-balanced composition of these high-level organic parts *in addition to* that of the lower-level simpler parts (4.741K).[31] That is, "in the ideal healthy body, *homoiomerous* structures, both those existing separately and those contributing to the structure of organic parts, have a proper balance of the four elemental qualities (*krasis*), while organic parts are normal in terms of size, conformation, number, and position."[32]

Second, as Peter Singer highlights, Galen sees "an indissoluble connection between this kind of correct balance or composition [of all parts] and the correct functioning of the parts of the body"; for the latter, the organs' or the parts' ability to perform their activities correctly is "an essential, if

---

elements, and humors in Galen, see R. J. Hankinson, "Philosophy of Nature," in *The Cambridge Companion to Galen*, ed. R. J. Hankinson (Cambridge: Cambridge University Press, 2008), 217–21. Cf. Nemesius, *On the Nature of Man* 4.44; 5.47.

28. This means "having parts like each other."

29. Cf. Peter N. Singer, "The Fight for Health: Tradition, Competition, Subdivision and Philosophy in Galen's Hygienic Writings," *British Journal for the History of Philosophy* 22 (2014): 976. Cf. Nemesius, *On the Nature of Man* 4.45.19–22.

30. Cf. Nemesius, *On the Nature of Man* 4.46.1–5.

31. Cf. *Hygiene* 1.4 (11–12K).

32. Ian Johnston, "General Introduction," in *Galen: Hygiene Books 1–4*, vol. 1, ed. and trans. Ian Johnston, LCL (Cambridge, MA: Harvard University Press, 2018), xxiv; see *The Art of Medicine* 3–4 (314–315K).

not the essential, feature of health."[33] In *Method of Medicine*, Galen states: "I see all men using the terms 'health' (*hygieia*) and 'disease' (*nosos*) in this way.... For they consider that person to be healthy in whom no function of any part [of the body] is damaged, and someone to be diseased in whom there is damage" (1.5 [41K]). As Ian Johnston notes, for Galen, function (*energeia*) is best understood in relation to capacity or faculty (*dynamis*) and action (*ergon*), which are also Aristotelian and Hellenistic Alexandrian notions in light of the anatomical and physiological contributions of Herophilus and Erasistratus.[34] These three are in a causal relationship, in that capacity or faculty (*dynamis*) is the potential to carry out a function; the function (*energeia*) is what is carried out, a movement (*kinēsis*) that comes from what is moved; and the action (*ergon*) is the observed application of the function.[35] For example, the arms have the capacity or faculty of stretching; their function is to stretch; their action is the actual process of stretching. Hence, "the body which is in the best state with regard to all activities will ... also be found to be the least liable to illness amongst all bodies. For the part which performs its activity (*energeia*) best is the product of a good-mixture (*eukrasia*) of the homogeneous parts and of a well-balanced constitution of the organic ones" (*The Best Constitution of Our Bodies* 4.741–742K). Nonetheless, Galen warns people not to determine the healthy and the diseased simply on the "basis of the strength or weakness of the functions"; instead, one "must apply the term 'in accord with nature' (*kata physin*) to those who are healthy and the term 'contrary to nature' (*para physin*) to those who are diseased, as health is a condition in accord with nature productive of functions and disease a condition contrary to nature injurious to functions" (*Hygiene* 1.5 [21K]).[36]

So, what is nature and what does it mean to be "in accord with nature" for Galen?[37] Nature for Galen has at least two meanings. First, like Hippocratic physicians, nature is basically "the nature of the body itself" but in

33. Singer, "The Fight for Health," 976, 979.
34. Johnston, "General Introduction," xxv; Singer, "The Fight for Health," 980.
35. Johnston, "General Introduction," xxv; see also Armelle Debru, "Physiology," in Hankinson, *The Cambridge Companion to Galen*, 265.
36. See also Galen, *The Art of Medicine* 21 (358K): "For all things are in accord with nature (*kata physin*) in those who are healthy while they are contrary to nature (*para physin*) in those who are diseased, to the extent that they are diseased."
37. On this see Jacques Jouanna, "Galen's Concept of Nature," in Eijk, *Greek Medicine from Hippocrates to Galen*, 287–311; Hankinson, "Philosophy of Nature," in Hankinson, *The Cambridge Companion to Galen*, 210–41.

three levels: "that which proceeds from the primary elements, which are all mixed; that which proceeds from the secondary elements, those that are perceptible, and which are also called homoeomers; and the third, in addition to these two, that which proceeds from organic parts" (*Opt. med.* 3.60). More importantly, nature is the organizing principle of the living being: "by the word nature (*physis*) we should understand the capacity or faculty (*dynamis*) residing in the very bodies that were organized by her [i.e., nature]" (*Hipp. Epid.* 6.5.1). Nature is in charge of the basic functions of the human body such as growth and everyday bodily maintenance.

Then, what is it for the body to work "according to nature" (*kata physin*)? It is "that which is produced by nature in the first place (*ho kata prōton logon hypo tēs physeōs*)," that is, "what nature intends as an aim (*skopōn*)," and not what results necessarily from other things (*PHP* 6.1.8–9). For example, Galen says eyebrows and eyelashes "are generated not in the manner of grass, but in the manner of plants which have been fashioned according to the original plan (*kata prōton logon*) of Nature; their status is different from that of necessary consequences of mixture" (*Temp.* 2.5 [619K]). The growth of the eyelids and eyelashes is the work of nature that is primary, essential, or primordial, whereas the growth of grass is the work of the labor that is secondary or accidental.[38] Therefore, embedded in Galen's notion of function is the Aristotelian teleological understanding of nature that "seeks to ascribe to the smallest part of the body both a role and a structure perfectly adapted to its function or functions."[39] Indeed, all the functions (*energeia*) and their causes work together with a view toward "producing a single result, the maintenance of the functional unity which is the living creature,"[40] as expressed in *On the Usefulness of the Parts*: "all the parts of the body are in sympathy, that is to say all of them cooperate in producing one effect" (1.13). This is Galen's teleological anatomy.[41] All parts of the body are "instruments" (*organa*) for diverse types of function, arranged hierarchically, but in a cooperative manner, and the entity that

38. Cf. Jouanna, "Galen's Concept of Nature," 293–94.

39. Debru, "Physiology," 266. Galen follows Aristotle's teleology that "nature does nothing in vain" (Aristotle, *Part. an.* 65829).

40. Debru, "Physiology," 267.

41. On Galen's teleological theory in *On the Usefulness of the Parts*, see Philip J. van der Eijk, "Galen on the Nature of Human Beings," in *Philosophical Themes in Galen*, ed. P. Adamson, R. Hansberger, and J. Wilberding, Bulletin of the Institute of Classical Studies, Supp. 114 (London: Institute of Classical Studies, University of London, 2014), 89–134.

oversees the whole thing is Nature (*physis*) or Demiurge (*dēmiourgos*).[42] Therefore, "to be according to nature" (*kata physin*), which had already been a familiar concept in the Hippocratic Corpus and classical philosophy, denotes the "conformity to a natural or normal order, by reinterpreting the expression within [Galen's] understanding of Nature-Demiurge, producing in a primordial way the different parts of the body in view of a precise aim: their usefulness."[43] It is a natural and normal order (= health) guided by Nature-Demiurge's creative design or purpose to "optimize its creation" in its function and usefulness (*chreia*).[44]

Identifying Nature as the Demiurge of Plato's *Timaeus*, the creator who fashioned and orders the world, influenced by Stoics' understanding of providence, and following Aristotle, whose "wonderment at the order he discovered in the bodies of animals, which he interpreted as evidence for a benign and intelligent creator,"[45] Galen also thinks that the evidence undeniably attests to the divine status of Nature-Demiurge, completely sufficient for all and in every aspect (cf. Hippocrates, *Nutrient* 15.P.141.24). The Nature-Demiurge's primordial plan of the structure and functioning of the human body has main qualities of providence/foresight (*pronoia*), wisdom/intelligence (*sophia*), art/skill (*technē*), power/capability (*dynamis*), and even justice (*dikaiosynē*).[46] And they manifest in the usefulness (*chreia*) of the parts of the human body. Their usefulness (*chreia*) is related to the soul, for the body is the soul's "instrument" (*organon*)[47] and has been crafted in whole and in part to adapt to the character and faculties of the soul (*UP* 1.2). Nature made a human the most "intelligent" (*sophos*) and "godlike" (*theios*) creature and, in place of any and every defensive weapon that other animals

42. Debru, "Physiology," 267.

43. Jouanna, "Galen's Concept of Nature," 295.

44. Rebecca Flemming, "Demiurge and Emperor in Galen's World of Knowledge," in *Galen and the World of Knowledge*, ed. Christopher Gill, Tim Whitmarch, and John Wilkins (Cambridge: Cambridge University Press, 2009), 63.

45. Mattern, *The Prince of Medicine*, 145.

46. E.g., *UP* 1.1113, 112, 128, 159, 172; 2.3, 439, 446. Cf. Flemming, "Demiurge and Emperor," 64; see also Jouanna, "Galen's Concept of Nature," 303–4; Michael Frede, "Galen's Theology," in *Galien et la Philosophie: Huit Exposés Suivis de Discussions*, ed. Jonathan Barnes, Entretiens sur L'Antiquité Classique 49 (Vandoeuvres and Geneva: Fondation Hardt, 2003), 105, 111.

47. Galen accepted the Platonic idea of the tripartite souls ruling and yet serving the body: the rational, spirited (irascible), and desiderative (concupiscible); they are seated in the brain, the heart, and the liver, respectively. See *PHP* 7.3.2; *QAM* 4.770–773K.

have (like horns or hoofs), gave humans hands, the "instruments necessary for every art and useful in peace no less than in war" (1.2). Thus, it was better for a human to make use of all weapons and all the arts through her hands and through reason, Nature's endowment (1.6). It was also best for a human to have fingers to lay hold on many different shapes and sizes of objects (1.7–8). How about fingernails? How are they useful? Unfortunately, Plato spoke dismissively of the usefulness of fingernails, describing them as failed versions of animals' claws; even Aristotle fell short in identifying their usefulness, saying that nails were formed for protection. However, Galen discerns the full extent of their usefulness: "Nature gave the finger tips a structure suited to laying hold of all kinds of objects. This is the reason why the finger tips were not composed of either nail or flesh alone, but of both together, having the best mutual arrangement" (1.7). Elsewhere in the same text, he further observes actual functions in apparently purposeless parts of the body. For example, he highlights the functional adaptiveness of the gall bladder, for "it fulfills the same role for yellow bile as the spleen does for black" bile (1.272–274).[48] In his *On the Usefulness of the Parts*, Galen provides "a complete account of the usefulness in all the parts" and sees "total optimisation in all the parts" (1.8).[49]

In the "Epode" to the book, these discoveries and his total trust in Nature's creative design and providence inspire Galen to "admire the skill of Nature more than ever" (2.439), who "exhibits the symmetry of the parts both on the outside, as sculptors do, and also deep below the surface" (2.442). For example, Nature used the very best proportions in establishing not only the size of the arm but also the size of the legs and the usefulness of each of their motions (2.443). "For if you imagine a man whose legs are only half the proper size, you will understand, I think, how hard to move and how heavy he will find the body above them, second, how unsteady he will be if he tries to talk, and third, how impossible it will be for him to run" (2.443). Thus, anyone with a detailed anatomical knowledge cannot fail to acknowledge the highest artistry "at work in marrying structure to function, and producing no superfluities" (2.441–443).[50] What Nature produces is not only exceedingly beautiful and symmetrical but also optimally designed to serve a necessary purpose (*chreia*); there is almost no failure or abnormality in Nature's creation (2.443–444):

48. Also Hankinson, "Philosophy of Nature," 227.
49. Flemming, "Demiurge and Emperor," 67.
50. Hankinson, "Philosophy of Nature," 234.

Who could be so stupid, then, or could there be anyone so hostile and antagonistic to the works of Nature as not to recognize immediately . . . the skill (*technē*) of the Creator? Who would not straightway conclude that some intelligence possessed of marvelous power was walking the earth and penetrating its every part? And yet, what part of the universe is more ignoble than the earth? . . . Even here there appears to be some intelligence reading us from the bodies above, and anyone seeing these is at once forced to admire the beauty of their substances, first and foremost that of the sun, after the sun that of the moon, and then of the stars. (2.446)

This is Galen's argument from design, his "natural theology,"[51] against those atomists such as Epicurus and Asclepiades who "call Nature an aimless workman," denying her foresight and providence. Galen's argument from design is also against Erasistratus and his followers who, "though they always commend her [Nature] for doing nothing in vain, do not really carry their point by demonstrating in every instrument that their praise is justified," as in the case of spleen (1.267; 1.232; cf. 1.391); these are the ones who "frequently suggest that the usefulness of the parts (of which they know nothing) does not exist" and leave matters to the operation of contingency or chance (2.450). Against these kinds of perspectives, later Christian authors will also make an extensive use of the argument of the *transcendent* God from design for divine providence and workmanship of the human body.[52] Since there is an organic link between what is healthy and thus normal, and what is useful (functional), symmetrical, and thus beautiful, implicit in Galen's totalizing and teleological argument is that there is an "image of physiological perfection."[53] Galen asks, "what goes along with health?" "Three things," he responds: "beauty (*kallos*), a good bodily state (*eueksia*), and soundness/completeness (*artiotēs*)" (*Hygiene* 6.1 [383K]).[54]

---

51. On Galen's natural theology and religious beliefs, see Frede, "Galen's Theology," 73–126; Fridolf Kudlien, "Galen's Religious Belief," in *Galen: Problems and Prospects*, ed. Vivian Nutton (London: Wellcome Institute for the History of Medicine, 1981), 117–30.

52. For example, Lactantius, *Opif.* 8–13; Gregory of Nyssa, *On the Making of Man* 30; Theodoret of Cyrrhus, *Prov.* 3.1–4.34; Nemesius, *On the Nature of Man* 4.46.16–17; 28.92.1–93.19; 42.122.1–4.

53. Robert Garland, *The Eye of the Beholder: Deformity and Disability in the Graeco-Roman World*, 2nd ed. (London: Bristol Classical Press, 2010), xx.

54. Cf. *Thrasybulus on Whether Health Belongs to Medicine or Gymnastics* 14 (829K); 15–16 (833–34K).

Although Galen also insists on the different levels of health in varying degrees of function, and thus in physiological imperfection (see below), this image of bodily perfection "is so consistent and so unwavering that it virtually denies the possibility of the accidental true-to-life deficiencies of authentic human anatomy";[55] and this ideal of bodily perfection manifests in broader Greco-Roman culture, especially in art. Indeed, it is no accident that Galen relates the work of the sculptor Proclitus to that of Nature as a pale comparison (*UP* 2.442). Thus health, physical perfection, and the beauty of the human (male) body all go together in the Greco-Roman context. The Neoplatonist Plotinus in the late third century confirms that relationship: "What is it that rouses the eyes of those who observe something beautiful, and which draws them to it and pulls them along and causes them to take delight in the spectacle? . . . Virtually everyone agrees that the proportion of parts in relation to one another and towards the whole combined with freshness of complexion produces beauty to the eye, and that in this as generally with everything else that which furnishes beauty is well-proportioned and well-measured" (*On Beauty* [*Enneads* 1.6.1]). This ideal of perfect body, functional health, and physical beauty manifested through symmetry would persist in Greco-Roman culture despite the increasing realism of the Roman era.

Finally, like his Hippocratic predecessors and despite his own argument for the optimalization of the human body, Galen also recognized the latitude (*platos*) of health in people:[56] "A healthy constitution is judged by functions that are in accord with nature, and that this is the best of health, or . . . perfect and excellent health, while that which is, as it were defective and is neither perfect nor excellent has, we may say a very wide range (*platos*)" (*Hygiene* 1.4 [12K]). This notion of health acknowledges the reality that there may be (only) a small group of people existing in a state of perfection—"exact, optimal, complete," "perfect and excellent health" (1.4 [12, 13K]). Indeed, most people would fall within a wide range of suboptimal, imperfect health, free from pain.

> For health is not . . . that which is indivisible alone, but also that which is inferior to this, but not so inferior as to have failed in its function. For we all . . . require health both for the performance or the activities of life, which are impeded, interrupted or terminated by sickness, and also

---

55. Garland, *Eye of the Beholder*, xx.
56. Cf. Singer, "The Fight for Health," 981–83.

for the sake of an untroubled existence. For when we are in pain, we are troubled by that significantly. Such a state, in which we are neither in pain nor impeded in the performance of life's activities, we call "health." If someone wishes to apply a different term to it, he will not gain anything from that—no more than those who introduce the notion of "permanent illness" (*aeipatheia*). (1.5 [18K]: trans. Singer)

This seems to be the condition Galen later describes as "nonnatural" (*ou physei*), that is, neither "in accord with nature" nor "contrary to nature"; it is a condition that has gone beyond "balance" but is not hindered with respect to function, like people with *leuke*, leprous warts or regular warts (6.2 [384–385K]). In this context, people should compare their health not to the optimal, perfect standard of health but to a standard calibrated by the activities they need to accomplish, and their ability to perform those activities, as well as according to their own constitutions. Say we have a fever, but it is so slight that we do not perceive it, and can go out and do our business, bathe, drink, eat, and conduct other necessary tasks. In that case, "it is the absence of interference with use (*chreias*) that especially defines health. Nor is weakness of the functions a sign of disease, strictly speaking, but [only] what is contrary to the nature (*para physin*) of each function. In fact, we all see badly if we compare ourselves to eagles and to Lynceus, and we do not hear properly if we compare ourselves to Melampus. . . . Indeed, in each part we would be deemed to come close to being disabled, if we were to compare ourselves to those who are preeminent in respect of that part" (1.5 [19–20K]). In this sense health is relative to our own natural (inborn) capacities, and also to our different stages of life, such as youth or old age, as the latter is not to be considered automatically pathological; there is a health proper to old age (1.5 [21K]).

Nonetheless, despite this recognition and his prescriptions for suboptimal lifestyles (1.12 [60–62K]), Galen still provides instructions for those who might achieve a level of "well-conditioned," optimal health, "the goal of all men"[57]—those elite men with the best natural constitution and the freedom to pursue the right activities in the prime of their lives (i.e., "the third seven-year period of life").[58] This man "who has the time to devote himself solely to the care of the body and to set aside all other things as secondary," should be engaged in exercises first, "followed by food and drink, and next

57. *The Best Constitution of Our Bodies* 4.740K.
58. *Hygiene* 2.1 (81–83K); cf. *Good Condition* 4.751K.

in order, sleep, and then sexual intercourse," all in moderation, following Hippocrates's regimen.[59] What types of exercise does Galen propose to that end? In *The Exercise with a Small Ball* he writes: "The form of exercise most deserving our attention is therefore that which has the capacity to provide health of the body, harmony of the parts, and virtue in the soul; and all these things are true of the exercise with the small ball" (4 [906K]).

Greco-Roman Philosopher Plutarch (c. 40–120 CE)

As we move to Galen's older contemporary, Middle Platonist Plutarch, in the mid to late first century and early second century CE, we see that Galen's precepts of health care for elite men are an appropriate connector to Plutarch's own precepts of health care for elite men in public life. Having written it sometime after 81 CE, Plutarch in his *Precepts of Health Care* (*De tuenda sanitate praecepta*) reveals his notion of health as he focuses on a suitable regimen that, together with surgery and pharmacology, was one of the three branches of ancient medicine. Already in the Hippocratic Corpus and certainly by the time of Plutarch, regimen (*diaita*) denoted not only diet and exercise but also the whole lifestyle and habitual behaviors of an individual or a people for the prevention of sickness and the maintenance of health, as well as for the treatment of diseases. In the process, Plutarch displays his knowledge of Hippocratic treatises and reflects a contemporary debate over regimen among physicians, gymnastic trainers, and philosophers.[60] The text starts with two characters, Moschion, a physician, and Zeuxipuus, a friend of Plutarch's, in a dialogue about Glaucus, the physician who attacked philosophy's legitimacy in addressing medicine and regimen and also dismissed the health-care advice of Zeuxipuus's companion, a philosopher (122C–D). Through the lips of Zeuxipuus, Plutarch defends philosophers' engagement with medicine/regimen as one area of the liberal arts, and champions those precepts of health care approved by the physician, Moschion.

One of the early pieces of advice given by the companion (and ridiculed by Glaucus) is about the food that doctors prescribe to the sick. The readers

---

59. *Hygiene* 2.2 (84K); *Epidemics VI* 4 (23K); *VI* 6 (2K).

60. Lieve Van Hoof, *Plutarch's Practical Ethics: The Social Dynamics of Philosophy* (Oxford: Oxford University Press, 2010), 212; Galen, in his *Thrasybulus on Whether Health Belongs to Medicine or Gymnastics*, argues that regimen belongs to medicine, not gymnastics.

should take the food and habituate it in time of health so that they do not get disaffected at the diet in time of illness (123B). Plutarch's point is that they "habituate the appetite to be obedient to expediency with all serenity" and "inculcate a fixed habit during the periods of soundest health, so as to make these things agreeable, familiar, and congenial to [their] nature," keeping in mind "how some men fall and act in times of sickness, being angry and fretful" (123C–D). This recalls the Hippocratic author's emphasis on the power of habit in diet as an important factor in the evaluation of an individual's health in the *Regimen in Acute Diseases* (35, 9). Many, says Plutarch, including Emperor Titus, perished because they could not keep the regimen prescribed in illness (*Precepts of Health Care* 123D). Moreover, the way to preserve health is "to guard against excess (*plēsmonas*) in eating and drinking, and against self-indulgence," that is, "to keep to moderation (*metriōn*) . . . and at the same time to avoid the extreme disagreeableness which makes one appear offensive and tiresome to the whole company" in various social contexts such as attending festivals, hosting visiting friends, entertaining high officials, or frequenting a symposium (123D–E). Here Plutarch taps into the long-held philosophical notion of moderation as a moral conduct, as it also relates to its medical counterpart, denoting human health and one's natural environment (e.g., *Airs, Waters, Places* 5, 10). Thus, his elite readers should eat moderately to avoid physical discomfort or pain at the banquet while fully participating in it. This, however, is a departure for Plutarch from a doctor's precept, which would prioritize health over social obligations (i.e., keep to moderation and avoid those social occasions). In the words of Van Hoof, "Plutarch's *Precepts of Health Care*, then, are not just concerned with the healthiest diet for an individual taken by himself, but also take into account that individual's social position and role: far from discouraging the reader from attending social occasions which, with their abundance of delicacies and alcohol, risk destroying his health, Plutarch teaches him how to reconcile the demands of health care and social decorum."[61] Indeed, for Plutarch, "the reason why health care is so important, is because health is a necessary condition for living the life his readers are living"; therefore, his readers should take care of their health in socially acceptable ways.[62]

The principle he draws is from Socrates, who did not forbid people to enjoy delicacies and wine but urged them to consume them "instead,

61. Van Hoof, *Plutarch's Practical Ethics*, 223.
62. Van Hoof, *Plutarch's Practical Ethics*, 223.

rather than on top, of one's usual diet" (124E–F).[63] Not all pleasure is self-indulgent and excessive, but the pleasure produced by love, eating, and drinking in "serenity and calmness" of our body, and health, is good (126D) because the key to good health and bodily pleasure does not depend on externals (i.e., food and drinks) but on that which is within our control (*to plēthos hypokeimenon*); therefore, for Plutarch, taking care of our bodies is of the utmost importance.[64] Then, his precepts continue in the following way: first, "we must not stuff and overload our body, and afterwards employ purgatives and injections, but rather keep it all the time trim, so that if ever it suffer depression, it shall owing to its buoyancy bob up again like a cork" (127D, 128F, 129F). If health is more about general knowledge and wisdom, philosophers are better suited to give precepts of health care than physicians, who are more concerned about technical knowledge. Second, "we ought to take special precautions in the case of premonitory symptoms and sensations" that precede illnesses (127D, 129E) in matters of food, exercise, sleep, dreams, and the emotions of the soul (129B); and failure to do that "will do us harm" (128A). Thus, "eating not unto satiety, laboring not unto weariness, and observance of chastity (*syntērēsin*), are the most healthful things" (130A; cf. Hippocratic *Epidemics VI* 4).

Then Plutarch dwells on the specific exercises suitable for men of letters (*philologois*): daily speaking (aloud) utilizing breathing (*pneuma*)[65] (*Precepts of Health Care* 130A–C); reading (130C–D); rubbing with oil (130D); and taking warm baths, although the latter is unnecessary if the body is already in good condition (131C). What follows is food, which should be focused on lighter, natural (*kata physin*) substances such as vegetables, poultry, and fish with less fat rather than heavier foodstuffs like meat and cheese (131E). In terms of drinks, they should drink water rather than wine (though wine "is the most beneficial of beverages"), for the former is mild and calm and the latter "intensifies the disturbances of the body" (132B–D). Finally, Plutarch recommends cultivated discussions (*symposia*) following dinner for controlling one's excessive appetites and the "soul's satisfaction" (133B, 133E). Note that all the activities mentioned are exercises for both the body and the soul. Throughout the *Precepts*, Plutarch reiterates the

---

63. Van Hoof, *Plutarch's Practical Ethics*, 224.

64. Van Hoof, *Plutarch's Practical Ethics*, 226.

65. Van Hoof, *Plutarch's Practical Ethics*, 233, notes Plutarch's suggestion of speaking as an exercise is linked with ancient theories of *pneuma*, which also stressed the importance of the *pneuma* of the soul. On *pneuma*, see p. 53 below.

importance of Aristotelian moderation (136A). At the "meta-level of health care as a whole: on the one hand, one should not neglect one's health out of gluttony, ignorance, or desire for honor; on the other one should not make health care the only or main aim in life, nor take pride in abstinence or familiarity with medical dietetics. What one does should cause neither pain nor repentance."[66] Regarding the latter point, his readers should continue to engage in active social and public life, following Plato, Democritus, and Theophrastus, as health cannot be gained by inactivity (135B, 135D–F); this, as we have seen, is an opposite position of Galen, who believed that the best way to live a healthy life was to solely pursue bodily care free from consideration of any activity (art, business, public life, etc.).[67] Once again, activities should be balanced by necessary recuperation because that is natural (*kata physin*). Overall for Plutarch, as health is "a necessary condition for action" (126B), "action is [also] the aim of health" (*to tēs hygieias telos*; 135C); hence, health care is a means to an end, that is, activity.[68] That is why his readers should have self-knowledge in their own constitution, regimen, and health care and be therefore self-sufficient from doctors (136E–F). This is a philosophical reframing of health and health care. In keeping health, as Plato notes, the soul and the body need to work together, the soul giving heed to the body when the latter needs it (137D). Then, "of the good gifts which fair and lovely Health (*hygieian*) bestows[,] the fairest is the unhampered opportunity to get and to use virtue (*aretēs*)," the highest goal of philosophy, "both in words and in deeds" (137E). Health and health care are ultimately important because they enable elite men to realize a higher philosophical objective—virtue. We will come back to the relationship between health, philosophy, and virtue later when we discuss the soul-body relationship in more detail.

Women's Health

We have analyzed, thus far, notions of health in the Greco-Roman world projected as a universally definable condition based on the basic biological structure common to all human beings. However, we have also seen how the standard of health and the pursuit of optimal health and health care belonged to (elite) men, and therefore were understood as fundamentally

---

66. Van Hoof, *Plutarch's Practical Ethics*, 240.
67. *Hygiene* 1.12 (62K) and 2.2 (84K).
68. Cf. Van Hoof, *Plutarch's Practical Ethics*, 245.

male concerns. When it comes to women's health, Greco-Roman medicine displays highly contextualized (i.e., patriarchal) yet again universalized understandings. In the Hippocratic Corpus (and Greco-Roman medicine in general), women's health is closely and explicitly tied to women's reproductive functions and reproductive organs, particularly the uterus (*hystera*).[69] While health is a balance because of the evacuation of blood due to menses, women's bodies were always in flux, and therefore frequently prone to states of excess and deficiency. Moreover, the process of achieving equilibrium is always in danger because of the ability of the uterus—the main organ responsible for maintaining balance—to move, tilt, close, or open up the body in response to smells and other factors.[70] Thus the health and stability of the womb are critical to women's health. To that end, first of all, health manifests in regular monthly menstruation (*Seven Months' Child* 9). If a woman is healthy, "her menstrual blood will pass like that of a sacrificial animal, and it will quickly congeal" (*Diseases of Women I* 6). And "in every healthy woman the amount of the menstrual flux to pass is equal to" more or less "two Attic cotyles" (a pint!) in two or three days, which indicate the womb's capacity to hold outside pregnancy (*I* 6). The assumption behind these numbers is that the amount that fills up the womb each month must be emptied completely, and this figure shows up later again in Soranus's *Gynaecology* (1.20) as a maximum rather than as the norm, indicating its large amount. Second, besides regular and heavy menstrual bleeding, sexual intercourse is also linked to women's health. According to *Generation* 4, if women have intercourse with men, they are more likely to be healthy because "intercourse makes the menses pass more easily by warming and moistening the blood." Third, conception and pregnancy are healthy because they show that the womb must be in place, open, and receptive (*Diseases of Women I* 37, 59). Giving birth after a full term also purges and cleanses the womb and thus makes the woman healthy again (60).

If women's health is fundamentally linked to their reproductive functions, can a woman be healthy and infertile? It seems it is possible at least in two categories, notwithstanding the social stigma. The author of *Diseases of Women I* first observes a woman who cannot conceive because her uterus, although healthy, is rather weak. The treatment is to strengthen the

---

69. The following draws on Helen King, "Women's Health and Recovery in the Hippocratic Corpus," in King, *Health in Antiquity*, 156–60.

70. King, "Women's Health and Recovery," 156.

feeble uterus "until it appears to be healthy" (12). The author also mentions a woman whose treatment for severe ulceration of the womb ends—she becomes healthy but is no longer fertile (9). Helen King poignantly points out, "if we take our definitions of health as universally valid, then it is possible to argue that women's health was defined by the Hippocratics in such a way that it could never be attained."[71] For anthropological data from rural Sri Lanka and premodern England and Wales all suggest that menstruation was "normally both scanty and infrequent."[72] This suggests two possibilities: either ancient Greek women consistently failed to meet the medical standard set for them and thus were persistently sick according to their cultural norms, or they did bleed to that extent and became very sick according to our cultural definitions.[73] Ironically, then, the very idea of health for women in the Hippocratic Corpus is a precarious one and implies medical intervention by either their or our cultural standard, as women are seen as physically more at risk of disease than men are due to their natural physical instability and imperfection.

Galen also believes that man is superior to woman following the Aristotelian model of women as "misbegotten men," and that male health is innately superior to female health (although the former is not more secure in relation to illness). He continues with his predecessors in linking women's health to the health of women's reproductive organ(s), womb, and hence to regular menstruation (although Galen believed the uterus was rather fixed).[74] But Galen, grounding his theory in cultural assumptions about women's domestic life, adds a teleological explanation for women's monthly evacuation that identifies it as in accordance with nature's design. "Does [Nature] not evacuate all women every month, by pouring forth the surfeit of the blood? For it is necessary, I think, that the female genus, who stays at home, neither leading a life of hard work nor coming into

71. King, "Women's Health and Recovery," 158.

72. King, "Women's Health and Recovery," 158. Helen King, "Medical Texts as a Source for Women's History," in *The Greek World*, ed. Anton Powell (London: Routledge, 1995), 210, says that the actual capacity of the nonpregnant uterus is two to three fluid ounces and that "heavy loss and regularity [in menstruation] is highly unlikely to reflect reality, and should instead be understood as part of a 'grand ideological statement' about the nature of the female."

73. Cf. King, "Women's Health and Recovery," 158.

74. This paragraph draws on Rebecca Flemming, *Medicine and the Making of Roman Women: Gender, Nature, and Authority from Celsus to Galen* (Oxford: Oxford University Press, 2001), 311–43.

contact with direct sunlight, and because of both these things generating excess, should have a natural remedy—the evacuation of the excess."[75] According to *On the Usefulness of the Parts*, Nature has instituted both the excess and its release to the goal of the continuation of the species, that is, the propagation of generations. A "well-purged woman" through monthly evacuation is unaffected by a host of diseases such as arthritis, pleuritis, phrenitis, melancholia, mania, and any other major and serious diseases.[76] Thinking of health as balance and excess as a main form of imbalance, Galen "opens the way for menstrual immunity" to most, if not all, of these kinds of diseases.[77] Again, a healthy uterus through regular menstruation and childbearing (through sexual intercourse) constitute the sum of women's health. In the words of Monica Green: "when the womb is healthy so is the woman, and when the woman is healthy so must be the womb."[78] On the other hand, Galen differed from the Hippocratic Corpus in that he departed from explicitly drawing a direct relationship between reproductive functions and female health and saw pregnancy and birth as rather dangerous experiences for women. Women's high mortality in childbearing throughout history in fact confirms Galen's perspective.

Disease in Greco-Roman Literature, Medicine, and Philosophy

In the previous section, we examined the representative visions of health in Greco-Roman literature, medicine, and philosophy. In this section, we will explore their main notions and theories of disease in their cultural context. Here we note one conceptual point from medical sociology and anthropology: the distinction between disease and illness.[79] As mentioned in the introduction, in (Western) biomedicine disease tends to denote an objectively measured or seen pathological condition determined by a physician, whereas illness tends to refer to the patient's subjective feelings about and experience of the significance and meaning of his or her pathological condition firmly

75. *On Venesection against Erasistratus* 5 (164–165K).

76. *On Venesection against Erasistratus* 5 (164–165K).

77. Flemming, *Medicine and the Making of Roman Women*, 338.

78. Cited in Flemming, *Medicine and the Making of Roman Women*, 333.

79. On this distinction see, for example, Kleinman, *Patients and Healers*, 72; Kleinman, *The Illness Narratives*, 3–5. See also King, "Introduction: What Is Health?" in *Health in Antiquity*, ed. Helen King (London: Routledge, 2005), 5; Lloyd, *In the Grip of Disease*, 1–2; Susan P. Mattern, *Galen and the Rhetoric of Healing* (Baltimore: Johns Hopkins University Press, 2008), 159–61.

embedded in a given culture and time.[80] While noting this distinction and therefore treating disease and illness separately in this section, we should also note that this distinction can be blurred at times and that the notion of disease (as well as health) is also, as we will see, culturally shaped.

Myth and Medicine

It seems that in classical Greece, people believed that health was the original state of humanity. In his myth *Works and Days*, archaic poet Hesiod describes the origin of disease as it was released from the first woman Pandora's jar along with hunger and hardship (*Erga* 102–104). The disease came upon humans "freely" or "of their own accord" (*automatoi*), and "silently" (*sigēi*), as Zeus took their voices away. In this particular account, the disease was not sent by god as punishment but came to humanity of its own accord, "in the nature of things," although Hesiod distinguishes it from another kind of disease, plague, as punishment from god later.[81] As we move to classical medicine, first Alcmaeon and then Hippocratic authors, the idea of disease arising from natural causes, "in the nature of things," and not as a result of divine justice or vengeance, is important. In Alcmaeon we have already witnessed the basic notion of disease as *monarchia*, a destructive prevalence of one power/element over the others as opposed to *isonomia*, their equality in mixture (*Fr.* 4; Aëtius 5.30.1). According to him, first disease arises from *monarchia* as noted, the disequilibrium particularly "by excess of heat or coldness"; second, it occurs from "surfeit or lack of nourishment" (foodstuffs) or from external factors such as "water of certain quality, local environment, exertions, hardship, or something similar to these"; finally, disease occurs in certain bodily parts such as "the blood, the marrow, and the brain" (Aëtius 5.30.1). This definition of disease by Alcmaeon accounts for "the elemental changes in the body with reference to the diet of an individual, [one's] physical activities, and the climatic and topographical conditions of [one's] place"; therefore, as in the definition of health, it also laid a foundation for Hippocratic medicine's notion of disease, which in turn became foundational for the subsequent Greco-Roman medicine.[82]

80. See Karin Nijhuis, "Greek Doctors and Roman Patient: A Medical Anthropological Approach," in *Ancient Medicine in Its Socio-Cultural Context*, ed. Ph. J. van der Ejik, H. F. J. Horstmanshoff, and P. H. Schrijvers, Clio Medica 27 (Amsterdam and Atlanta: Edition Rodopi, 1995), 1:53–54.

81. See R. M. Frazer, "Pandora's Diseases, *Erga* 102–104," *GRBS* 13 (1972): 236.

82. Kouloumentas, "The Body and Polis," 876.

The Hippocratic Corpus

The Hippocratic authors theorize disease as the disruption of the normal state against nature (*para physin*). In Hippocratic treatises, the etiology of disease as disequilibrium of bodily constituents more or less follows that of Alcmaeon, with further development. First of all, according to *Diseases I* 2 *and Affections* 1, all diseases arise from the state of the body that is contingent on its internal constitutions and functions. Particularly "bile and phlegm produce diseases when, inside the body, one of them becomes too moist, too dry, too hot, or too cold." The author of *Diseases IV* names the four humors (phlegm, blood, bile, and water in this case) as the cause of all diseases except those caused by violence (1). Too much or too little of these humors (whether the two or the four), which cause diseases, arises from the effects of nourishment (foodstuffs and drinks)[83] and also from exertions, wounds, smell, sound, sight, and venery, and from heat and cold (*Affections* 1; *Diseases I* 2). One of the earliest Hippocratic treatises, *On Ancient Medicine*, focuses on food and dietary regimen first as the root of all diseases and then as the origin and foundation of medical art (*technē*). According to the text, the original state of humanity was disease; this is in contrast to Hesiod's myth. In the beginning human beings ate the same raw and uncooked foods as wild animals and suffered violently from them, falling into pain and disease, and even death (3.3–4). Original human nature (*physis*) then was weak; and due to their sickness and suffering, humans experimented with cooking (e.g., grinding, baking, boiling, blending, etc.) and learned to prepare a diet that was properly adapted to human nature (*physis*) (3.5). As Jacques Jouanna notes, this necessary adaptation of diet to human constitution, whether one is sick or in good health, "forms the basis of a long history of dietetics that this medical writer retraces, beginning with the first discovery of the diet of people in good health, before coming to successive discoveries of diets for patients."[84] "For they considered that if foods that are too strong are ingested, the human constitution will be unable to overcome them, and from these foods themselves will come suffering, diseases, and death, while from all those foods that the human constitution can overcome will come nourishment, growth, and health" (3.5; trans. Schiefsky). Implied in this statement is that people had different constitutions (*physeis*) and there was no single food or type of food suitable

---

83. *Affections* 1; *Diseases IV* 2.
84. Jouanna, "Dietetics in Hippocratic Medicine," 145–46.

to all *physeis*; thus, "even among the earliest humans the process of discovering proper foods for health was dependent on calibrating to individual *physeis* and understanding fully the properties of all foods."[85] Only then came the discovery of the diet most appropriate for sick people, which was the birth of medicine (5).

Second, the author of *Humors* more or less confirms the constitutional etiology, but also adds the environmental conditions that are contingent on changes of climate and seasons, on wind, air, and water quality, and on residential location: some diseases "are the result of the physical constitution, others of regimen, of the constitution of the [year], of the seasons. Countries badly situated with respect to the seasons engender disease analogous to the season" (12).[86] For example, "when it produces irregular heat or cold on the same day, diseases in the country are autumnal, and similarly in the case of the other seasons. Some spring from the smells or mud or marshes, others from waters, stone, for example, and diseases of the spleen; of this kind are waters because of winds good or bad" (12). As mentioned earlier, change (*metabolē*) is the basis of Hippocratic pathology, and according to *Airs, Waters, Places*, "one should be especially on one's guard against the most violent changes of the seasons" (11). While the author of *Breaths* attempts to explain the single cause of disease as air (2), the author of *Airs, Waters, Places* speaks of not only general diseases due to the seasons but also of local diseases dependent on the orientation of cities. In cities exposed to hot and humid winds from the south, diseases with moisture and phlegm are dominant, such as infertility for women, asthma for children, and dysentery, diarrhea, and eye inflammation for men (3). However, in cities exposed to the cold and dry winds of the north, diseases with bile prevail, such as pleurisies, eye inflammations, nose bleeding for men, infertility and unhealthy menstruation for women, and dropsies for children (4). Cities facing to the west are the unhealthiest (6). As we have seen earlier, this text also correlates the orientation of places to the seasons—a city facing east, the healthiest city, resembles the spring by its moderate climate (5), whereas a city facing west, the unhealthiest city, is like autumn because of its great contrasts of temperature (6). Thus, there is a correlation between local diseases and seasonal diseases.[87]

---

85. Ralph M. Rosen, "Towards a Hippocratic Anthropology: *On Ancient Medicine* and the Origins of Humans," in *Ancient Concepts of the Hippocratic: Papers Presented at the XIIIth International Hippocrates Colloquium, Austin, Texas, 11–13 August, 2008*, edited by Lesley Dean-Jones and Ralph M. Rosen (Leiden: Brill, 2016), 250.

86. See also diseases in each season in *Aphorisms* 3.1, 20–23; cf. *Nature of Man* 8.

87. Jouanna, *Hippocrates*, 148.

In the same text, water quality and seasonal diseases are correlated as well. Stagnant waters are the unhealthiest as they provoke a sickly state in those who drink them (7), but rainwater is best for health (8). Those who drink stagnant water are affected by seasonal diseases and their permanently sickly state: dysentery, diarrhea, and quartan fevers in summer and other diseases in winter (7). Also, the seasons change the water's nature as stagnant waters are hot in summer and cold in winter (7); this change determines a particular disease. As in the case of foods and human constitution, proper waters for health depend on adjusting to individual *physeis*, too. Therefore, speaking of using spring waters, the author says that those in health and strength can drink any water at hand without distinction, but for those suffering from a disease, the most suitable water can best attain health in the following way (7): "those whose digestive organs are hard and easily heated will gain benefit from the sweetest, lightest and most sparkling waters. But those whose bellies are soft, moist, and phlegmatic, benefit from the hardest, most harsh and saltish waters, for these are the best to dry them up" (7). In this way, "the patient's natural constitution determines the choice of the water to be administered according to the principle of contraries, which is found throughout the Hippocratic Corpus"; the text identifies both the natural constitution and environmental influences as explanatory factors of pathology.[88]

Plato

In *Timaeus* Plato, similar to Alcmaeon and the Hippocratic physicians, offers three theories about the origin of disease. First, disease arises from disorder in the fundamental bodily elements when one or more of the four bodily elements (earth, air, fire, and water) increases (excess) or decreases (deficiency) unnaturally (*para physin*), or when elements change places, or the wrong variety of an element is present (82a1–82b7).[89] Excess (*pleonexis*) and deficiency (*endeia*)[90] in *Timaeus* bring about not only diseases (*nosoi*) but also disorder (*staseis*) in that one or more elements overstep a boundary, which creates alterations of every variety and countless diseases and destruction (82ab–b7). Second, disease affects the simple *homoiomer-*

---

88. Jacques Jouanna, "Water, Health and Disease in the Hippocratic Treatise *Airs, Waters, Places*," in Eijk, *Greek Medicine from Hippocrates to Galen*, 169.

89. Cf. Prince, "Metaphysics of Bodily Health," 909.

90. Lloyd, *In the Grip of Disease*, 154, notes the Hippocratic terms for these notions: *plerosis* (repletion) and *kenosis* (depletion).

ous parts (e.g., bone, muscle, and sinews) when the basic substances do not interact and generate one another (82c1–82ef.). "When all the substances become reversed and corrupted" (82ef.), they no longer preserve their natural order (*taxis*) but, "being at enmity with themselves (*echthra*)[,] they have no enjoyment of themselves, and being at war (*polemia*) also with the established and regular constitution of the body, they corrupt and dissolve it" (83a). As G. E. R. Lloyd notes, here disease as hostility and warfare is contrasted to order in the body.[91] Finally, the third kind of disease comes from air (*pneumatos*, 84d2–85a1), phlegm (*phlegmatos*, 85a1–b5), and bile (*cholēs*, 85b5–86a8). Whenever the lungs, which dispense air to the body, fail to keep their outlets clean, the air, unable to pass, forces and distorts the vessels of the veins and causes "countless diseases (*nosēmata*) of a painful kind" (84e1). Then when white phlegm is blocked inside because of the air in its bubbles, it is blended with black bile; spreads over to the most divine part of the body, the head; and causes "the sacred disease" (i.e., epilepsy). Lastly, bile causes all the inflammations due to the burning and inflaming of the body. Lloyd again notes that here the disturbances of order (*taxis*) and proportion (*symmetria*), which are the natural state of the body, result in disorder (*ataxis*) (85e4).[92] These are the diseases that affect the body, but *Timaeus* also discusses the diseases of the soul resulting from a bodily condition. We will examine it later under the heading of the disease of the soul.

Back to Medicine: Galen

Corresponding to his notion of health, disease (*nosos/nosēma*) for Galen has a constitutional or structural classification to begin with. First, just as health is a balance, so disease is an imbalance (*dyskrasia*). An imbalance may arise from a change in the balance of one or more of the four elemental qualities (hot, cold, wet, and dry). This in turn occurs within the body itself (i.e., the process is entirely internal to the body) or from an inflow of material into the affected body from without, which alters the balance of qualities.[93] Galen mentions eight possible imbalances of the four elemental qualities involving *homoiomerous* parts or the whole of the body:

91. Lloyd, *In the Grip of Disease*, 155.
92. Lloyd, *In the Grip of Disease*, 155.
93. *Galen: On Diseases and Symptoms*, translated and introduced by Ian Johnston, LCL (New York: Cambridge University Press, 2006), 70.

four "simple" imbalances, in which there is a dominance or isolation of only one elemental quality (hot, cold, wet, or dry), and four "compound" imbalances, in which there is a dominance or isolation of one of the four possible combinations (hot and wet, hot and dry, cold and wet, or cold and dry).[94] Second, some diseases affect the organic parts and structures, having to do with abnormalities of morphology and composition such as size of parts, number of parts, conformation (form), or position such as cavities and dislocations.[95] Finally, Galen mentions diseases involving dissolution of continuity such as bone fractures and ulcers applying to both *homoiomerous* and organic structures.[96]

As in the case of health, there is an intimate connection between constitution (*kataskeuē*) and function (*energeia*) in Galen's definitions of disease. In *On the Differences in Diseases*, he defines disease as "either some constitution contrary to nature (*kata physin*) or a cause of damaged function" (6.837–838K). He again combines the constitutional and functional definition of disease in *On the Differentiae of Symptoms*: "Now disease (*nosos*) is spoken of as being any constitution contrary to nature (*para physin*) by which function is harmed primarily" (7.43K).[97] A bit later in the same text, Galen nuances the definition of disease by differentiating it from a condition or disposition (*diathesis*), a cause, and a symptom (*symptōma*):

"A disease is a condition (*diathesis*) of a body primarily impeding function." Accordingly, those conditions preceding this are not yet diseases. And even if some other conditions were to coincide with them like some accompanying shadows, these too we shall not call disease, but symptoms. And for us in the same way not everything in the body that would be contrary to nature (*para physin*) will be what one must immediately call a disease, but [only] what is primarily harmful to function [is called] a disease and what precedes this [is called] a cause of disease, but not yet a disease. If some other condition involving the body follows the disease, this will be termed a symptom. (7.50K)

94. Ian Johnston, introduction to *Galen: Method of Medicine I: Book 1–4*, ed. and trans. Ian Johnston and G. H. R. Horsley, LCL (Cambridge, MA: Harvard University Press, 2011), lxv.

95. Johnston, introduction to *Galen: Method of Medicine I: Book 1–4*, lxv.

96. Cf. *Galen: On Diseases and Symptoms*, 71.

97. See also 7.47K: "disease is either a constitution of the body contrary to nature, or a cause of damage to function. Or to put it more succinctly, disease is a condition contrary to nature which impedes functions."

A condition or disposition (*diathesis*) is an abiding temperament of the body, which is responsible for the proper or improper functioning of its various systems.[98] As Robert Hankinson notes, Galen in this text holds that any proper analysis of physical functioning involves: (1) the conditions of the physical parts; (2) their proper activities; (3) the causes of the conditions; and (4) the symptoms that follow the alterations in the bodies.[99] Of these four, only the first and the second can be called disease. Yet Galen himself acknowledges that a clear line of demarcation among them is difficult, if not impossible, as shown in the following statements: "A symptom is anything that should befall the animal which is contrary to nature. Consequently, a disease will be referred to under the designation of the class 'symptom,' for it is in a way itself a symptom" (*On the Differentiae of Symptoms* 7.53K). As Galen earlier recognized latitude of health as balance judged by function, he also admits relative degrees of disease (and symptoms) as imbalance and thus damage to function. According to Ian Johnston, there are three divisions of damage of function: (1) privation (complete loss) of function; (2) deficient (reduced, diminished) function; and (3) defective (abnormal) function.[100] For example, if we consider sight and hearing among the five sensory modalities (sight, hearing, smell, taste, and touch), blindness (*typhlotēs*) would be an example of privation of function, dim-sightedness (*amblyopia*) a deficient function, and false vision (*parorasis*) a defective function; likewise, deafness (*kephotēs*) would be an example of privation of function, hardness of hearing (*baryekoia*) a deficient function, and false hearing (*parakousis*) a defective function (7.56K). To the last category of defective function also belong tremor, spasm/convulsion, palpitation, shivering, rigor, and agitation, as he gives special attention to them in *On Tremor, Palpitation, Convulsion, and Rigor* (7.584–642K).[101]

Another way Galen conceptualizes disease is through the notion of the affected place (*locus affectus, peponthōs topos*), which is "a way of thinking about disease [which] focuses on the internal spaces of the body, and attempts to identify which of these places is affected by each disease."[102]

98. Hankinson, "Philosophy of Nature," 231.
99. Hankinson, "Philosophy of Nature," 231.
100. *Galen: On Diseases and Symptoms*, 73.
101. *Galen: On Diseases and Symptoms*, 73–74.
102. Glenda Camille McDonald, "The 'Locus Affectus' in Ancient Medical Theories of Disease," in *Medicine and Space: Body, Surroundings, and Borders in Antiquity and the Middle Ages*, ed. Patricia A. Baker, Han Nijdam, and Karine van 't Land (Leiden: Brill, 2012), 63. This paragraph draws on the insights from her chapter.

According to his *On the Affected Parts,* "an activity (*energeian*) is never damaged without an affection of the active organ. An organ itself is at any rate affected in some form when there is pain (*odynē*) or an unnatural (*para physin*) swelling, and still more, if its activity (*energeia*) is impaired (*blaptētai*)" (1.2 [8.29K]). In other words, when a part of the body shows unnatural (*para physin*) symptoms like pain or swelling, or when the function (activity) of that part is impaired, that part of the body is affected, that is, it undergoes, suffers, or experiences a disease. This notion of affection (*pathos; pathēma*) presupposes Galen's insistence that a good physician must use anatomical study to acquire a detailed knowledge of the structure and function of different parts (1.1, 6 [8.16K]); "for," says Galen, "I always searched in which [primarily] affected part or in which part involved by sympathy the activity was damaged (*tēs energeias egeneto blabē*). When I was convinced that I had found the [affected] part, then I further looked for its condition (*diathesin*)" (3.4 [8.146K]). As Glenda McDonald notes, this statement reflects Galen's view "that diseases are primary or sympathetic (i.e., transferred) affections of particular parts of the body" that perform specific activities or functions (*energeia*).[103] Then a physician can identify diseases by carefully examining a patient's symptoms, that is, by looking for changes in the organs (i.e., affections) in the form of color, size, form, irregularities in excretions, or damages to the function of the affected part (see 1.3, 5 [8.44–5K]).[104] When the disease is located in the organ itself, primary affection takes place; and when an organ suffers through transference of affection from a different primarily affected organ, then sympathetic affection occurs (3.7 [8.166K]). For Galen, it is important to locate the exact affected organ because he believes that remedial treatment should be applied directly to the affected part. For example, "those who[se] spinal nerves become affected require application of the medicine to the vertebral column" (1.6 [8.56–59K]).

In sharp contrast, however, Galen's main rivals, the Methodists,[105] opposed the theory of *locus affectus.* Themison, a "founder" of Methodism according to Celsus and Pliny the Elder, saw no need for a complex system or theories of disease featuring characteristics to search for so as to determine a disease's hidden causes.[106] Methodists, challenging the more traditional, Hip-

---

103. McDonald, "The 'Locus Affectus,'" 65.

104. Cf. McDonald, "The 'Locus Affectus,'" 66–67.

105. Methodists are also called Methodics in some scholarship.

106. Ido Israelowich, *Society, Medicine, and Religion in the* Sacred Tales *of Aelius Aristides* (Leiden: Brill, 2012), 60.

pocratic humoral theories of disease, argued for a new epistemology of medicine in which all diseases shared some general and visible characteristics, described as "common features" (*koinotēs*); once physicians properly identified these "commonalities" based on the patients' condition, the decision on how to treat the disease easily followed. There was no need for repeated observation, logical demonstration, or anatomical investigation into the origin of these commonalities.[107] So, what are the commonalities of all diseases? Methodist Caelius Aurelianus of the fifth century CE, in his *On Acute and Chronic Diseases*, explains that as bodies are made of atoms and pores, all diseases result from the following three common features or states: stricture (*strictura*; where the atoms are pressed too tightly together), looseness (*solutio*; where the atoms are too dispersed), or a mixed state (*complexio*) in combination of stricture and looseness (e.g., 1.7.52; 2.5.24; 2.12.90).[108] Once Methodist physicians differentiate between these features or states based on the bodily excretions, they choose their treatments according to the principles of opposites: remedies with relaxing properties used for cases of stricture; remedies with stringent properties used for cases of looseness; and a combination of those for cases caused by the mixed features (e.g., 1.7.52; 2.5.24; 2.12.90).[109] They also divide the treatment of all diseases in three common and universal stages: a beginning and initial increase; a middle period when the level of the disease is constantly leading up to the highest stage; and a final stage of decline.[110] Thus the focus of the Methodists is not a single or a few affected parts either in diagnosis or treatment; the common features are active in the whole body at once, and thus diseases affect the whole body, for example, through fever. Then, treatments must be applied to the whole body, too. In the case of phrenitis, Caelius explains thus: "Now we hold that in *phrenitis* there is a general affection of the whole body [not just brain and eye as by Galen], for the whole body is shaken by fever. And fever is one of the signs that make up the general indication of *phrenitis*, and for that reason we treat the whole body" (1.8.55). In the end, while Methodists emphasized a commonality of symptoms and thus an identical form of treatment, Galen preceded from an observation of the individual constitution and condition of the patient that required restoration.[111]

107. Nutton, *Ancient Medicine*, 195.

108. Caelius Aurelianus, *On Acute Diseases and on Chronic Diseases*, ed. and trans. I. E. Drabkin (Chicago: University of Chicago Press, 1950), xix.

109. McDonald, "The 'Locus Affectus,'" 71–72.

110. Drabkin, introduction to *On Acute Diseases and on Chronic Diseases*, by Caelius Aurelianus, xix.

111. Nutton, *Ancient Medicine*, 197.

Women's Diseases

Just as women's health was distinctive in contrast to men's health in Greco-Roman medicine, so were women's diseases, largely pertaining to their reproductive organ(s) and functions in addition to general diseases. Within the Hippocratic Corpus, *Diseases of Women I & II*, *Sterility*, *Nature of Women*, *Barrenness*, *Places in Man 47*, and a part of *Aphorism V* are devoted to women's diseases. As one can imagine, the primary cause of diseases in women is the condition of the uterus. *Places in Man* states: "the uterus (*hai hysterai*) is the cause of all these diseases [i.e., diseases of women]; for however it changes from its normal position—whether it moves forward, or whether it withdraws—it produces diseases" (47). As Jouanna notes, although the modern term "hysteria" denotes a nonspecified neurosis in women, its original meaning was supposed to be a particularly female complaint associated with the uterus.[112] Given the crucial importance of regular monthly menstruation, an irregular menstruation or issue of blood (*rhoos*; *flux*) was a source of disease. For example, "When menstruation is too copious, diseases ensue; when it is suppressed, disease of the womb occurs" (*Aphorism V* 57). Also, "if a flux (*rhoos*) starts from a woman's uterus, her blood flows copiously, congealed clots are expelled, and she has pain in the loins, flanks, and lower belly. . . . Chills and acute fever set in, and weakness develops. . . . This condition arises most often after abortions, but it may also follow when the menses fail to pass for a long time and then suddenly break out" (*Diseases of Women II* 3, 110). *Diseases of Women I* covers various menstrual disorders, including suffocation caused by the moving uterus ("uterine suffocation"), which causes the sufferer epilepsy-like symptoms such as unconsciousness, grinding of teeth, and difficulty in breathing, and bilious and phlegmatic menses (1–9; *Diseases of Women II* 201).

Women were also liable to suffer from the diseases that threatened to undermine their indispensable role in the generation and survival of the human race—their fertility. Barrenness or sterility in a society in which giving birth was essential to one's identity as a full, mature woman (*gynē*) could have been the most important reason to seek a doctor's help. Many particular causes of infertility could be envisioned in relation to the condition of the uterus itself, which might be too dense, too cold, too watery, too hot (*Aphorism V* 62; *Diseases of Women I* 12, 17, 18, 20); they may also be due to the deviation, obstruction, turning, or closure of the cervix, the

112. Jouanna, *Hippocrates*, 184.

ulceration or the presence of a cancer in the cervix or the uterus, the inflammation or suppuration of the uterus, etc. (e.g., *Diseases of Women I* 13,
16, 17, 20, 24, 63–66).[113] Female sterility is explained by averring that the
womb is unable to attract, receive, or retain the male sperm; that it prevents normal menstrual discharge; or that it does not provide a suitable
environment for the coagulation of the seed and the development of the
embryo (e.g., *Diseases of Women I* 18).[114]

Galen mentions having observed many women who called themselves
hysterical (*histerikas*) whose symptoms varied from muscle contractions
to nonresponsive lethargy, to nearly complete asphyxia (*Loc. affect.* 6.5
[8.414K]); widows who used to menstruate, have intercourse, and bear
children well, and who are deprived of all those things, are particularly
susceptible to that (*Loc. affect.* 6.5 [8.417K]). In contrast to the Hippocratic authors who imagined a very actively mobile uterus ("a wandering
womb"), Galen, like his older contemporary Soranus of Ephesus, knew
of Herophilus's discovery of the broad ligaments anchoring the womb to
the pelvis and rejected the notion that the uterus moved around the body
like an animal wreaking havoc (*Loc. affect.* 6.5 [8.430K]).[115] However, he
thought the uterus could become withdrawn in some direction or inflamed
and thus recommended the Hippocratic and folk practice of fumigating the
vagina with a sweet-smelling aroma to attract the uterus to its proper place
(*Meth. med. Glauc.* 1.15 [54K]). Galen attributed hysterical conditions to the
noxious buildup of female seed or menstrual fluid and, like the Hippocratic
authors, explained a wide variety of women's diseases by attributing them
to the disruption of the menstrual cycle.[116] Rebecca Flemming notes Galen's arresting commentary on Hippocratic *Epidemics I* as the epitome of
his perspective on women's distinctiveness. *Epidemics I* reads: "In women
these [a whole range of symptoms] too and pain from the womb." Galen
comments: "Everything previously mentioned also occurs in women, as
in males, in so far as they are human beings and have all the same organs
as males; but in so far as there exists in them a special organ, the womb,
thus they fall prey also to the diseases of it, especially those in which it is
affected together with the gullet" (*Hipp. Epid. I* 2.60). Males are the abso-

113. Jouanna, *Hippocrates*, 174–75.

114. Jouanna, *Hippocrates*, 175.

115. Mattern, *The Prince of Medicine*, 233; for Galen on women's pathology, see also
Flemming, *Medicine and the Making of Roman Women*, 331–47.

116. Mattern, *The Prince of Medicine*, 233.

lute, standard human beings, and females are the relative, qualified sort; insofar as they are human, women are the same as men; however, there is also something more—therefore less—to them, the uterus, which does not exist for men. This signifies an intrinsic gap and subsequent hierarchy between men and women that manifest in all other areas of life.[117] Thus, he also stressed women's susceptibility to certain diseases not tied directly to menstruation but to their naturally colder and wetter (and thus inferior) constitution when compared to men.[118] Thus women are more susceptible to rigors on account of their greater coldness and to seasonal *dysenteria* (intestinal flux) due to their greater wetness (cf. *Aphorism V* 69). Just as women's health was dictated, articulated, and regulated by men in the Greco-Roman world, so also were women's diseases diagnosed, treated, and controlled by men (male doctors and relatives) but silently and shamefully suffered by women.

Disease of the Soul

Thus far, we have focused on Greco-Roman concepts of diseases of the body. However, Greco-Roman writers understood the diseases of the soul to have originated from the diseases of the body, and thus saw close connections between the two. Therefore, we will briefly examine how Hippocratic and Galenic medicine conceptualized the soul and the diseases of the soul. The Greek word *psychē* does not appear frequently in the Hippocratic treatises, and its usage varies from one treatise to another.[119] According to *Nature of Man*, *psychē* is the life principle that leaves the body at death and is equated with blood (6). In *Regimen*, *psychē* refers not only to the male and female seed as the vehicle of life but also to the mental component of the human being in that it is the agent of intelligence (*phronēsis*), memory, and character, depending on its particular mixture of fire and water, and is associated with sense perception and dreams (*I* 28–29; *I* 6;

---

117. Flemming, *Medicine and the Making of Roman Women*, 343.

118. This was certainly shared by the Hippocratics and Soranus. See Helen King, "Producing Woman: Hippocratic Gynaecology," in *Women in Ancient Societies: An Illusion of the Night*, ed. Léonie Archer, Susan Fischler, and Maria Wyke (New York: Macmillan, 1994), 102–14; King, "Medical Texts," 204.

119. This part on soul and body in the Hippocratic Corpus largely draws from Beate Gundert, "Soma and Psyche in Hippocratic Medicine," in *Psyche and Soma: Physicians and Metaphysicians on the Mind-Body Problem from Antiquity to Enlightenment*, ed. John P. Wright and Paul Potter (Oxford: Clarendon, 2000), 13–35.

IV 86–87). While its essence is heat, it needs moisture for nourishment and travels inside the body along the same passages as blood (*I* 10; *II* 56, 62). In *Breaths* the soul and blood are explicitly related—intelligence is located in the soul, whose condition depends on the particular state of the blood (14). In *Sacred Disease*, although the author does not necessarily link soul and brain, we see the brain as the source of mental activity and sense perception, judgment, and emotions (7, 14). Intelligence (*phronēsis*) and coherent movement depend on an unimpeded flow of air to the brain and its unblocked distribution from there to the cavity, the lung, vessels, and eventually the rest of the body (7). Then *psychē* "corresponds to the expressions for mind (*nous, dianoia, gnōmē*)" when used in contrast to the body (e.g., *Sacred Disease* 16).[120] *Psychē* thus signifies not only life principle but also mental principle in terms of function or agent, and as the latter it corresponds to the brain or blood as the site of mental events.[121]

As one can see, then, the Hippocratic treatises give somatic explanations for mental and psychic phenomena, and the health of those phenomena depends on the health and integrity of bodily elements, parts, and their functions (e.g., blood and brain). If the latter malfunctions or something goes wrong in some way, it causes not only disease of the body but also disease of the soul (in mental and psychic functions). For instance, the author *of Sacred Disease* painstakingly provides a "natural" explanation for the various symptoms of epilepsy, particularly disturbance of intelligence and coherent movement, against those who attribute "sacred" character and divine origin to this disease. Rather, epilepsy is caused by the brain, which has not been sufficiently cleared before or after birth; it is caused by flows of phlegm descending from the head in two veins coming from the liver and the spleen; then the cold phlegm coagulates the blood and impedes the passage of air (6–9), and the disturbance in the brain then not only causes epileptic seizures but also leads to madness (17). In *Glands*, when the brain is harmed and irritated, the mind (*nous*) is deranged and the person suffers apoplexy with speechlessness and suffocation (12); derangements (*paraphrosynē*) and delirious states (*mania*) are other affections of the brain (15). According to *Aphorisms*, "if the tongue is suddenly paralysed, or a part of the body suffers a stroke, the affection is melancholic (*melancholikon*)" (*VII* 40). On the other hand, psychological factors influence somatic processes as well. In melancholic affection, apoplexy of the whole body, convulsions, or blind-

---

120. Gundert, "Soma and Psyche," 33.
121. Gundert, "Soma and Psyche," 33.

ness determines the melancholy humor in *Aphorisms* (*VI* 56). In *Humors*, the appropriate member of the body responds by its action to each of the following emotional states. Fear, shame, pain, pleasure, and passion, for example, might encourage the sweating of the skin and palpitations of the heart. Furthermore, sense perceptions have a positive effect on the bodily fluids. Sight, sound, thought, and speech (and reading and singing) all move the soul; and as the soul moves, it grows warm and dry and consumes the fluid of the body, making the person lean (*Regimen II* 62). Therefore, both body and soul are involved in various pathological processes and treatments in the Hippocratic Corpus and have a reciprocal relationship affecting each other as two distinct, yet related, aspects of human nature.[122]

The aforementioned statement similarly applies to Galen's works. However, as we transition to Galen's notion of the soul and its relation to the body and its diseases,[123] it is important to keep in mind, in Hellenistic philosophies (Epicureanism and Stoicism) and physicians (Herophilus and Erasistratus), "that all *psychē* is *sōma* but not all *sōma* is *psychē*; that only what is spatially extended, three-dimensional, and capable of acting or being acted upon exists; that the soul meets these criteria of existence; that this corporeal *psychē*, like the rest of the body, is mortal and transient; that the *psychē* is generated with the body; that neither exists before the body nor exists eternally after its separation from the body—that is, the soul does not exist independently of the body in which it exists."[124] This was the physicalist position, supported particularly by the discoveries of the sensory and motor functions of the nerves, and of the relation of sensory and motor nerves to the brain, and as such, it challenged Platonic and Aristotelian theories of the soul and its relation to the body (e.g., the soul's preexistence and immortality; the tripartite soul).

Galen, for all his attacks on Erasistratus and the Stoics (for their materialism), and his allegiance to Platonism, followed the physicalists in *On the Differentiae of Symptoms*, where he makes a distinction between natural activities and psychical activities: the former includes appetite, "digestion, distribution of nutrients, the generation of blood, the pulse, and the excre-

---

122. Gundert, "Soma and Psyche," 30–32.

123. On this, see Luis Garcia Ballester, "Soul and Body, Disease of the Soul and Disease of the Body in Galen's Medical Thought," in *Le Opere Psicologiche di Galeno: Atti del Terzo Colloquio Galenico Internazionale Pavia, 10–12 Settembre 1986*, ed. Paola Manuli and Mario Vegetti (Naples: Bibliopolis, 1988), 117–52.

124. H. von Staden, "Body, Soul, and Nerves: Epicurus, Herophilus, Erasistratus, the Stoics, and Galen," in Wright and Potter, *Psyche and Soma*, 79.

tion of residues," as these are the capacities that belong to nature, but the latter is confined to cognitive (such as imagination, reason, and memory) and voluntary sensory (*aisthēsis*) and motor (*kinēsis*) activity (five senses such as sight, smell, taste, hearing, and touch), as in the theories of Herophilus, Erasistratus, and some Stoics (7.55–56K).[125] Thus the capacities of the soul are not innate in their organs or instruments (as those of nature are), but "flow to the soul's instruments from its source [*archē*, i.e., brain], similar to the light of the sun [radiating out]" (*Loc. affect.* 8.66.8K). Galen further explains the cognitive activities as the authoritative, hegemonic (*hēgemonikon*) activities that belong to the ruling part of the soul, that is, the brain[126] (cf. *Symp. diff.* 7.55K). In this brain-centered model, *psychē* carries, in addition to sense perception and motor functions, a wide range of mental functions that are distinctively cognitive.

In *On the Doctrines of Hippocrates and Plato*, however, Galen provides a rather different concept of the soul. Galen adopts the tripartite soul (rational, spirited, desiderative) from Plato's *Republic*, and from *Timaeus* the location of the three sources or centers (*archai*) of the soul in the head, the heart, and the liver, respectively.[127] Here Galen sees the physical parts, brain, heart, and liver, as sources of distinct bodily networks, that is, those of nerves, arteries, and veins, respectively. The works or functions (*erga*) of the source (*archē*) in the head are, by themselves, the higher cognitive activities such as "imagination, memory and recollection, knowledge and thought and reasoning, and in its relation to the other parts of the animal to guide the sensation of the sensory parts and the motion of the parts that move voluntarily" (7.3.2). The work of the *archē* in the heart is, by itself, the proper tension (*tonos*) of the soul, to help maintain the right mixture or temperament (*eukrasia*), to be constant "in the things that reason commands, and in states of passion (*pathos*) to provide the boiling, as it were, of the innate heat"; in its relation to other things, it is to be a source of warmth for the individual parts and of pulsing motion for the arteries (7.3.3). Here it seems that Galen implicitly identifies *archē* (source/center) as a *dynamis* (capacity/faculty), and thus "the remaining faculty (*dynamis*), seated in the liver, has as its work all the things that have to do with nutrition in the animal, the most important of which in us and in all

125. Staden, "Body, Soul, and Nerves," 107.

126. On Galen and the hegemony of the brain, see Jessica Wright, "Brain and Soul in Late Antiquity" (PhD diss., Princeton University, 2016), 86–128.

127. See n. 47 above.

sanguineous animals is the generation of blood" (7.3.3). It is therefore to "supply the substances whose mixture forms the body's blend or temperament (*krasis*)";[128] also, it is to the third capacity that the enjoyment of pleasure and its excesses (intemperance and licentiousness) belongs. Later in the text, while Galen is repeatedly agnostic about the substance (*ousia*) and origin of the soul, he consistently stresses that the *psychē* uses *pneuma* as its "first instrument (*organon*) in its relations with the physical organism and its functions" and various bodily parts as its secondary instruments (*PHP* 7.3.14–21, 27, 30). *Pneuma* is an ethereal, vaporous "stuff" that is the vehicle of perception, motion, and life itself in two separate forms in the human body: vital *pneuma* produced in the heart's left ventricle, and psychic *pneuma* in the brain, mainly through the refinement of the former. As the soul's "first instrument," the brain's psychic *pneuma* travels throughout the body through the nerves and mediates sensory and motor signals between the brain and the sense organs; it therefore allows the body's mental faculty to process movement and sensation (as external sights and sounds are received from the eyes and ears) and thus enables cognitive functions within the mind.[129]

In *That the Soul Follows the Mixture of the Body (QAM)*, Galen declares that the capacities (*dynameis*) of the soul depend on the mixtures or temperaments (*kraseis*) of the primary physical constituents of the bodily organs (5.1K). There he identifies the soul with the form (*eidos*) of each of the three main organs and explains form in terms of the mixture of bodily elements.[130] As Teun Tieleman notes, Galen therefore brings together the Platonic "tripartition-cum-trilocation with the Aristotelian definition of the souls as the form of the body."[131] Since the proper mixture (*eukrasia*) is different in different parts of the body, the brain, the heart, and the liver each has its own proper mixture (*eukrasia*); hence, the brain, the heart, and the liver will each have an effect on the capacities (*dynameis*) that reside in these organs; and the same capacities make use in turn of the organs in order to carry out the manifold functions of the living being, as we have already encountered Galen speaking of the body as an instrument (*organon*) that the soul makes use of in *On the Usefulness of the Parts* (1.2). For Galen,

128. Staden, "Body, Soul, and Nerves," 109.
129. On the role of *pneuma* in Galenic medicine, see J. Rocca, *Galen on the Brain* (Leiden: Brill, 2003), chap. 6; Wright, "Brain and Soul in Late Antiquity," 102–15.
130. Teun Tieleman, "Galen's Psychology," in Barnes, *Galien et la Philosophie*, 150.
131. Tieleman, "Galen's Psychology," 150.

this interactive relation between bodily mixture and the soul's capacities is foundational; bodily factors influence our mental and emotional functioning, including our moral character, for good or ill.

Here we have two apparently different (if not incompatible) models of the soul in Galen. One is the physicalist, more unified brain-centered model with the soul-nature contrast, and the other is the Platonic-Aristotelian "tripartition-cum-trilocation" model of the soul with the more interactive soul-body relation. Galen may have harmonized this tension by opining that it makes no difference whether one calls the capacity/faculty (*dynamis*) from the liver "desiderative" (*epithymētikē*; a Platonic term), "natural" (*physikē*), or "nourishing" (*threptikē*; an Aristotelian term), "just as it makes no difference if [one] calls it 'soul' (*psychē*) or 'capacity' (*dynamis*)" (*Meth. med. Glaucon* 9.10 [635K]). Also, in *On My Own Opinions* he says his vocabulary is adapted to his audience: "Among Platonists I refer to that which governs plants as 'soul' (*psychē*; e.g., Plato and Aristotle) but among physicians as 'nature' (*physis*; e.g., Hierophillus)."[132] In fact, the work of the brain-based rational soul rather corresponds to the combination of cognitive activities and perceptive and voluntary motor functions of the brain-centered unified psychological model elsewhere. And the part of the work of the heart-based spirited soul involving passion is reacting in a "physiological way appropriate to its character by 'pulsing' activity in a response to a rational command" sent from the brain.[133] For the third part, it could be interpreted that the liver acts in a natural way but also responds to commands from the brain via the nervous system; in this way, the functions of the two nonrational parts could be explained simply as "natural" or as physiological reactions activated by the rational messages of a brain-based control center.[134] However, that is not how Galen reconciles them after all, and it remains a challenge to reconcile the last part concerning the liver, as seen in the words of Staden: "Galen's Platonic view that the lower functions too—not only heat and pulse, but also the vegetative, nutritive, reproductive, desiderative functions (especially desires for food, drink, and sexual pleasure)—are activities of the *soul*, and that human beings share this desiderative kind of soul with plants, is hardly compatible with the Galenic views traced above, and especially not with this distinction between *physis*

---

132. Quoted in Staden, "Body, Soul, and Nerves," 110.

133. Christopher Gill, *Naturalistic Psychology in Galen and Stoicism* (Oxford: Oxford University Press, 2010), 116.

134. Gill, *Naturalistic Psychology*, 116.

and *psychē*—that is between human capacities, activities, and instruments that belong to 'nature' and 'soul' respectively."[135]

As for mental and psychical diseases,[136] like the Hippocratic Corpus, Galen also provides somatic explanations. For Galen, many kinds of diseases that affect reason, memory, or other functions of the rational soul arise from a disorder in the brain, the ruling part of the soul (*hēgemonikon*).[137] In *On the Causes of Symptoms II*, he lists three primary classes of symptoms of those diseases. First is destruction of function; second is damage; the last is a turning aside to a difference of form. First, destruction occurs in dementias (*mōrosis*) and amnesias (*lēthē*), as a person completely forgets both letters and skills (7.200K). Such diseases occur as a consequence of the cooling of the actual body of the brain, just as the apoplexies and epilepsies occur "through an abundance of phlegmatous humour gathered together in the cavities of the brain itself" (7.201K). Second, "the moderate damages, like the 'numbness' (*narkē*) of reason and of memory, occur in response to a more slight cooling, either through one of the cold medicines being taken into the body, . . . being applied to the head, or when a cold humour has been gathered in the brain" (7.201–202K). Finally, "all the deliria (*paraphrosynē*), which are defective movements of the authoritative capacity," occur on account of abnormal humors or through a *dyskrasia* of those humors in the brain (7.202K)—*phrenitides* when accompanied by fevers, and *manias* when without them; only melancholic derangements

135. Staden, "Body, Soul, and Nerves," 110. Other scholars have also criticized this discrepancy and difficulty of Galen's seeming incompatible notions of soul and its relation to body. See, for example, Tieleman, "Galen's Psychology," 141–44. However, Christopher Gill has attempted to ease the tensions in Gill, *Naturalistic Psychology*, 113–17.

136. Many recent works are available on this topic of mental and psychical diseases in Galen. See, for example, Peter N. Singer, "Galen's Pathological Soul: Diagnosis and Therapy in Ethical and Medical Texts and Contexts," in *Mental Illness in Ancient Medicine: From Celsus to Paul of Aegina*, ed. Chiara Thumigur and Peter Singer, Studies on Ancient Medicine 50 (Leiden: Brill, 2018), 381–420; L. Michael White, "The Pathology and Cure of Grief (λύπη): Galen's *De indolentia* in Context," in *Galen's* De indolentia: *Essays on a Newly Discovered Letter*, ed. Clare K. Rothschild and Trevor W. Thompson (Tübingen: Mohr Siebeck, 2014), 221–50; Vivian Nutton, "Galenic Madness," in *Mental Disorders in the Classical World*, ed. William V. Harris, Columbia Studies in the Classical Tradition 38 (Leiden: Brill, 2013), 119–27; and Brooke A. Holmes, "Disturbing Connections: Sympathetic Affections, Mental Disorder, and the Elusive Soul in Galen," in Harris, *Mental Disorders in the Classical World*, 147–76.

137. On this see Mattern, *The Prince of Medicine*, 254–55.

have a colder humor as a cause. Phrenitis does not simply arise due to hot humors but occurs after inflammation involving the brain and cerebral membranes (7.202K).[138] In other cases of fevers, deliriums (*paraphrosynē*) arise when such a humor is increased in the brain (7.202K). What is in common in all of these—particularly in the melancholic derangements—is fear and despair as stated by Hippocrates (*Aphorisms* 6.23). All such sufferers are despairing with no reason, and many fear death or something not worthy of fear; others even become suicidal (*On the Causes of Symptoms II* 7.203K). Such fears arise through the black bile taking possession of the *archē* of the rational soul; so, when either the black bile itself takes hold of the brain or when some melancholic vapor arises, some kind of darkness is stirred up by the melancholic humor within, just as in the disease called flatulence or hypochondriasis (7.203–204K). In *Method of Medicine*, Galen further explains the psychic affections—sudden and strong fears, which people call terrors, and the extreme pleasures that are the opposite to those fears, which people call excessive joys—as creating dissipated and weakened *pneumas* (12.5 [841K]). "Grief, anguish, outbursts of strong anger and anxiety, and in like manner frequent episodes of insomnia also cause harm by dissipating the capacity" (12.5 [842K]). The *psychē* moves itself by itself and the body in all these, but the patients obviously fail in strength and suffer dissipation of the capacity and even die due to the destruction of *pneuma* (12.5 [842K]). Moreover, alteration of the tension (*tonos*) of the *pneuma* through physical pressure can also result in headache, loss of sensation and motion, and even loss of consciousness (*Loc. affect.* 4.3).[139] Thus, the soul and body again have a reciprocal relationship in Galen, as they affect each other in determining a person's pathologies.

Galen in fact pays special attention to the affections and errors of the soul and their proper treatments in a treatise called *The Affections and Errors of the Soul*. There he first distinguishes between affections (*pathē*) and errors (*harmartēmata*) of the soul: the former arise from an "irrational power in us that is not amenable to reason," and the latter arise from false opinion (5.2); he then devotes the rest to the cure (*therapeia*) of the emotions and of the intellectual/cognitive errors in the context of both Stoic (Chrysippus) and Platonic-Aristotelian traditions. The genre here is Stoic-Epicurean therapy

138. Aretaeus of Cappadocia also explains *phrenitis* as an organic disorder. "It is the delirium caused by any acute infectious disease affecting 'the head and the senses.'" See Sotiris Kotsopoulos, "Aretaeus the Cappadocian on Mental Illness," *Comprehensive Psychiatry* 27, no. 2 (1986): 177.

139. Jessica Wright, "John Chrysostom and the Rhetoric of Cerebral Vulnerability," *Studia Patristica* 81 (2017): 118.

of the soul (emotions),[140] following the Stoic-Epicurean idea that emotions are sicknesses that need to be cured or removed. Hence, the goal is Stoic passionlessness (*apatheia*): we should "identify all affections (*pathē*) as being in need of rectification," "free (*eleutherōsai*) [ourselves] from the affections, and remove the affections (*pathē*) of the soul (*psychē*)," that is, those universally acknowledged affections such as spirit (*thymos*), anger, fear, envy, and excessive (*sphodra*) desire (5.5; 5.7–8). Particularly, Galen names anger as a kind of madness (*mania*) and thus a sickness (*nosēma*) of the soul along with other affections of the soul—grief, rage, desire, and fear—that overlap with the four basic Stoic emotions (*pathē*): grief/distress, desire, pleasure, and fear.[141] On the other hand, Galen also likens the goal to moderation (*metriopatheia*), which is the Platonic and Peripatetic ideal; and his distinction between the affections and errors and his corresponding therapies of the irrational and rational souls themselves reflect Platonic tripartite psychological structure and therapy (5.2; 5.27–29). Especially, his distinction between the spirited soul and the desiderative soul corresponds to the two Platonic methods of treatment, training (*paideusis*) for the former and discipline or correction (*kolasis*) for the violent latter (5.27–29). What processes does this therapy actually entail? Galen recommends the three-step therapeutic process. First, recognize the need for self-knowledge (including self-deception) and have a reliable guide or advisor for the correct diagnosis of one's soul. Second, the person appointed for this task should be a mature, older, good, and upright man to guide one's self-criticism. Finally, carry out one's self-criticism in a steady and sustained way, including training (*paideusis*) and correction (*kolasis*) (5.3–7, 30–32). In this process, we see a clear relationship between the psychological model and ethical development, especially the latter depending on the innate individual nature and the correct habituation over time, making virtuous action pleasurable.[142] We have earlier encountered the importance of habit (in diet) in Plutarch's *Precepts of Health Care*. Here, while the exercise of self-control (*enkrateia*)—which involves the cultivation of fine habits—is the means, "the maximization of temperance" (*sōphrosynē*) is the end (5.32–34).

Galen then takes a specific affection—the insatiable appetite for food—as an example. Plutarch earlier advised Aristotelian moderation in food intake and its pleasure for his elite audience through physical and mental

---

140. On this, see chap. 4 and its last section, "Therapy for the Soul (Medico-Philosophical Treatises)."

141. Cf. Gill, *Naturalistic Psychology*, 255.

142. Cf. Gill, *Naturalistic Psychology*, 255.

exercises so that they could engage in active social and public life. Galen's approach as a doctor, despite his own claim to be a philosopher, is both to highlight the negative physiological effects of insatiability on the body (e.g., causing bad humor in the veins) and its negative psychic effects (e.g., the insatiable desire for material objects), and to advise self-sufficiency against "the life of indulgence" through consistent training (5.46–50). The goal here is not moderation but the removal of the affection of insatiability through the same three-step process (5.55–56). As far as the soul's rational errors are concerned, another three-step treatment is necessary: first, "investigate the question whether matters which are not evident admit of proof by argument"; then inquire into the method of proof with men of the "highest credentials in terms of veracity, natural intelligence, and education in logical theory" for a considerable time; finally, take the ultimate inquiry, which, by virtue of the good (or the goal of life), will make one happy or blessed (5.61). It is a pursuit of the method of logical proof for the ultimate happiness, and one should look to physicians like Galen who could withstand the scrutiny of all philosophers and practitioners of the liberal arts rather than "sophistic" and boastful philosophers to see to the education or reeducation of men with the goal of leading them toward virtue (5.102, 103). One should look to physicians "precisely insofar as they are in a position to ameliorate the moral and intellectual qualities of souls which, in view of their relation of dependence upon the temperaments of the bodily organs, will be responsive to the changes in the dietary regime, environment and tenor of life which medical science will ultimately impose."[143] If Plutarch reframed health and health care philosophically (away from the doctor) and entrusted them to the philosopher in the *Precepts of Health Care*, Galen here and elsewhere[144] proposes to entrust to the doctor the treatment of both intellectual errors and psychic (emotional) and moral sickness rather than to the philosopher.

Disease and Environmental Factors in Roman Public Health

Earlier we considered the environmental causes and factors in disease in Hippocratic pathology. Now we examine Roman public health (or lack thereof), hygiene, and disease in the first two centuries of the imperial pe-

---

143. Pierluigi Donini, "Psychology," in Hankinson, *The Cambridge Companion to Galen*, 196; QAM 5.804–806, 807–808K.

144. QAM 4.773–776, 804–806, 807–808K.

riod, especially around the time of Galen in the late second century. Study in this area is severely limited by scholars' reliance on a relatively small amount of (textual and material) data. However, despite the challenge, the following snapshots of Roman society are still helpful, if only for their approximate value. In terms of Roman physical well-being (health), a good place to start is life expectancy and longevity. Based on the census returns from Roman Egypt in the first three centuries CE, Walter Scheidel notes a mean life expectancy at birth of twenty-two years for women and twenty-five-plus years for men—with a big caveat that these data are not representative of the Roman world, given that they are heavily dominated by documents from the Fayyum Oasis (south of Cairo);[145] the Fayyum was unhealthy even by the low standards of Egypt, and Egypt was at the bottom of the scale of longevity and physical well-being in the empire.[146] Egypt was "exceptionally densely populated, hot, annually inundated, and until quite recently a hotbed of both endemic and epidemic disease and consequently subject to very high death rates."[147] While Roman evidence from outside Egypt also points to low life expectancy, the mean life expectancy of Roman emperors who had natural deaths is from twenty-six to thirty-seven years, and of Roman elite (*honestiores*) in Italy, it is from twenty-five to thirty years.[148] As one can see, in general, longevity was not much better for Rome's elite than for the general population, and the health of elites did not significantly improve until the eighteenth century in England and China.[149]

In Italy, the main reason for this low longevity was heightened vulnerability to fatal seasonal infections such as malaria.[150] The city of Rome expe-

---

145. Another caveat may be the impact of infant mortality on life expectancy; the considerable rate of infant/child death brought down the average life expectancy. However, if one survived infancy/childhood, one would likely live into one's fifties or even later.

146. Walter Scheidel, "Physical Well-Being," in *The Cambridge Companion to the Roman Economy*, ed. Walter Scheidel (Cambridge: Cambridge University Press, 2012), 322; Walter Scheidel, "Roman Wellbeing and the Economic Consequences of the 'Antonine Plague,'" Version 3.0, Princeton/Stanford Working Papers in Classics, January 2010, 3.

147. Scheidel, "Physical Well-Being," 322.

148. Scheidel, "Physical Well-Being," 322; Scheidel, "Roman Wellbeing," 4. Cf. Ann G. Carmichael, "Public Health and Sanitation in the West before 1700," in *The Cambridge World History of Human Disease*, ed. Kenneth F. Kiple et al. (New York: Cambridge University Press, 1993), 193.

149. Scheidel, "Roman Wellbeing," 4–5.

150. Scheidel, "Roman Wellbeing," 7.

rienced a disproportionately high mortality rate among men and women under fifty in August, September, and October, most likely because of various types of malaria, from relatively benign tertian fever and quartan fever to the more malignant tertian fever.[151] According to Caelius Aurelianus, the physician Asclepiades in the first century BCE testified to the prevalence of quotidian fever ("typical of primary falciparian infection") in Rome (2.10.63–64; 4). Two centuries later, Galen also observed the commonplace of semitertian fever in Rome: "just as other disease thrives in other places, this one abounds in that city [Rome]."[152] Furthermore, combinations of falciparian malaria[153] and other common seasonal infections such as gastrointestinal disorders and respiratory diseases intensified and contributed to persistence of high seasonal mortality rates.[154] While Pliny the Elder in the first century CE spoke of "novel diseases" introduced to Rome and Italy, including lichen and elephantiasis (lepromatous leprosy) (*Nat.* 26.1–9), based on numerous case studies, Galen called Rome "this populous city, where daily ten thousand people can be discovered suffering from jaundice, and then thousand from dropsy."[155] In general, Scheidel would expect higher mortality in warm, high-density, low-altitude areas—thus Britain to have been healthier than Italy, and Italy healthier than Egypt. He would also "expect the majority of the population to have been concentrated in the more disease-prone parts of the empire—in coastal lowlands, along rivers, in areas that favored Mediterranean farming—with unfavorable consequences for morbidity and mortality."[156] Hence, with lower population densities and consequently lighter disease loads, the inhabitants of northwestern Europe were on average taller (and healthier) than Mediterranean populations and had a diet richer in dairy (and meat).[157]

Another main reason for low longevity was a chronic lack of housing and proper sanitation. According to Alex Scobie's classic study of Roman

151. Walter Scheidel, "Disease and Death in the Ancient City of Rome," Version 2.0, Princeton/Stanford Working Papers in Classics, April 2009, 4.

152. Quoted in Scheidel, "Disease and Death," 4.

153. On this see Jerry Stannard, "Diseases of Western Antiquity," in Kiple, *The Cambridge World History of Human Disease*, 267–69.

154. Scheidel, "Disease and Death," 5.

155. 11.328 Kühn, cited in Scheidel, "Disease and Death," 7.

156. Scheidel, "Roman Wellbeing," 13.

157. Scheidel, "Physical Well-Being," 326.

urban housing and hygiene, the enormous gulf between the elite and non-elite was felt most of all in housing.[158] While the imperial family and the rest of *honestiores* monopolized Rome's most prestigious hill away from flood-prone low regions near the Tiber, lower-status Romans (*humiliores*) lived in crowded apartments (*insulae*) or single rooms or complex shop-dwellings (*tabernae*), and the destitute (*egeni*) barely managed to live in shanties, shacks in slums, or even tombs.[159] The shanties and shacks often had structural defects such as leaking roofs, inadequate toilet and washing facilities, and inadequate space because of overcrowding.[160] Extremely inadequate sanitation, that is, the improper disposal of human excreta and sewage, posed a major threat to public health, as the mere exposure of sewage or its improper disposal immediately set the stage for potential disease transmission.[161] According to archaeological evidence from Pompeii, Ostia, and Rome, very few dwellings were directly connected to street drains, and wastes were barely discharged into sewers beneath the streets.[162] On top of that, people typically relieved themselves in streets, doorways, and tombs (despite the existence of public or sometimes private latrines); animals were herded through the streets and butchered there for sale; and sick, dying, and dead slaves were also found there, which attracted stray dogs and vultures as scavengers of the dead.[163] With a very high risk of food and water contamination in these environmental conditions, diseases were rampant. The most common were cholera, dysentery, gastroenteritis, infectious hepatitis, leptospirosis (causing jaundice), and typhoid, attested by Celsus (*Med.* 4.18.1) and Galen;[164] and babies, young infants, and undernourished adults were particularly susceptible to these infections.[165] Moreover, while the supply of water through aqueducts in Rome (and elsewhere) was adequate, its quality and purity were hardly safeguarded when

---

158. Alex Scobie, "Slums, Sanitation, and Mortality in the Roman World," *Klio* 68 (1986): 399–433.

159. Scobie, "Slums, Sanitation, and Mortality," 401–3.

160. Scobie, "Slums, Sanitation, and Mortality," 404.

161. J. A. Salvato, *Environmental Sanitation* (New York, 1958), 186, quoted in Scobie, "Slums, Sanitation, and Mortality," 407.

162. Scobie, "Slums, Sanitation, and Mortality," 409, 411. See also Carmichael, "Public Health and Sanitation," 193.

163. Scobie, "Slums, Sanitation, and Mortality," 417, 419–20.

164. See his case studies in Mattern, *Galen and the Rhetoric of Healing*, appendix B.

165. Scobie, "Slums, Sanitation, and Mortality," 421.

it reached *insulae* or *tabernae*, and for that matter, public baths.[166] Despite the frequent recommendations of baths for various ailments and diseases (including fevers, tuberculosis, dysentery, cholera, or bowel troubles) by (Roman) medical writers and physicians like Celsus, Pliny the Elder, and Galen, or even because of their recommendations of the baths,[167] Roman public baths "might not have been as sanitary as is commonly assumed," and "the risks of becoming infected with a wide range of contagious and infectious diseases in such establishments would have been great."[168] While the elite were not exempt from the general conditions of Rome's poor sanitation, the poor were certainly hit by them much more severely and their lives were dangerous due to fire, food shortage, and housing collapse, as well as the constant risk of the rapid spread of diseases.

Only about 10–12 percent of the population of the Roman Empire of 60–75 million lived in cities (of five thousand or more), and about 14–18 percent in smaller towns (of one thousand residents), including army camps.[169] Thus, at least 70 to 75 percent of the total population lived in villages and rural towns and worked on farms. Galen was an urban elite and considered the rural countryside a foreign environment and peasants a strange species living in poverty and isolation.[170] They ate a kind of food indigestible to city people; during famine, they were forced to eat grasses and green twigs and suffered from skin disease, dysentery, and fever;[171] they also suffered ear pains due to cold, sickle wounds, and snakebite that was often self-treated or left untreated.[172]

Finally, there were dreadful epidemic outbreaks in Rome. Scheidel notes a relative paucity of records—only five occasions—between the mid-second century BCE and the late second century CE (then ten occasions between 284 and 750 CE). Galen witnessed one that he called "anthrax" in his young adulthood in Pergamum.[173] Of these, the most dramatic crisis was the so-called Antonine Plague that was brought from the eastern Mediterranean by the Roman army and spread throughout the empire from

---

166. Cf. Galen, *Simp. med.* 3.8 (11.554K).
167. E.g., Celsus, *Med.* 2.17.2, 7; 3.2.5; 3.6.14; 3.22.1, 6; 4.18.1; 4.23.3; 4.25.2.
168. Scobie, "Slums, Sanitation, and Mortality," 426.
169. Scheidel, "Roman Well-Being," 12.
170. Mattern, *The Prince of Medicine*, 23.
171. *On Good and Bad Humors* 1 (6.749K).
172. *Simp. med.* 6.4 (11.866K); 10.2.22 (12.299K).
173. Mattern, *The Prince of Medicine*, 21.

165 CE to late 180 and beyond.[174] As it reached Rome in 166, Galen left Rome for Pergamum, his homeland, but in 168/169 he was summoned to attend the emperor Marcus Aurelius and his son Lucius Verus in Aquileia in northern Italy, where plague harshly hit the army.[175] Egypt, "the most densely populated and one of the most heavily urbanized provinces," may have been among the most heavily affected regions of the empire.[176] Galen's descriptions of an ulcerated rash and corneal ulceration with black diarrhea would correspond to smallpox, and he considered the plague one of the diseases caused by an excessive accumulation of black bile.[177] In Rome, he lost most of his slaves and his good friend Theuthras (*Ind.* 35). This plague was also attested by Aelius Aristides, who in fact suffered and recovered from it by following a diet prescribed in his dream by Asclepius; according to Aristides, it infected nearly all his neighbors and household in Smyrna and attacked the livestock (*Or.* 48.38–44). If it was indeed a case of smallpox visiting a previously unexposed "virgin" population, about 25 percent of the empire's population (perhaps 15 million people) may have died in this "great plague."[178]

Illness as a Cultural Construct

Galen did not consider the "great plague" a form of divine punishment or anger, but many of his contemporaries undoubtedly did in their cultural and religious contexts. In this section, we investigate how Greeks and Romans constructed and interpreted the experiences of their own illness as patients and sufferers, which were part of everyday lived reality, and how they produced their meanings in their cultural, religious, and moral framework. As a reminder, illness according to Arthur Kleinman is the personal, psychosocial, and cultural responses of the patient to certain

---

174. For detailed studies on this plague and its manifold impact, see R. P. Duncan-Jones, "The Impact of the Antonine Plague," *JRA* 9 (1996): 108–36; Walter Scheidel, "A Model of Demographic and Economic Change in Roman Egypt after the Antonine Plague," *JRA* 15 (2002): 97–114; Walter Scheidel, "Roman Wellbeing and the Economic Consequences of the 'Antonine Plague'"; also R. J. Littman and M. L. Littman, "Galen and the Antonine Plague," *AJP* 94 (1973): 243–55.

175. *On My Own Books* 1.19.15K; 2.19.17–18K.

176. Scheidel, "A Model of Demographic and Economic Change," 98.

177. *Meth. med.* 5.12 (367K); 5.11–12 (360–363K); Mattern, *The Prince of Medicine*, 201.

178. Scheidel, "A Model of Demographic and Economic Change," 100–101.

physical or psychological symptoms, whereas disease is a malfunctioning of biological or psychological processes.[179] This distinction enables us to perceive "the symbolic meanings and religious significance attached to physical suffering"[180] through the construction of illness narratives. Illness narratives "tell us about the way cultural values and social relations shape how [people] perceive and monitor [their] bodies, label and categorize bodily symptoms, interpret complaints in the particular context of [their] life situation; [people] express [their] distress through bodily idioms that are both peculiar to distinctive cultural worlds and constrained by our shared human condition."[181]

With this idea of illness and the illness narrative in mind, we begin with an understanding of illness as a form of divine punishment or displeasure. Starting from the Homeric poems and continuing to Greek tragedy, and also to the inscriptions from the sanctuary of Asclepius at Epidaurus, there had been a link between human action, divine punishment, and illness.[182] In the case of pestilence, when plague fell upon the Achaean army laying siege to Troy in the *Iliad*, people automatically believed that the wrath of Apollo sent it; then the diviner Calchas revealed in a prophecy that its cause was the sin committed against Chryses, a priest of Apollo, by the leader of the Achaeans, Agamemnon, who refused to return his daughter; and that its solution was for Agamemnon to return Chryses's daughter to him without ransom and an expiatory sacrifice to be offered in honor of Apollo.[183] Then the great plague attacked Athens at the beginning of the Peloponnesian War in the mid-fifth century BCE when the city was under siege by Sparta. The inhabitants resorted to supplication in the sanctuaries and oracles but to no avail. The end of a plague (*loimos*) was to be brought about only by driving out the *miasma* ("pestilence" with the connotation of religious defilement) or by its purification according to a religious ritual.[184] During the Roman Republic, Livy recorded a plague at Rome in 174 BCE whose impact was so massive that religious measures were taken to combat it with the belief that it was impiety that had brought about plague.

179. Kleinman, *Patients and Healers*, 72; Kleinman, *The Illness Narratives*, 3–5.

180. Anne Elizabeth Merideth, "Illness and Healing in the Early Christian East" (PhD diss., Princeton University, 1999), 21.

181. Kleinman, *The Illness Narratives*, xiii.

182. Nutton, *Ancient Medicine*, 291.

183. Homer, *Iliad* 1.60–70; Jouanna, *Hippocrates*, 101–2.

184. Thucydides, *The History of the Peloponnesian War* 2.47.4–2.53.4.

According to Ancus Martius, plague was Rome's punishment for neglecting the worship of the gods, leading religious leaders to consult the Sibylline Oracles and to call for observation of a day of prayer.[185] Then, soon after the Antonine Plague first struck the empire during the reign of Marcus Aurelius (168/169 CE), Salus, the goddess of health who protected against disasters, was a dominant motif in numismatics. Thus, in times of great plagues, both Greeks and Romans repeatedly resorted to religious constructions as both their causes and solutions and sought religious meanings and significance in and out of them.

Moreover, as Vivian Nutton notes, with few exceptions, all the human errors that incurred illnesses can be classified as direct offenses against the gods (i.e., religious offenses), such as ritual impurity (e.g., eating forbidden food or not wearing clean cloth), failure to fulfill a vow or offer sacrifice, perjury, theft, and damage to sanctuaries and their possessions.[186] Written in plain Greek and dedicated to a variety of local deities, a group of propitiatory inscriptions from Lydia and Phrygia in Asia Minor dating from the second and third centuries CE records this broader pattern of sin, punishment, and repentance: (1) a person confessing a sin committed by oneself (or a relative); (2) a deity forcing the sinner to confess by punishing either the individual or a relative with sickness, accident, or death; (3) the specific details of the illness such as blindness, diseases of the breasts (by female deities in particular), mental illness (madness), and diseases of the legs, buttocks, and arms (in order of the most common); blindness was the most common illness in antiquity and was also believed to be a form of divine punishment; (4) some afflictions are chronic while others are the result of an accident; (5) sometimes the afflicted organ is represented in a panel above the text; (6) the sinner (or a relative) expiates the sin through appropriate rituals and incantations involving a priest or priestess, warns others not to underrate or dismiss the divine power, and praises the god.[187] As Angelos Chaniotis notes, despite the variety of diseases mentioned in the propitiatory inscriptions, "it seems that diseases which could not easily

---

185. Livy, *The History of Rome* 41.21.5–11; Duncan-Jones, "The Impact of the Antonine Plague," 113.

186. Nutton, *Ancient Medicine*, 291; cf. Ramsay MacMullen, *Paganism in the Roman Empire* (New Haven: Yale University Press, 1981), 32.

187. Angelos Chaniotis, "Illness and Cures in the Greek Propitiatory Inscriptions and Dedications of Lydia and Phrygia," in Ejik, Horstmanshoff, and Schrijvers, *Ancient Medicine in Its Socio-Cultural Context*, 2:324.

be treated by secular medicine (ocular and mental disorders) were most commonly attributed to divine punishment."[188]

Some dedications from the same area mention healings after the divine punishment as they are labeled as vows (*euchē*) to the same gods: "To Zeus Peizenos. Diogenes had made a vow for the ox, but he did not fulfill it; for that reason his daughter Tatiane was punished in her eyes. But now they propitiated and made the dedication."[189] Another case involves Stratonike, who did not repay a loan belonging to the deity Axiottenos and was punished by the god in her right breast; she then repaid the sum with all the interest and with praises to Axiottenos.[190] Another example comes from Glykia, who, having been punished by the goddess Anaeitis from Metro (with a disease) in the buttocks, asked her what to do and then dedicated a stone.[191] The most commonly healed body parts mentioned in vows and dedications are legs, eyes, and breasts; thus, there is a certain parallel between the illnesses mentioned in the propitiatory inscriptions and those mentioned in the dedications.[192] These propitiatory inscriptions and vows in fact attest to a long tradition in Greek (religious) culture, as we can find the same pattern of sin, illness, repentance, and healing in the earliest collection of Asclepius's healing miracles at Epidaurus in the fourth century BCE. For instance, Hermon of Thasus's blindness was cured by Asclepius. However, he didn't bring the thank offerings afterward, and the god made him blind again. When he came back, presumably making the thank offering, and slept again in the temple, the god made him well.[193] Here's another case:

> Cephisias . . . laughed at the cures of Asclepius and said: "If the god claims he has healed lame people, he is lying. For, if he had the power to do so, why has he not cured Hephaistos? But the god did not conceal that he was inflicting penalty for the insolence. For Cephisias when riding was struck

188. Chaniotis, "Illness and Cures," 2:329.

189. Chaniotis, "Illness and Cures," 2:324.

190. Chaniotis, "Illness and Cures," 2:325.

191. R. A. Kearsley, "Magic, Medicine and Cults," in *New Documents Illustrating Early Christianity*, vol. 6, *A Review of the Greek Inscriptions and Papyri Published in 1980–81*, ed. S. R. Llewelyn (Grand Rapids: Eerdmans, 2007), 192.

192. Chaniotis, "Illness and Cures," 2:329.

193. Emma J. Edelstein and Ludwig Edelstein, *Asclepius: Collection and Interpretation of the Testimonies*, 2 vols. (Baltimore: Johns Hopkins University Press, 1945), T. 423, I 22.

by his bullheaded horse which had been tickled beneath the saddle, so that instantly his foot was crippled and he was carried into the temple on a stretcher. Later on, after he had entreated him earnestly, the god made him well.[194]

While Asclepius was the popular healing god, the local gods of Lydia and Phrygia were just as competent in both inflicting diseases on sinners and healing them in the ancient Greek context, which considered divine punishment and healing "natural." That is why the author of *Sacred Disease* in the Hippocratic Corpus is at pains to convince his readers that epilepsy, traditionally considered a "sacred disease," is not "sacred" at all but has real, "natural," that is, rational and physiological, causes and cures rather than religious purifications or incantations by "superstitious" healers who invoked the gods as cures. In typical cases, the ill sought divine help simply because they thought the god's (or gods') anger was responsible for their illness according to their cultural understanding, and therefore both illness and healing were religious experiences for them as the god(s) relieved sinners from their sins. And these understandings were also shared by the ancient Israelite culture and by early Christian communities, as examined in the next chapter.

The account of Galen's contemporary Aelius Aristides of his illness also makes good sense in this cultural context and tells us about how his cultural values and social relations shaped how he perceived and monitored his body and labeled and categorized his bodily symptoms.[195] One of the most famous (or infamous) illness narratives of this period was his *Sacred Tales* (*Hieroi Logoi*). His extravagant fixation on his own bodily illness and suffering is well-known. Modern scholars and commentators' unflattering critique of (and little sympathy for) his graphic representation of his sick self describes his work as that of a neurotic, psychosomatic hypochondriac.[196] However, the *Tales* was famous for its literary quality and authori-

---

194. Edelstein and Edelstein, *Asclepius*, T. 423, II. 36.

195. Cf. Kleinman, *The Illness Narratives*, xiii.

196. See a compilation of those descriptors in scholarly literature in Helen King, "Chronic Pain and the Creation of Narrative," in *Construction of the Classical Body*, ed. J. Porter, 269–86 (Ann Arbor: University of Michigan Press, 1999), 277; Glen W. Bowersock, *The Greek Sophists in the Roman Empire* (Oxford: Clarendon, 1969), 72; Peter R. L. Brown, *The Making of Late Antiquity* (Cambridge, MA: Harvard University Press, 1978), 41.

tative self-representation in its reception in antiquity.[197] Born in 117 CE in upper Mysia in Asia Minor of wealthy parents, Aristides had a successful oratorical career before he fell ill in 143; unable to recover, he spent 145–147 in incubation at the temple of Asclepius in Pergamum, receiving divine visitations, advice, and prescriptions in dreams and visions. In 147 he recovered significantly enough to resume his career, but continued his close relationship with Asclepius until his death in 171, especially when he fell sick with plague (smallpox) in 165.

The six orations of the *Sacred Tales*, written at the end of his life, are not simply a record of but are a representation of his pained existence throughout his chronic illness and recovery by divine power. The significant beginning of the *Tales* records the repeated failure of the doctors, first in Rome and then in Smyrna in 144, to understand and identify the "complex nature of [his] disease" (*Or.* 48.69), and thus their failure to treat and alleviate his manifold pain and symptoms—coughs, vomiting, swellings, shivering, fever, headache, pain in the abdomen and groin, and intermittent inability to eat, breathe, or walk properly (48.37–45). No amount of purging or bloodletting provides him with relief; and there's no physician like Galen to rescue him by identifying and treating the affected areas of his body. It is then that the god of medicine Asclepius steps into the Asclepieion[198] in Pergamum to open up another way to understand his mysterious illness—through dreams. Both Aristides and his physicians share a common understanding of dreams in their culture, that they are believed to exist independently as a means of communication between gods and humans and are an appropriate means of medical therapy.[199] The Hippocratic *Regimen IV* was devoted to dreams as an integral part of the regimen; in it the author advised patients to seek the proper interpretations from authoritative and expert physicians (rather than dream interpreters). Galen himself

197. As attested by Philostratus, Sopater, Libanius, and Synesius. Cf. Israelowich, *Society, Medicine, and Religion*, 26–29, 128.

198. The temple of Asclepius.

199. On the role of dream in medical diagnosis and remedies and the role of dream and its interpretations in the *Sacred Tales*, see Steven M. Oberhelman, "Dreams in Greco-Roman Medicine," *ANRW* II.37.1 (1993): 121–56; Steven M. Oberhelman, "Galen, on Diagnosis from Dreams," *Journal of the History of Medicine and Allied Sciences* 38, no. 1 (1983), 36–47; Patricia Cox Miller, *Dreams in Late Antiquity: Studies in the Imagination of a Culture* (Princeton: Princeton University Press, 1994) 184–204; Israelowich, *Society, Medicine, and Religion*, 71–86. See also C. A. Behr, *Aelius Aristides, and* The Sacred Tales (Amsterdam: Hakkert, 1968), 171–252.

attributed his recovery from his own serious illness to a radical therapy of self-bloodletting from the arteries (arteriotomy) inspired by a series of dreams (perhaps by Asclepius) and told a story of another patient saved by similar means as directed in a dream by Asclepius at Pergamum.[200] Later he told the emperor Marcus Aurelius: "I had declared myself [Asclepius's] servant ever since he had saved me from a deadly condition of an abscess" (*On My Own Books* 2.19.18–19K).[201]

Coming back to Aristides, dreams have a pivotal role in his medical treatment. He deciphers the mysterious suffering of the abdomen through a dream (*Or.* 47.4ff.). Once he is able to interpret the dream, he is led to follow the appropriate therapeutic response. For example, dreaming of a trapped bone indicates bloodletting (47.28), dreaming of self-anointing means no bathing (47.29), and dreaming of a fig prescribes vomiting or fasting (47.54–55; cf. 47.9, 26, 40).[202] As Brooke Holmes observes, "by determining how to act on the sick body, Aristides, not unlike his contemporaries committed to elaborate regimens of self-care underwritten by physicians, gains control over it."[203]

Nonetheless, although Aristides succeeds in deciphering his illness through dream and views himself as having direct relations with Asclepius, he often has to depend on others to interpret his dreams.[204] In a dream that both Aristides and one of the two temple wardens (*neōkoroi*), Philadelphus, have in the winter of 146, the god prescribes drinking wormwood diluted in vinegar at a religious festival in which a host of men, wearing white garments, assemble before the god in the Sacred Theater in the Asclepieion; in it Aristides is directed to deliver a hymn to the god during this festival (48.29–36). Aristides then summons a physician, Theodotus, and the other temple warden, Asclepiacus, for consultation of the dream. While Theodotus "marveled how divine they [the dreams] were," Asclepiacus informs Aristides that his colleague Philadelphus had a very similar dream about Aristides. Then Philadelphus joins them and narrates his own dream

---

200. *On Treatment by Venesection* 23 (314–315K); *Meth. med.* 14.8 (971–972K).

201. For more on how Galen took dreams seriously in his medical practice, particularly those of Asclepius, see Oberhelman, "Galen," 37–39; Mattern, *The Prince of Medicine*, 172–74; Israelowich, *Society, Medicine, and Religion*, 73–75.

202. Cf. Brooke A. Holmes, "Aelius Aristides' Illegible Body," in *Aelius Aristides between Greece, Rome, and the Gods*, ed. William V. Harris and B. Holmes (Leiden: Brill, 2008), 91.

203. Holmes, "Aelius Aristides' Illegible Body," 91–92.

204. Cf. Israelowich, *Society, Medicine, and Religion*, 102.

concerning Aristides, and with this confirmation of his dream, a convinced Aristides complies with Asclepius's prescription and drinks wormwood (48.36). On other occasions, Asclepiacus, as a temple warden, consistently helps Aristides interpret his dreams and find the exact necessary remedy prescribed by Asclepius. For instance, Aristides received a prescription of a "royal ointment" through dream for tension in his neck and ears (49.21). When he asked Asclepiacus about it, Asclepiacus went to the Temple of Hygieia (Health) to get the ointment, which immediately relieved Aristides from the tension upon its application (49.22). These examples of cooperation in deciphering dreams and following the god's prescriptions indicate a certain synergy between the rational medicine and temple medicine in the shared cultural context on the one hand and the superiority of Asclepius (temple medicine) as the "better physician" (*iatros*) on the other (at least from the patient's perspective).[205] In fact, Aristides had more than one hundred dream therapies from the god, which were comprehensive and thus helped him to recuperate both physically and psychologically:[206] "But if someone would like to know how the god acted upon me, it is time to search for the parchments and the dream reports themselves. In them, he will find remedies of all kinds and a few confidential talks and lengthy orations and various phenomena and prophecies and oracles of all kinds on diverse subjects, partly in prose, partly in meter, all of them devoted in binding and indescribable gratitude to the god" (48.8).

Indeed, the precise reason why Aristides turns to Asclepius was that he thought the god was the better physician rather than that the god provided an alternative to medicine.[207] At the outset of the *Tales*, Aristides declares that he has "decided to submit to the god, as to a doctor and to do in silence whatever he wishes (*hōsper iatrōi tōi theōi sigēi poein*)" (47.4). On several

205. "These dreams appeared to me while the doctor had arrived and had prepared himself to help, as much as he knew how. But when he heard the dreams, being a sensible man, he also yielded to the god. And we recognized the true and proper doctor for us, and we did what he commanded" (*Or.* 47.57). Cf. Israelowich, *Society, Medicine, and Religion*, 103, 113.

206. Florian Steger, "Aristides, Patient of Asclepius in Pergamum," in *In Praise of Asclepius: Aelius Aristide, Selected Prose Hymns*, ed. Donald A. Russell et al. (Tübingen: Mohr Siebeck, 2016), 137.

207. On the theme of Aristides's glorification of Asclepius as the best physician, see Georgia Petridou, "The Curious Case of Aelius Aristides: The Author as Sufferer and Illness as 'Individualizing Motif,'" in *Autoren in religiösen literarischen Texten der späthellenistischen und der frühkaiserzeitlichen Welt*, ed. Eve-Marie Becker and Jörg Rüpke (Tübingen: Mohr Siebeck, 2018), 208–16.

occasions, the god prescribes remedies that oppose contemporary medical practices, such as prescribing a certain drug (*pharmaka*) after dining that doctors think should not be taken with food, particularly when one was eating only one meal a day (49.27). However, Aristides follows the god's regimen, and, as always, it proves effective against the doctors' concerns and thus testifies to the superiority of the medical method of Asclepius. Again, on other occasions such as the following one, the god, after his epiphany to Aristides, prescribes him a regimen deemed dangerous, suspicious, or inadequate by his physicians:

> It was the middle of winter and the north wind was stormy and it was icy cold, and the pebbles were fixed to one another by the frost, so that they seemed like a continuous sheet of ice, and the water was such as is likely in such weather. When the divine manifestation was announced, friends escorted us and various doctors, some of them acquaintances, and others who came either out of concern or even for the purposes of investigation. There was also another great crowd, for some distribution happened to be taking place. . . . But being still full of warmth from the vision of the god, I cast off my clothes, and not wanting a massage, flung myself where the water was deepest. Then as I was in a pool of very gentle and tempered water, I passed my time swimming all about and splashing myself all over. When I came out, all my skin had a rosy hue and there was a lightness throughout my body. There was also much shouting from those present and those coming up, shouting that celebrated phrase, "Great is Asclepius!" Who could tell what came next. During the rest of the day and night till bed time, I preserved the condition which I had after the bath, nor did I feel any part of my body to be drier or moister. None of the warmth left me, none was added, nor again was the warmth such as one would have from a human contrivance, but it was a certain continuous body heat, producing the same effect throughout the whole of my body and during the whole time. My mental state was also nearly the same. For there was neither, as it were, conspicuous pleasure, nor would you say it was like human joy. But there was a certain inexplicable contentment which regarded everything as less than the present moment, so that when I saw other things, I seemed not to see them. Thus I was wholly with the god. (48.19–23)

This scene captures Aristides's perception of and response to Asclepius as both a doctor and a god, which results not only in a therapeutic experience but also in a mystical union with a deity in the most intimate and yet public way. Throughout the rest of the day and the night Aristides retained warmth, and a perfect balance of elemental qualities in his body, a balance that could not have been attained by human contrivance. This state of a perfectly balanced mixture of the four elemental qualities (wet, dry, hot, and cold) in the human body is what Hippocratic authors, Plato, and Galen called a "proper mixture" (*eukrasia*) in their treatises (e.g., *Nature of Man* 6; Galen, *Hygiene* 1.4 [11–12K]). This perfect bodily and mental state transposes then into almost a sort of initiation into a mystery cult by Aristides with a privileged relationship with the god, his savior (*sōtēr*).[208] Aristides's self-identification as a sacred servant of Asclepius (*therapōn, diakonos, therapeutēs*; e.g., *Or.* 48.23) and his understanding of and total but reasoned submission to this dual role of Asclepius, even through enduring physical pain and suffering, are crucial for him to gain consistent encouragement and confidence from the god throughout his sick journey (*kathedra*). Furthermore, this is crucial for him eventually to gain psychosomatic healing and a renewed sense of calling for his oratory as the physician-god restores him to life and to his oratory with divine sanction (50.38–39). This is a "double redemption" for Aristides, as it were, which was based upon his complete obedience to the doctor-god while still consulting his team of doctors. In this way, Aristides is able to weave both medical and religious discourses into his illness narrative with the transformed meaning of his illness and divine intervention, including the nature of dreams and their interpretations.[209]

As Ido Israelowich underscores, Aristides's conceptualization of being ill was an "intrinsic aspect of a broader conceptual scheme of a religious nature"; and his narrative of "his religious experience owed much to social institutions and to some of the most distinctive cultural phenomena in the Greek world of his age," referred to as "Second Sophistic."[210] The age of the Second Sophistic in the early imperial era exhibited a characteristic of

208. Cf. Georgia Petridou, "Becoming a Doctor, Becoming a God: Religion and Medicine in Aelius Aristides' *Hieroi Logoi*," in *Religion and Illness*, ed. Annette Weissenrieder and Gregor Etzelmüller (Eugene, OR: Cascade Books, 2016), 310n11.

209. We can see how Aristides's illness narrative shares the following common narrative elements listed by Ron Loewe in the introduction: "mystery (disease is unexpected or difficult to diagnose), betrayal by one's own body, conflict with medical professionals or medical bureaucracies, the failure of medical science to heal, the need for self-reliance, and generally, but not always, a return to good health." Loewe, "Illness Narratives," 43; see above, p. 6.

210. Israelowich, *Society, Medicine, and Religion*, 117, 180.

"hypersensitivity in literature and bodily care," one in which the suffering body became a focal point of significant cultural concern, and a "new subjectivity—the self as sufferer"—emerged.[211] In that context, in words of Judith Perkins, "if there is a pathology, it belongs to the culture rather than to the psychology of any individual."[212]

The other representative of that culture is Marcus Aurelius, the Stoic Roman emperor contemporaneous with Aristides and also Galen's imperial patient, whose constant ill health was well-known even in his time.[213] Despite his strict Stoicism, scholars have noticed his preoccupation with his own bodily ailments and those of his tutor Cornelius Fronto, with whom, alongside hypochondriac Aristides, Marcus forms a trio.[214] While health news was the natural basis of epistolary intimacy among other Roman elites, in over eighty of the extant letters of friendship between Fronto and Marcus Aurelius, they, especially the former, freely share their health concerns and illness in great detail. Fronto complains of a host of symptoms: pains in his shoulder, elbow, knee, ankle, arm, finger, eyes, and groin; ailments from sleeplessness; crippled hands, a sore throat, diarrhea, heartache, grief, etc. Marcus Aurelius's illnesses seem less serious than those of his teacher: casual fever, chest pain, tiredness, eating difficulties, digestive discomfort, and discomfort while walking. Consider the following correspondence. Fronto writes: "I am anxious to know, my Lord, how you are keeping. I have been seized with pain in the neck" (*Ad. M. Caes.* 5.27). The caesar responds: "I think I have got through the night without fever. I have taken food without repugnance, and am doing very nicely now. We shall see what the night brings. But, my master, by your late anxiety you can certainly gauge my feelings when I learnt that you had been seized with pain in the neck" (5.28). Then Fronto writes back: "My neck, Lord, hurts terribly, but the pain in my foot has disappeared" (5.29). The caesar responds: "If the pains in your neck get better, even in two days' time, it will help on my convalescence more than anything, my master. I have had a bath and to-day even a little walking and taken a little more food,

---

211. Bowersock, *Greek Sophists*, 74; Judith Perkins, *The Suffering Self: Pain and Narrative Representation in the Early Christian Era* (London: Routledge, 1995), 7.

212. Perkins, *The Suffering Self*, 193.

213. Cassius Dio, *Roman History* 71.6; 72.36.2–4.

214. Bowersock, *Greek Sophists*, 73; J. E. G. Whitehorne, "Was Marcus Aurelius a Hypochondriac?" *Latomus* 36 (1977): 413–21, concludes that Fronto is a hypochondriac but Marcus Aurelius is not, tempered by his concern more for Fronto's physical condition.

but not as yet without discomfort" (5.30). Then Fronto states: "The pains in my neck are no easier, but my mind was set at rest as soon as I knew that you had been able to take a bath and relish your wine" (5.32). In these exchanges, these two elite men not only openly display their ill health and incapacities but also insist on how the health of one directly and tangibly influences the health of the other. And thus, each man "depend[s] on each other to validate the other's state of health."[215]

Interestingly, as Rebecca Flemming notes, their sick role—in fact, the sick roles of all three elite men, including Aristides—while exhibiting "weak" behaviors in ways characteristic of and appropriate to women of that culture, "appears less a challenge to, than a privilege of maleness," a sharp contrast to women's tendency to suffer in silence.[216] In the context of the Second Sophistic, their illness narratives, rather than the works of neurotic men, reflect an "intensification" of a way to form oneself "as a subject who [has] the proper, necessary, and sufficient concern for his body," as Michel Foucault writes about the penchant of this period.[217] As a matter of fact, since, according to Perkins, the pathology belongs to the culture, this intensification of the concern for their bodies displays their illness and suffering as a focus on the self. Then there is a strong irony in that their ability to articulate their illness symptoms and their acting out of the sick role in active relationship with the body is done on their own terms where they could still control their and others' medical care and access; in fact, both men "maintain the male prerogatives in illness" in contrast to women's inability and silence.[218] For Marcus, there is certainly a contrast between his own "illness narrative" in his correspondence and his *Meditations*. In *Meditations*, where he is writing to himself (not to others), Marcus hardly mentions his own illness; and when he does, he speaks about "cheeriness in sickness as well as in other circumstances" (1.15) and about receiving antidotes "against the spitting of blood and vertigo" through dreams and setting his heart on (Stoic) philosophy, not busying himself "with physical phenomena" (1.17). Indeed, he adopts the attitudes of Epicurus, who said:

215. Annelise Freisenbruch, "Back to Fronto: Doctor and Patient in His Correspondence with an Emperor," in *Ancient Letters: Classical and Late Antique Epistolography*, ed. Ruth Morello and A. D. Morrison (Oxford: Oxford University Press, 2007), 241.

216. Flemming, *Medicine and the Making of Roman Women*, 74, 76.

217. Michel Foucault, *The Use of Pleasure*, vol. 2 of *The History of Sexuality*, trans. Robert Hurley (New York: Pantheon, 1986), 108.

218. Flemming, *Medicine and the Making of Roman Women*, 74.

"In my illness my talk was not of any bodily feelings, nor did I chatter about such things to those who came to see me, but I went on with my cardinal disquisitions on natural philosophy. . . . Nor did I . . . let the physicians ride the high horse as if they were doing grand things, but my life went on well and happily" (9.41). Here he is reflecting a traditional Stoic-Epicurean attitude that the internal space remains completely undamaged by any external events, including illness (cf. 3.16). In these seemingly contradictory writings, then, Marcus Aurelius exhibits male frailty and privilege in illness on the one hand, and a philosophical virtue of *apatheia*, another male virtue, on the other.

Conclusion

This chapter has investigated how Greco-Roman authors constructed notions and theories of health, disease, and illness in their rich historical and cultural contexts. Health as a mixture or balance within and without the body; a sense of proportion and proper function in the body; disease as isolation or imbalance within and without the body in relation to the soul and the environment; disease as an impediment to function; the religious and gendered meaning-making of the experience of illness—all were part of the common cultural currency in Greco-Roman medicine, philosophy, and literature. Indeed, a consistent undercurrent throughout this chapter has been how medicine (and philosophy) relates to religion; in construction of health, disease, and illness, Greco-Roman medicine always related to religion even when it put forth natural and physiological definitions and explanations, as nature itself was deemed divine and a standard by which one's health and disease were measured (either according to or against nature). Thus, while acknowledging the significance of the development of medical theories and practices in Galen's work and the Hippocratic Corpus, for instance, the Greco-Roman world embraced a medical discourse, which obscured a closed boundary between science and religion and between rational medicine and temple medicine.[219] The second undercurrent is the relation between medicine and philosophy. While there was a synergy and positive interaction between them (especially in the case of Galen), doctors and philosophers also competed for opportunities to guide elites to their own precepts of health care and therapies of the soul. These precepts all shared the common goal of helping a person to attain

219. Cf. Israelowich, *Society, Medicine, and Religion*, 179.

virtue. Third, Greco-Roman medicine recognized the reciprocal relationship between the soul and the body in impacting and shaping both health and diseases in one another. Finally, Greeks and Romans constructed their illness narratives within their cultural frameworks, conferring personal and socioreligious meanings on their illness experiences. The Israelite/Jewish and Christian texts, discussed in the next chapter, will also adopt and engage with these larger cultural paradigms and reconfigure them according to their emerging theological visions.

HEALTH, DISEASE, AND ILLNESS
IN THE BIBLE AND EARLY CHRISTIANITY

In light of the topics covered in the previous chapter, this chapter offers an overview of health, disease, and illness in the Hebrew and Christian Bibles, as well as how early Christians living in the Greco-Roman context developed their own theories of health, disease, and illness. Compared to Greco-Roman medicine and philosophy, the concepts of health, disease, and illness in the Hebrew and Christian Bibles are primarily religious in accordance with Hebrew monotheism. However, they also evolve to adopt the notions from Greco-Roman medicine and philosophy by the time of Second Temple Judaism. After a brief presentation of representative (but not exhaustive) biblical views on health, disease, and illness, I will focus on analyzing early Christian texts as they interacted with ideas from Greco-Roman medicine and philosophy. I will show how early Christian authors adopted and appropriated Greco-Roman medico-philosophical principles for their purposes, and developed their own illness narratives incorporating various etiologies.

Health in the Hebrew Bible, Second Temple Literature, and the Christian Bible

While there is no direct statement on health in the Hebrew Bible, the first chapters of Genesis establish a fundamental biblical notion of health that "God is the Creator of life and thus the Giver of health and well-being."[1]

---

1. Gerhard F. Hasel, "Health and Healing in the Old Testament," *AUSS* 21, no. 3 (1983): 191.

Health given by Creator God is holistic—it is comprised not only of physical well-being but also of spiritual, mental, emotional, relational, and sociopolitical well-being[2]—and thus is expressed with the Hebrew word *shalom*, meaning the "wholeness," "completeness," and "peace" of a person in relation to and in harmony with God, the person himself or herself, fellow humans, and the world.[3] In its 237 occurrences in the Hebrew Bible, the word *shalom* carries three main meanings: most frequently it means a sense of physical well-being and material prosperity, as when Joseph was sent by his father, Jacob, to check out the *shalom* of his brothers and the flocks (Gen. 37:14). It is also used in a second sense with reference to social or political relationships, as that of Tyre and Israel (1 Kings 5:12e; cf. Jer. 38:22), and thirdly, in a moral sense of integrity or straightforwardness, as in the psalmist's instruction:

> Mark the blameless, and behold the upright,
>> for there is posterity for the peaceable [lit. the man of *shalom*]). (37:37)[4]

A common denominator in all three meanings is that *shalom* "defines how people, situations, things, should be, i.e., how such entities are to be healthy and wholesome."[5] Therefore, *shalom* is "the perfection to which the creation [after the Fall] longs to return, that world of being in which every individual and people are full and complete, free from injustice, oppression, pain, and sickness"; it is "that place and state where God, humanity, and the environment are one in harmony and peace."[6] Isaiah 32:16–17 associates *shalom* with justice, righteousness, quietness, and trust. In the beatific vision of Isaiah 2:2–4 and Micah 4:1–5, many peoples and nations go up to the temple mountain of Jerusalem, letting go of their weapons and

---

2. Cf. R. K. Harrison, "Healing, Health," in *The Interpreter's Dictionary of the Bible: An Illustrated Encyclopedia* (New York and Nashville: Abingdon, 1962), 2:541.

3. Hasel, "Health and Healing," 192; see also Nigel Allan, "The Physician in Ancient Israel: His Status and Function," *Medical History* 45 (2001): 378.

4. Roger W. Uitti, "Health and Wholeness in the Old Testament," *Consensus* 17, no. 2 (1991): 56. Unless otherwise indicated, all Bible quotations come from the New Revised Standard Version.

5. Uitti, "Health and Wholeness," 56.

6. Uitti, "Health and Wholeness," 56. John Wilkinson mentions five subcategories of *shalom*: justice and obedience are conditions of attaining *shalom*; strength, fertility, and longevity are blessings of *shalom*. See John R. Wilkinson, *The Bible and Healing: A Medical and Theological Commentary* (Grand Rapids: Eerdmans, 1998).

wars and living in peace, and the people enjoy the return of Solomon-like prosperity (1 Kings 4:25e). Since *shalom* is God's creational design, disease, death, or chaos, in whatever form, is not an established part of the divine order of things—neither in the prelapsarian world nor in the eschaton. The postlapsarian but antediluvian period, in which longevity is one of the manifestations of health, tells the story of people who reach as many as 777 or even 969 years of life (Gen. 5:27, 31). This is a stark contrast to the life spans of postdiluvian patriarchs, the oldest of whom saw 110 to 120 years (Joseph in Gen. 50:26; Moses in Deut. 34:7), numbers that are still very high.[7]

Following the exodus, God promised the Israelites health and immunity from the diseases that God had afflicted the Egyptians with: "I will not bring upon you any of the diseases that I brought upon the Egyptians; for I am the LORD who heals you" (Exod. 15:26b). Here, God is the giver of not only health (and healing)[8] but also diseases. This immunity was conditional on their consistent obedience to God's law codified in the Mosaic law (Exod. 15:26a). Indeed, individual and corporate health and related blessings such as human fertility, agricultural fecundity, prosperity, longevity, and security could only be maintained by "a punctilious observance of the divine commands" throughout life (cf. Deut. 28:2–14).[9] With this inextricable association between obedience and health (and healing),[10] the Israelites were also warned that their persistent disobedience of God's law would bring about, along with other forms of punishment, an epidemic of various diseases (Deut. 28:58–61).[11] For our topic in particular, Israelites were to observe the Sabbath law, which was linked with creation (Gen. 2:1–3; Exod. 20:11; 31:17), a gift of the Creator for humanity. Just as YHWH had worked six days and had rested on the seventh day, so was Israel to work six days and spend the seventh day in restful imitation. According to Deuteronomy 5:14–15, Israelites were also to keep the Sabbath day as a way to remember and commemorate their former servitude in Egypt and YHWH's deliverance of them. Thus, it was, on the one hand, a weekly reminder of their religious-spiritual responsibility to worship God

---

7. Cf. Hasel, "Health and Healing," 192; Harrison, "Healing, Health," 541–42.

8. See Richard S. Hess, "The COVID-19 Virus, Illness, and Biblical Interpretation in Its Ancient Context," *Canon and Culture: A Journal of Biblical Interpretation in Context* 14, no. 2 (2020): 68–69.

9. Harrison, "Healing, Health," 545.

10. See Hess, "The COVID-19 Virus," 68–69.

11. On epidemics as God's judgment and punishment, see Hess, "The COVID-19 Virus," 64–66.

as the Creator and Redeemer and, on the other hand, an opportunity to meet the social and humanitarian responsibility to care for their household servants, to look after their domestic animals, and to attend to their own physical, mental, spiritual, and emotional health and renewal. Therefore, the Sabbath stood as "an agent of restoration, health, and wholeness for Hebrew society as a whole."[12] Moreover, the law of the sabbatical year extended the Sabbath rest to the land, which was to lie fallow every seventh year after the harvest, during which time orchards and vineyards would remain untended (Lev. 25:1–7). As Gerhard Hasel notes, this is significant in its interest in "ecological conservation, [and] the continued health of the land by preserving natural resources and permitting the land to rejuvenate itself in a seven-year cycle."[13]

Leviticus devotes significant attention to health and disease in the context of ritual cleanliness and uncleanliness and the Holiness Code, laying special emphasis on ritually required physical wholeness "as a necessary reflection of the desired spiritual well-being pictured in social hygiene and sanitary regulations." These regulations make up 213 of the total 613 biblical commands.[14] Thus, for example, the law excludes from the sphere of holiness anyone who has a physical "blemish" or "deformity" (e.g., a blind or lame man); hence, priests with any physical "defect" or "imperfection" such as a mutilated face or limb or an itching disease were excluded from ministry (i.e., offering sacrifices) but could still eat consecrated food (21:16–23); and men with genital mutilation were also excluded from the assembly (21:20; Deut. 23:1). Regarding the dietary rules, the distinction between "clean" and "unclean," which goes back to the flood (Gen. 7:2), was unique in the ancient Near East;[15] the Israelites were to eat only "clean" creatures as opposed to "unclean" ones (Lev. 11:2–19). As Averell Darling notes, the division was linked to creatures' feeding habits and their manner

12. Uitti, "Health and Wholeness," 54.

13. Hasel, "Health and Healing," 194.

14. Donald J. Wiseman, "Medicine in the Old Testament World," in *Medicine and the Bible*, ed. Bernard Palmer (Exeter, UK: Paternoster, 1986), 19. While these regulations were mainly seen as hygienic and have been thus praised as unique in the ancient world until recently, contemporary biblical scholars and medical practitioners approach them as primarily theological. However, these laws may still contain some valid health benefits such as isolation of contagious patients and disinfection by washing. See Averell S. Darling, "The Levitical Code: Hygiene or Holiness," in Palmer, *Medicine and the Bible*, 98–99.

15. Harrison, "Healing, Health," 543; Hasel, "Health and Healing," 195.

of progressing over the earth if they were earth-born.[16] Clean were herbivores that chewed the cud and walked on divided hooves; unclean were all others, whether herbivores or carnivores. Among birds, grain eaters were clean but predators were unclean. In the sea, all creatures with both fins and scales were clean but all others were unclean. "Thus, we have all predators, scavengers and 'bottom-feeders' excluded."[17] Darling answers the question, "did these dietary prohibitions have a hygienic benefit?" with a qualified yes.[18] While the issue of defilement is ceremonial rather than hygienic, there was "some gain in choosing for [one's] diet those animals and birds that were least likely to be infested with harmful organisms."[19]

In the wisdom literature, Proverbs likens the sage's words as "life to those who find them, / and health to one's whole body" (4:22 NIV) and promises health to the body for those who fear the Lord and shun evil (3:7–8). In the Second Temple literature, written in early second century BCE, the Wisdom of Ben Sira (Ecclesiasticus) considers bodily health among the highest goods in life:

> Better off poor, healthy, and fit
> > than rich and afflicted in body.
> Health and fitness are better than any gold,
> > and a robust body than countless riches.
> There is no wealth better than health of body. (30:14–16)

As for the notion of health in the Christian Bible, we can find no direct ideas or sources and can only indirectly deduce them from Jesus's numerous healing accounts in the Gospels, the Synoptics in particular. The fact that the (Synoptic) Gospels identify healing as central to Jesus's earthly ministry, and a sign of God's power and kingdom, indicates that Jesus inherited the Hebrew notions of health/wholeness as both God's gift and will for humanity, and of healing as a process that attends to the health and care of the whole person—physically as well as in that person's religious-ethical life, in accordance to God's will.

16. Darling, "The Levitical Code," 90.
17. Darling, "The Levitical Code," 90.
18. Darling, "The Levitical Code," 91.
19. Darling, "The Levitical Code," 91. Harrison, "Healing, Health," 544, notes that God's instruction for the covenant community through Moses "was the first of its kind to recognize that infection could be transmitted by both food and water."

Disease and Illness in the Hebrew Bible, Second Temple Literature, and the Christian Bible

In both the Hebrew Bible and the Christian Bible, including the Apocrypha, references to diseases and illness (and healing) far outweigh references to health. C. Frevel defines illness in these texts as "a disruption of the God-given integrity of human life and a disruption of the God-given, inherently good order";[20] this is the opposite of *shalom*. In the Hebrew Bible, disease and illness are seen as God's actions, as mentioned in Exodus 15:26b, intended to punish people for their disobedience or other religious failings, described as "sin."[21] In this context, diseases and illness can be seen as God's pedagogical tool as well—temporarily punishing individuals[22] or a people as a whole[23] for their sins, so as to bring them to repentance and redemption. Hence, while diseases and illness are the expressions of a disrupted relationship between God and human beings first and foremost, they also have a social dimension since, when ill, one is cut off from one's family and community, that is, from the cultic community of worship.[24] Therefore, when sick, one is deprived of fellowship in life and worship as well as being cut off from one's environment; in some instances, one might even be considered dead to one's fellow man and enemies while still living.[25] Not unlike that described by the psalmists, Job's illness is both a physical and social suffering that is reinforced by the attitudes and behaviors of his familiar circle of family and friends:[26]

---

20. C. Frevel, "Krankheit/Heilung," in *Handbuch theologischer Grundbegriffe zum Alten und Neuen Testament*, ed. Angelika Berlejung and C. Frevel (Darmstadt: Wissenschaftliche Buchgesellschaft, 2006), 284.

21. E.g., Pss. 38:3, 5–9; 41:3–5; Job 4:7–9; 1 Kings 17:17–18. Cf. Hess, "The COVID-19 Virus," 64–66.

22. Num. 12:10; Ps. 32:10; 2 Chron. 21:15–20. For example, Miriam was stricken with ṣāra'at (a skin disease) for her jealousy of Moses and claiming an equal role with Moses as God's instrument, and was healed by God at Moses's pleading (Num. 12:1–15).

23. E.g., plagues and childlessness in Exod. 23:25–26; plagues in Lev. 26:25; Num. 14:12; Deut. 28:21–22; 1 Kings 8:37; madness, blindness, and mental confusion in Deut. 28:27–29, 35.

24. Note Hector Avalos's definition of illness in the Israelite health-care system: "illness is any condition that . . . renders a person physically or mentally unfit to execute a social role defined as 'normal' by the society . . . here ancient Israel": *Illness and Health Care in the Ancient Near East: The Role of the Temple in Greece, Mesopotamia, and Israel*, HSM 54 (Atlanta: Scholars Press, 1995), 250.

25. E.g., Pss. 38:11; 88:8, 18; Job 2; cf. Ps. 22:14–18.

26. See also Pss. 27:10; 31:11; 35:13–14; 38:10–15; 41:6–7, 9; 55:14–15; 69:8; 88:8, 18.

> "He [God] has put my family far from me,
>   and my acquaintances are wholly estranged from me.
> My relatives and my close friends have failed me. . . .
> My breath is repulsive to my wife;
>   I am loathsome to my own family. . . .
> All my intimate friends abhor me,
>   and those whom I love have turned against me."
>   (Job 19:13–19)

As we can see, physical illness and suffering are often intertwined with emotional suffering, in "that the illness and its social consequences blend into one another"; the psalmists' and Job's laments are, in this way, versions of illness narratives.[27]

H. Mowvley especially highlights a link between illness and Sheol: "Illness is potential death. There is no essential difference between illness and death, but only a difference of degree, since in both cases the same powers are at work."[28] Therefore, illness is "the incursion of death itself into the realm of life. It is all of a piece with the conception of the threat of returning chaos";[29] and it is sin that unleashes this incursion of death, curse, and chaos through illness (and other calamities) (e.g., Pss. 38:3; 25:18); thus sin and its consequences are not two separate things but belong together as parts of the same process since there is no escape from sin's consequences.[30] And since all laws governing the created world and human life are dependent on God, it is God who declares human sin to be sin and it is God who sends the illness: "If you do not diligently observe all the words of this law that are written in this book, fearing this glorious and awesome name, the LORD your God, . . . he will bring back upon you all the diseases of Egypt, of which you were in dread, and they shall cling to you. Every other malady and affliction, even though not recorded in the book of this law, the LORD will inflict on you until you are destroyed" (Deut. 28:58–61).

---

27. E.g., Pss. 6:5; 18:4–5; 25:18; 38:1–6; 88:3–7. Angelika Berlejung, "Written on the Body: Body and Illness in the Physiognomic Tradition of the Ancient Near East and the Old Testament," in *Religion and Illness*, ed. Annette Weissenrieder and Gregor Etzelmüller (Eugene, OR: Cascade Books, 2016), 158.

28. E.g., Pss. 106:3–4; 88:3–7; cf. 2 Sam. 22:5; Hos. 13:14. H. Ringgren, *The Faith of the Psalmists* (London: SCM, 1965), 63, quoted in H. Mowvley, "Health and Salvation," *Baptist Quarterly* 22 (1967): 103.

29. Mowvley, "Health and Salvation," 104.

30. Mowvley, "Health and Salvation," 105.

If illness is an incursion of the forces of death and chaos, healing is then deliverance from those forces, or salvation.[31] Thus the psalmist declares:

> Some were sick through their sinful ways,
> and because of their iniquities endured affliction;
> they loathed any kind of food,
> and they drew near to the gates of death.
> Then they cried to the LORD in their trouble,
> and he saved them from their distress;
> he sent out his word and healed them,
> and delivered them from destruction. (107:17–20; cf.
> Jer. 17:14)

Just as God is the sender of diseases and illness, so it is God who sends forgiveness and healing:

> Bless the LORD, O my soul,
> and do not forget all his benefits—
> who forgives all your iniquity,
> who heals all your diseases,
> who redeems your life from the Pit,
> who crowns you with steadfast love and mercy. (Ps. 103:2–4;
> cf. Job 5.17–18)

Therefore, the "sick role" in Israel involves "assuming the posture of a penitent before God and pleading with God for forgiveness and healing."[32] Hence, in these texts "the experience of illness and healing is integrated into an overall framework of religious existential understanding."[33]

In Leviticus, illness is associated with ritual impurity, which requires cultic prohibitions. Leviticus 13–14 mentions ṣāra'at, a generic term for dreaded surface ailments (not leprosy!) that affect and discolor skin, cloth, leather, and the walls of houses that are generally nonpermanent and may or may not be infectious.[34] The priests, not medical professionals, are to

---

31. Mowvley, "Health and Salvation," 107.

32. Allen Verhey, "Health and Healing in Memory of Jesus," *ExAud* 21 (2005): 27.

33. Bernd Janowski, "'Heal Me, for I Have Sinned against You!' (Psalm 41:5 MT): On the Concept of Illness and Healing in the Old Testament," in Weissenrieder and Etzelmüller, *Religion and Illness*, 178.

34. Stanley G. Browne, "Leprosy in the Bible," in Palmer, *Medicine and the Bi-*

examine the symptoms of the person or object and declare the person or it ritually unclean or clean (Lev. 13:17; 14:3); priests would also offer atoning sacrifices for one's uncleanness so that he or she could be clean (i.e., cured) from the disease (Lev. 14:18–20, 28). However, they are not healers. Prophets[35] could be God's agents of healing, as in the case of Elijah restoring the life of a son of a widow in Zarephath (1 Kings 17:17–23) and Elisha instructing the Syrian army commander Naaman to bathe in the river Jordan for his *ṣāraʿat* (2 Kings 5:1–14). When Isaiah served as a consultant to gravely ill Hezekiah, he prophesied that Hezekiah would live an additional fifteen years and provided him with medical treatment (2 Kings 20:1–7). Physicians are mentioned only a few times, rather negatively, as in the case of Asa, the king of Israel, who was condemned because he sought healing from physicians instead of God for his lingering illness (2 Chron. 16:12; cf. Job 13:4). Only YHWH was to be addressed as Israel's healer and physician, especially as God declares in Exodus 15:26b: "I am the LORD who heals you."[36]

There are two exceptions or challenges to the conventional sin-illness (and forgiveness-healing) paradigm in the Hebrew Bible. One is Job's illness/suffering, which comes not from God but from Satan (though with God's permission); here illness can be permitted by God through Satan not to punish the sinner but to test the righteous person.[37] Moreover, Job's protest and lament and God's rebuke of Job's friends reject (at least partially) the traditional theology of retribution employed by his friends.[38] The other exception or challenge is the suffering servant in Isaiah, whose vicarious illness/wounds, endured for the guilty, rather bring about healing to the many (Isa. 53:4–5).

---

ble, 104. See also Elinor Lieber, "Old Testament 'Leprosy,' Contagion and Sin," in *Contagion: Perspectives from Pre-Modern Societies*, ed. Lawrence I. Conrad and Dominik Wujastyk (Aldershot, UK: Ashgate, 2000), 99–136. Lieber suggests that *ṣāraʿat* throughout the Hebrew Bible is applied to "the skin signs of certain non-contagious and relatively harmless diseases, particularly chronic psoriasis," but in Lev. 13 it could also be applied to "those of highly contagious chronic conditions such as bejel" (132, 136).

35. On the role of Israelite prophets as health-care consultants, see Avalos, *Illness and Health Care*, 260–77.

36. Cf. Num. 12:13; Hos. 5:13; 6:1; 1 Sam. 6; Pss. 103:3; 147:3; Job 5:18; Isa. 30:26; Jer. 33:6.

37. Cf. L. P. Hogan, *Healing in the Second Temple Period* (Freiburg, Switzerland: Universitätsverlag; Göttingen: Vandenhoeck & Ruprecht, 1992), 21.

38. E.g., Job 6:28–30; 7:5; 10:1–12; 13:3; 19:6–20; 23; 27:1–6; 38–40; 42:7–8.

There is a continuation and further development in the Second Temple literature concerning both the sin-illness paradigm and God as the only healer. On the one hand, recounting the exodus from Egypt, the Wisdom of Solomon carries on the traditional idea that God punishes his enemies (the impious Egyptians) with sickness and affliction while protecting his people (the righteous Israelites). While "no healing was found" for the Egyptians, the Israelites "were not conquered even by the fangs of venomous serpents, / for [God's] mercy came to their help and healed them" (16:9–10). When God chastised the Israelites, allowing serpents to bite them to remind them of God's word, they were quickly delivered from the effects of the venom (16:11). "For neither herb nor poultice cured them," the author opines, "but it was your word, O Lord, that heals all people" (16:12). On the other hand, Sirach, written around the turn of the second century BCE, for the first time endorses physicians and medicine (*pharmaka*) as God's creation and instrument for healing (38:1–8, 12–14); and both the physicians and the patients need to pray for healing (38:9, 14); nonetheless, it still attributes illness to sin so that the patient is to pray and to "give up [one's] faults and direct [one's] hands rightly, / and cleanse [one's] heart from all sin" (38:10). Given its date and the provenance of its Greek version in Alexandria, this positive shift toward physicians and medicine shows the major influence of Greek culture, and Greek (Hippocratic) medicine in particular, with its significant Hellenistic development in Alexandria through physicians such as Herophilus and Erasistratus.

In Tobit, purportedly written during the Assyrian captivity but most likely composed in the early second century BCE, illness is seen for the first time as a result of the work of demons in human life. When sparrows' droppings fell in the eyes of Tobit, a devout Jew, the consulted physicians could not cure it, but rather made him completely blind (2:10; cf. Mark 5:25)—yet there is no divine disapproval of Tobit's consultation of physicians. In contrast to Exodus 15:26a, his blindness is not the result of his disobedience to God but rather, "God makes use of natural causes to demonstrate his powers."[39] Then a narrative moves to Sarah, who had been given in marriage seven times but had never been able to consummate any of her marriages; on each occasion the demon Asmodeus had killed the bridegroom.[40] While Sarah prays for her release from her affliction, Tobit

---

39. Allan, "The Physician in Ancient Israel," 382.

40. Allan, "The Physician in Ancient Israel," 383n37, notes that this is the first appearance of the demon Asmodeus, which was associated with the Persian sky demon

also prays for his release from his life. Their prayers are answered, and the angel Raphael, whose name means "God heals," is sent to heal them both—"Tobit, by removing the white films from his eyes, so that he might see God's light with his eyes; and Sarah, daughter of Raguel, by giving her in marriage to Tobias, son of Tobit, and by setting her free from the wicked demon Asmodeus" (3:17). How does Raphael heal them? He uses the entrails of a fish to restore Tobit's sight (11:8) and, at the same time, uses fumigation to exorcise demons (6:7; 8:1–3). Nigel Allan notes that the magical use of vile-smelling smoke to expel an evil spirit was common in the ancient world, especially in Egypt, and that fish gall as a medical prescription for blindness was well known in the ancient world as well.[41] Now, with the marriage of Tobias and Sarah consummated, Tobit's sight restored, and Raphael's identity revealed, people praise God for all God has done (chap. 12). In this story, "the helplessness of the physicians serves to emphasize that health is solely within the gift of the Lord, by whatever means he chooses to employ, and it is significant that those [magical and medical] means, although foreign to Israel, were well known and accepted in a narrative written for the edification of devout Jews."[42]

Similarly, in 1 Enoch, another example of Second Temple literature written in the first half of the second century BCE, fallen angels cause human ailments (6–11), and also disclose to humanity forbidden heavenly secrets, charms, and enchantments with *pharmaka* (i.e., drugs, potions, or spells) (7:1; 8:3; 9:6). It is once again Raphael as God's agent of healing who announces the eternal judgment on the fallen angels and the subsequent healing of the earth, including all diseases and wounds of people (10:4–14). In the first century CE, the Jewish historian Josephus also shows the influence of both a belief in the connection between demons and illness and a belief in the Hellenistic medical tradition. According to his *Jewish Antiquities*, God granted Solomon knowledge of "the art used against demons for the benefit and healing of" humans; Solomon thus "composed incantations by which illnesses are relieved, and left behind forms of exorcisms with which those possessed by demons drive them out, never to return" (8.44–46). Josephus goes on to attest to an exorcism performed in the presence of the emperor Vespasian using these Solomonic exorcistic

Aeshma Deva, in Jewish writings, and became prominent in talmudic, Christian, and later Jewish literature.

41. Allan, "The Physician in Ancient Israel," 383–84.
42. Allan, "The Physician in Ancient Israel," 385.

formulas (8.46–47). At the same time, Josephus further highlights Solomon's divinely given detailed knowledge of all kinds of creatures, herbs, and trees that he studied "philosophically and revealed the most complete knowledge of their several properties" (8.44). Similar to the physicians described by Sirach and similar to Celsus and Pliny and the tradition defended by them, the "welfare of human beings is dependent on knowledge of herbs and other natural items with medicinal potential, to be administered by those, like Solomon, who are knowledgeable in such matters."[43] Thus, Josephus describes the medical interests and skills of the Essenes in this Solomonic tradition while noting that they followed a way of life based on the Pythagorian teachings: "They display an extraordinary interest in the writings of the ancients, singling out in particular those which make for the welfare of the soul and the body; with the help of these, and with a view to the treatment of diseases, they make investigation into medical roots and the properties of stones" (*Jewish Antiquities* 8.136).

The gospel tradition in the Christian Bible considers physicians a given in the society at the time. Jesus justifies his association with tax collectors and sinners with a proverbial saying: "Those who are well have no need of a physician, but those who are sick; I have come to call not the righteous but sinners" (Mark 2:17; cf. Matt. 9:12; Luke 5:31). And yet, at least an indirect challenge to and criticism of the adequacy of the physicians' therapy are preserved when Mark and Luke report the substantial and prolonged suffering of the woman with menorrhagia, who despite being under the care of many doctors, and spending all she had, only got worse over time (Mark 5:26; Luke 8:43); by contrast, she is healed instantly by Jesus. Luke also reports Jesus's prophetic insight when he says to his hometown audience in Nazareth that they will reject him even as they acknowledge his healing power: "Doubtless you will quote to me this proverb, 'Doctor, cure yourself!'" (Luke 4:23).

However, the gospel tradition carries on the (relatively recent) belief in the demon-illness connection and introduces the belief in Jesus as God's agent of healing[44] par excellence through his healing ministry.[45] Of the

43. Howard Clark Kee, *Medicine, Miracle, and Magic in New Testament Times*, SNTSMS 55 (Cambridge: Cambridge University Press, 1986), 23.

44. Jesus's healings in the Gospels always include cures.

45. On the topic of Jesus's healing, see, for example, Frederick J. Gaiser, *Healing in the Bible: Theological Insight for Christian Ministry* (Grand Rapids: Baker Academic, 2010), chaps. 11–15; Graham H. Twelftree, *Jesus the Miracle Worker: A Historical and Theological Study* (Downers Grove, IL: InterVarsity Press, 1999), especially chaps.

twenty-eight healing and exorcism stories and summaries mentioned in the Gospels (including three resurrection stories), five have to do with demonic possessions and Jesus enacting deliverance through exorcisms, as in the cases of a "demoniac" near Gerasa (Mark 5:1–20; Matt. 8:28–34; Luke 8:26–39) and the Syrophoenician woman's daughter (Mark 7:24–30; Matt. 15:21–28). Here, as Gary Ferngren notes, exorcism and healing should be differentiated in Jesus's ministry[46] as Mark 1:34 does: "And he [Jesus] cured (*etherapeusen*) many who were sick with various diseases, and cast out (*exebalen*) many demons." However, Amanda Porterfield cautions not to read into this distinction a Cartesian dichotomy between mind and body, or "to suggest that the gospel writers viewed demonic possession as a spiritual problem wholly separate from the natural problems to which human beings were heir."[47] Jesus's exorcisms (which occur only in the Synoptics) are expressions of God's rule/kingdom and a direct assault against the dominion of Satan;[48] they free the possessed and isolated persons to be themselves (identity/integrity) and to be with others (community).[49] However, in three cases in the Synoptics, demonic possession is accompanied by physical impairment: that of a mute man (Matt. 9:32–34), that of a mute and blind man (Matt. 12:22–23; Luke 11:14–16), and that of an epileptic boy (Mark 9:14–29; Matt. 17:14–21; Luke 9:37–43a). In the first two cases, a demoniac, who is mute, is brought to Jesus and speaks after Jesus casts the demon out of him, and a demoniac who is blind and mute speaks and sees after Jesus cures (*therapeuei*) him; while the physical im-

---

11–15; Morton Kelsey, *Healing and Christianity: A Classic Study* (Minneapolis: Augsburg, 1995), especially chap. 4; on Jesus's healing from the perspective of psychosocial or medical anthropology, see Justin Meggitt, "The Historical Jesus and Healing: Jesus' Miracles in Psychosocial Context," in *Spiritual Healing: Scientific and Religious Perspectives*, ed. Fraser Watts (Cambridge: Cambridge University Press, 2011), 17–43; John J. Pilch, *Healing in the New Testament: Insights from Medical and Mediterranean Anthropology* (Minneapolis: Fortress, 2000).

46. Gary B. Ferngren, "Early Christian Views of the Demonic Etiology of Disease," in *From Athens to Jerusalem: Medicine in Hellenized Jewish Lore and Early Christian Literature*, ed. S. S. Kottek, M. Horstmanschoff, et al. (Rotterdam: Erasmus, 2000), 185–86; Gary B. Ferngren, *Medicine and Health Care in Early Christianity* (Baltimore: Johns Hopkins University Press, 2009), 46.

47. Amanda Porterfield, *Healing in the History of Christianity* (Oxford: Oxford University Press, 2005), 22.

48. Cf. John T. Carroll, "Sickness and Healing in the New Testament Gospels," *Int* 49, no. 2 (1995): 137.

49. See Verhey, "Health and Healing," 34.

pairment is not directly attributed to demonic possession in either case, the close association is there. In the third case, of an epileptic boy brought to Jesus, Jesus cures the boy through exorcism, and there is a direct demonic etiology of epilepsy. In addition, while there is no formal demon-possession in the Lukan story of a crippled and bent-over woman, Jesus attributes her illness (*astheneias*) directly to Satan and sets her free on Sabbath (Luke 13:10–17).

A majority of the healing stories (seventeen out of twenty-eight) only describe medical symptoms/conditions, not causes, in "naturalistic" ways, and Jesus's healings; those conditions, which include blindness, muteness, deafness, fever, skin diseases ("leprosy"), paralysis, dropsy, a uterine hemorrhage, dysentery, and a withered hand, belong to the category of ordinary diseases or congenital conditions.[50] Of those seventeen healing stories/summaries of blindness, skin diseases ("leprosy"), and infirmities, only two stories—that of the paralyzed man in the Synoptics (Mark 2:1–12; Matt. 9:1–8; Luke 5:17–26) and that of the crippled man at Bethesda in John (John 5:1–15)—reveal the connection between personal sin and illness. In the former, Jesus heals the paralytic man with the words, "Son, your sins are forgiven" (Mark 2:5) and "I say to you, stand up, take your mat and go to your home" (Mark 2:11); in the latter Jesus heals the crippled man with the words, "Stand up, take your mat and walk" (John 5:8), and later warns him: "Do not sin anymore, so that nothing worse happens to you" (John 5:14). Nevertheless, the intention of the former seems more to stress Jesus's authority to forgive, as the ensuing controversy indicates; and the point of the latter healing story seems more to heighten the Jewish persecution of Jesus due to his healing on the Sabbath. While illness and sin can be closely related in the Gospels, so too are experiences of healing and forgiveness. Following in the tradition of the Hebrew Bible, in which people's health involves physical, psychological, spiritual, and relational dimensions (a kind of holistic health known as *shalom*),[51] stories in the Gospels feature the healing of the whole person; and yet, the gospel accounts of Jesus's healing as a whole tend to deny a direct one-to-one relationship between sin and illness as one of cause and effect. In fact, Jesus, in healing a crippled and bent-over woman on the Sabbath (Luke 13:10–17) and a man born blind (John 9:1–41), clearly rejects the people's assumption that their

50. Ferngren, "Early Christian Views," 186; Ferngren, *Medicine and Health*, 46.
51. Cf. Carroll, "Sickness and Healing," 139.

illnesses resulted from (personal or parental) sin. Most importantly, Jesus's own suffering and death most clearly breaks with the idea of retribution, as all parts of the Christian Bible (NT) affirm that he died not for his own sins but for the sins of all others.

Jesus's healing stories in Matthew and Luke are explicitly presented as the fulfillment of Scripture, particularly of Isaiah. In Matthew, two healing passages are followed by direct quotations from the Servant Songs of Isaiah. As the messianic Son of David and the Servant of God, Jesus drove out evil spirits and healed "all who were sick" (8:16) so as to fulfill Isaiah 53:4: "He took our infirmities and bore our diseases" (Matt. 8:17). Again in chapter 12, Jesus was followed by many and healed "all their sick," fulfilling Isaiah 42:1–4, in which God chose the Servant, he who would lead justice to victory, and put his Spirit upon him (Matt. 12:15–21). In Luke, Jesus himself frames his public ministry, including his acts of healing, with Isaiah 61:1–2 in Nazareth (Luke 4:16–20), declaring the Scripture's fulfillment in the audience's hearing (4:21); and in response to the question from John the Baptizer whether he is the promised Messiah, Jesus defines and describes his ministry by speaking of himself as the one who enables the blind to see, the lame to walk, lepers to be cleansed, the deaf to hear, and the dead to be raised (7:22; cf. Matt. 11:4–5; Isa. 29:18; 35:5–6; 61:1).[52] Jesus's words follow Luke's summary that he "cured (*etherapeusen*) many people of diseases, plagues, and evil spirits, and had given sight to many who were blind" (7:21). Jesus's healing ministry, with its tangible signs of the eschatological kingdom, confirms that he is the one who is to come and that disease or illness is not an established part of the divine order of reality.

On the other hand, Jesus's healings, particularly of the "lepers" and the woman with a heavy menstrual flow, upset the existing Jewish religious and cultic system devoted to preserving boundaries and separateness through purity laws. The Synoptic account of Jesus healing a man with "leprosy" (Mark 1:40–45; Matt. 8:1–4; Luke 5:12–16) and the Lukan account of Jesus healing ten "lepers" (Luke 17:11–19) opened a way for them to be reintegrated into their respective communities. Jesus calls the unclean woman,

---

52. C. Wahlen, "Healing," in *Dictionary of Jesus and the Gospels*, ed. Joel B. Green, Jeannine K. Brown, and Nicholas Perrin (Downers Grove, IL: IVP Academic, 2013), 368, notes that although "these passages in Isaiah do not mention resurrection, the Messianic Apocalypse from Qumran describes the messianic era in terms of these same passages, adding that the Lord 'will make the dead live'" (4Q521 II, 1, 8, 12).

who had been afflicted with an issue of blood for twelve years, "daughter" and affirms her healing faith as she is healed, and thus he also opens a way for her to be reintegrated into her community (Mark 5:25–34; Matt. 9:20–22; Luke 8:43–48). Jesus's healings on the Sabbath result in controversies and persecution but restore the Sabbath law toward healing, life, and freedom;[53] they echo "creation, not only in displaying divine power and authority, but also in returning individuals to total health on the day that was the first full day of life for human beings in the beginning."[54] In Mark and Matthew, Jesus carries on his healing ministry beyond the land of Israel to the gentile land (Mark 7:31–37; Matt. 15:21–28). And a return of a Samaritan "leper," who, although a foreigner, was the only one of the ten healed "lepers" to praise God (Luke 17:17–18), together with Jesus's healing of the Roman centurion's slave (Luke 7:1–10), is paradigmatically significant in Luke, as the healing of gentiles highlights Luke's vision of their inclusion in God's salvation. As a result of Jesus's healings, then, people who were once excluded from participation in the people of Israel due to their ritual impurity (due to their illness or physical impairment) are declared clean and included in the covenant people by Jesus's own authority.[55] Thus his healings (including exorcisms) function "as clues and foretastes of a new situation in which the purpose of God will finally be accomplished in creation and his people will be vindicated and at peace"[56]—*shalom*.

In Acts (and already in Matthew and Luke) the apostles also carry on Jesus's healing ministry, as in the story of Peter healing the crippled man at the temple gate (3:1–10), which is followed by Luke's summary of a mass healing and exorcism by Peter and the other apostles in Jerusalem (5:12–16). In a similar fashion, the story of Paul healing the crippled man at Lystra (14:8–10) is followed by Paul's healing of Publius's father and "the rest of the sick" in Malta (28:8–10). We can note here how the geographic and

53. Matt. 12:9–14; Luke 6:6–11; 13:10–17; John 9:1–41.

54. Wahlen, "Healing," 369. See also the comments of Peder Borgen on Jesus's healing on the Sabbath: "The act of Jesus was the action of the Creator, who in this case made God-given ordinance of the Sabbath subordinate in relation to the healing of this sick man. Since disease was part of the revolt of the creation against its creator, the healing meant that God's plan for humanity was being put into action." Peder Borgen, "Miracles of Healing in the New Testament: Some Observations," *ST* 35 (1981): 103.

55. Bethany McKinney Fox, *Disability and the Way of Jesus: Holistic Healing in the Gospels and the Church* (Downers Grove, IL: IVP Academic, 2019), 41.

56. Kee, *Medicine, Miracle, and Magic*, 79.

thematic spread of the gospel is followed by miracles of healing and exorcism by the two respective pillars of the early Christian church: Peter and Paul.[57] As in the gospel tradition, here in Acts there is an understanding that disease/illness is not from God but mostly from natural causes, as it is with ordinary conditions such as lameness, paralysis, dysentery; and there is also a distinction between healing the sick and casting out demons (5:16; 8:17; 19:11–12). Although disease/illness has been a constant factor in human existence since the Fall, it is not God's will for humankind.[58] However, at the same time, illness and death are also at times the divinely appointed consequences of the antagonists' sinful actions. For example, Elymas, a Jewish sorcerer, who attempted to hinder Paul's ministry, was struck with blindness at Paul's curse (13:6–12); King Herod Agrippa I, who had arrested Christians and accepted people's praise for himself rather than praising God, was immediately struck by the angel of the Lord and died soon afterward (12:21–23). Here, as one can see, Acts also makes a clear distinction between the apostles who perform miracles/healings and their antagonists (here non-Christian wonder workers), as in the case of Elymas and also Simon the sorcerer, who sought to buy the power of the Holy Spirit (8:18–23).

In the Pauline epistles, Paul himself never mentions the healing aspect of his ministry, at least not directly, even though he lists healing (but not exorcism) as one of the spiritual gifts (*charismata iamatōn*; 1 Cor. 12:9, 28, 30). In fact, as far as his own "thorn" (*skolops*) was concerned, healing was not necessarily the norm or a barometer of one's faith. Much like Job, we see a divine passive at work as he was "given" a thorn (*skolops*) in his flesh, "a messenger of Satan," to torment him, to keep him from becoming conceited due to his vision of paradise. His *skolops* is a chronic, recurring illness[59] that is torturously debilitating but not debarring him from his apostolic work. God's response to his repeated prayer for healing is, "My grace is sufficient for you, for my power is made perfect in weakness"

---

57. For more healings by Peter and Paul, see Acts 9:32–35 and 19:11–12, respectively.

58. Judith L. Hill, "Health, Sickness and Healing in the New Testament: A Brief Theology," *Africa Journal of Evangelical Theology* 26, no. 2 (2007): 171.

59. There is a plethora of ancient and contemporary suggestions as to what Paul's "thorn" might have been, ranging from a physical ailment, such as eye disease, malaria, migraine, or epilepsy, to religious opposition or persecution, mental or emotional illness such as depression, neurosis, or even spiritual temptation. I take it as some form of physical illness (perhaps malaria) based on the most natural reading of the text in a Pauline context. Cf. Colin J. Hemer, "Medicine in the New Testament World," in Palmer, *Medicine and the Bible*, 78–80.

(*astheneiai*) (2 Cor. 12:9a). This new revelation, radically different from Job, sets everything in perspective for him, and he will rather boast all the more gladly about his weaknesses (*astheneiais*), so that Christ's power may rest on him. He continues, "That is why, for Christ's sake, I delight in weaknesses (*astheneiais*), in insults, in hardships, in persecutions, in difficulties. For when I am weak (*asthenō*), then I am strong" (2 Cor. 12:9b–10 NIV). Here, healing is not divine favor, but rather God's power (grace) in weakness, illness, hardships, or persecutions; Paul's endurance of his thorn "demonstrates his participation in Christ's crucifixion."[60] This new shift in attitude will be noteworthy for the subsequent early Christian attitude concerning illness and pain. Furthermore, while Paul does *not* denigrate the physical dimensions of life, he seems to subordinate them to the higher, spiritual purposes of God.[61] For him the kingdom of God is primarily spiritual: "For the kingdom of God is not food and drink but righteousness and peace and joy in the Holy Spirit" (Rom. 14:17; cf. 1 Cor. 15:50). His body may be failing and may suffer now, but his inner spiritual life is renewed every day: "For this slight momentary affliction is preparing us for an eternal weight of glory beyond all measure, because we look not at what can be seen but at what cannot be seen; for what can be seen is temporary, but what cannot be seen is eternal" (2 Cor. 4:17–18). Nonetheless, it seems that "doctors and medical means of healing are also useful and do not contradict Paul's belief in miraculous healing," as seen in the Pauline tradition (Col. 4:14; 2 Tim. 4:11; cf. Philem. 24).[62] "Paul" in that tradition suggests a medicinal remedy (wine) to Timothy to treat his stomach issue and chronic illness (1 Tim. 5:23). Therefore, in Pauline theology, "healing is real and may come through a miracle, but it may also come through time and medicine—or it may not come at all."[63]

Finally, in James we see a connection between sin and illness, though it may be tentative. Toward the end of the treatise, in the context of prayers, the author of James exhorts the troubled to pray, the cheerful to sing praise, and the sick (*asthenei*) to pray.[64] The latter case is special since it involves not only calling the elders of the church to pray for the sick person with

60. Susan R. Garrett, "Paul's Thorn and Cultural Models of Affliction," in *The Social World of the First Christians: Essays in Honor of Wayne A. Meeks*, ed. O. L. Yarbrough and L. M. White (Minneapolis: Fortress, 1995), 96.

61. Cf. Hill, "Health, Sickness and Healing," 174.

62. Hill, "Health, Sickness and Healing," 177.

63. Hill, "Health, Sickness and Healing," 177.

64. On the health-care system in James from the medical anthropological per-

the anointing of oil but also the mutual confession of sins within the church so that those participants may be healed (5:13–16)—and if the sick person has sinned, that person will be forgiven. Here the care of the church is expressed in its belief that prayer can be conjoined with appropriate medical care. The use of olive oil, the medicinal use of which was "remarkably widespread," works together with prayer to heal.[65] In addition, both sickness and healing here seem to be more inclusive—both physical and spiritual; and while the "if" clause makes the connection between sin and illness as well as between healing and forgiveness, that connection is not absolute or linear in this context. Still, all healing ultimately comes from God—it is the Lord who raises up the formerly sick person (5:15b).

Health in Early Christianity

The notions of health in early Christianity in the second through the fifth centuries show how early Christians incorporated common concepts of health in Greek medicine and philosophy into a Christian framework of creation and providence. First, Christian authors adopted the well-established medico-philosophical model that health consists of a balance (*krasis*) of the elementary qualities (of heat, cold, wet, and dry) and an equilibrium of bodily constitution.[66] For the Cappadocians, health is the ultimate balance between those elements and humors: blood, phlegm, yellow bile, and black bile.[67] Nemesius, bishop of Emesa in the late fourth century, who wrote *On the Nature of Man* (*De natura hominis*), confirms the composition of the human body out of the four elements and the four humors (blood, phlegm, yellow bile, and black bile) (4.44); for him "the right mixture (*eukrasia*) of the warm, cold, dry and moist [elements] is health, and the symmetry (*symmetria*) of the limbs with a good complexion constitutes the beauty of the body" (2.25). Augustine further identifies health (*sanitas, salus*) with unity—health of the body is when there is a balanced order among its parts; health of the soul is when there is an agreement between its condition and

---

spective, see Martin C. Albl, "'Are Any among You Sick?': The Health Care System in the Letter of James," *JBL* 121, no. 1 (2002): 123–43.

65. Hemer, "Medicine in the New Testament World," 82.

66. E.g., Athanasius, *C. Gent.* 27.54–62; John Chrysostom, *De statuis hom.* 10.4.

67. Susan R. Holman, "The Social Leper in Gregory of Nyssa's and Gregory of Nazianzus's 'περί Φιλοπτωχίας,'" *HTR* 92, no. 3 (1999): 291; cf. Gregory of Nyssa, *Hom. op.* 30.5; *On Virginity* 22.

its action. A healthy person is one who has a duly ordered life with a proper balance between body and soul (*Civ.* 19.13.1).

In order to maintain health, diet must be plain and balanced. Clement of Alexandria recommends no sauces or cooked food for good digestion and lightness of body, while stomach problems are caused "by the un-happy arts of cookery and pastry making" (*Paed.* 2.2). In addition, he also recommends, for a healthy body and courageous soul, exercises such as walking, playing ball, and fencing before meals (*Strom.* 2.6.25.4). In fact, his *Paedagogus* is filled with his discussion and suggestions of different as-pects of the hygiene of life (regimen) for his relatively wealthy audience of Alexandria: besides simple dietary rules, he recommends measured sleep on a hard bed, hygiene, bathing, cleanness of clothing, and even sexual life, "all of which are intended to promote a simple life that conforms to nature" (*kata physin*).[68] As Jacqueline Lagrée notes, all of these are "valid for the equilibrium of both body and soul since they educate one in simplicity and in caring only for what is necessary."[69]

Second, and more prominently, early Christian authors argue for a func-tional notion of health by adapting Aristotelian-Galenic teleological anat-omy to the Christian understanding of creation and providence. A number of patristic authors in the fourth and fifth centuries devoted works to this theme especially as defenders of Christian providence against a certain non-Christian disbelief in providence. Lactantius wrote first about this theme during the first two years of the Great Persecution (303–313 CE); in *The Workmanship of God* (*De opificio Dei*) (c. 304 CE), he attempts to vindicate divine providence from the attacks of both ancient and current schools of philosophy, relying on the Stoic view of providence. Refuting Epicureans in particular, and in conversation with Cicero's *De natura deo-rum*, Lactantius stresses that the body is the sign of divine providence and that a human being is the special object of that providence; as a way to argue for this point, he engages in a teleological or functional account of the parts of the human body in detail. To begin with, God, "our Artificer and Parent," made a human being "naked and unarmed because he could be armed by his abilities and clothed by his reason" (*Opif.* 2.7). In fact, the

68. Jacqueline Lagrée, "Wisdom, Health, Salvation: The Medical Model in the Works of Clement of Alexandria," in Kottek, Horstmanschoff, et al., *From Athens to Jerusalem*, 237. For examples of the hygiene of life, see Clement of Alexandria, *Paed.* 3.6.50; 3.9.46; 3.10.49; 3.11.

69. Lagrée, "Wisdom, Health, Salvation," 237.

human's upright stature indicates the priority of the human's rationality (*mens*) as the ruler of the body and its position overlooking the body and the whole of creation (8.2–5; 16.4ff.).[70] Then he speaks of God's "delicate workmanship of the eyes" to affirm the reliability of the senses (8.9); God "fashioned two identical eyeballs and controlled them so completely that they could not only be directed and turned entirely, but also be moved and bent partially" (8.17). He filled the eyeballs "with a clear, liquid substance, and in the middle of these are held enclosed sparks of light which we call pupils" (8.17). Through these eyes as windows the mind sees, and "there is mingled and joined into one the vision or object seen of the two lights" (8.17). Lactantius indulges in the utilitarian and aesthetic "perfection of doubleness in unity" in human body parts (10.9–11);[71] two eyes, two nostrils, two ears, two lips, two hands, and two feet not only have power for their proper utility and function of seeing, hearing, and moving, etc., but also confer marvelous beauty. This combination of utility and beauty, which we have already seen in Galen, is a consistent and convincing argument for Lactantius throughout the treatise. Compare what he says about hands and fingers with Galen's statements on the same parts in chapter 1 of *On the Usefulness of the Parts*. The Master Artificer fashioned the hands with a slightly concave surface with which to grasp whatever, and fingers of varying length, and the perfect number of five so as to maximize their utility. What about fingernails? Like Galen, Lactantius sees the full extent of their usefulness: "The form of the nails, rounded and tightly grasping the ends of the fingers with curved protection lest the softness of the flesh falter in the function of holding," enhances holding and "furnishes beautiful ornament" (*Opif.* 10.22). Elsewhere in the same text, Lactantius further observes the functions of the inner organs that serve the nourishment of the body (such as stomach, intestine, and bladder), those that serve the nourishment of the soul (such as lungs and windpipe), and the reproduc-

---

70. Thomas Halton notes that the notion of the uniqueness of the human upright posture has a long history in Greek and Latin literature and the Hebrew Bible (Eccles. 7:29), including in Xenophon, Aristotle, Cicero, Seneca, and Galen. Early Christian authors such as Clement of Alexandria and Basil the Great also mention that. Gregory of Nyssa and Theodoret of Cyrrhus note this uniqueness as part of God's providential design of the human body. See Theodoret of Cyrus, *On Divine Providence*, translated and annotated by Thomas Halton, Ancient Christian Writers 49 (New York and Mahwah, NJ, 1988), 187–88.

71. Peter A. Roots, "The *De Opificio Dei*: The Workmanship of God and Lactantius," *ClQ* 37, no. 2 (1987): 476.

tive organs (11–12).[72] His teleological anatomy ends with God's plan and wisdom shown through the utility and beauty of the external, lower parts of the body such as legs, knees, feet, and toes (13.1–8).

Gregory of Nyssa, in *On the Making of Man* (*De hominis opificio*), the work intended to complete Basil the Great's *Hexameron*, also utilizes the argument of teleological anatomy to highlight the uniqueness and sovereignty of humanity. Like Galen and Lactantius, Gregory wonders why a human being lacks natural weapons and covering like animals (such as horns or hooves), and, like them, he also responds that "he who was promoted to rule over the rest of the creatures should be defended by nature with arms of his own so that he might not need assistance from others for his own security" (7.1). What seems to be a deficiency in human nature is a means of obtaining dominion over the subject creatures; for example, our slowness of foot brought the horse to supply our need, and our nakedness helped us to manage the sheep that produce wool for our clothing (7.2–3). For Gregory, human beings' upright posture is a mark of royal dignity, which is qualitatively different from the stooped position of animals; and thus the Maker's formation of hands is the very mark of this dignity and a special property and instrument of the rational nature of humanity, which reflects the image of God (8.1–2; 9.1). In other words, the human body is weaker than that of other animals because it is intentionally designed as an instrument for reason. The rational part of the soul, the mind (*nous*), pervades all sense organs and permeates the entire body, suggesting that rationality is endemic to human nature (14–15).

For Gregory, human biological life consists of three aspects. The first relates to the essential part of life: the brain, the heart, and the liver; the second is the sense organs through which a person enjoys "participation in the pleasures of life"; finally, the sex organs serve to supply future succession of life; the stomach and lungs act as support organs for these three aspects of human physical existence (30.2). In terms of the interaction of the elements of the material world and their utilization of the body, Gregory explains that the body is a perfect balance of heat and cold, and of moisture and dryness, the four fundamental elements. If a thing is too dry, it "does not admit the action of the sense"; if it is too moist, the impression of the form "would not remain in most substance" (30.7). The skillful workmanship of nature also took the hard (bones, joints) and the soft substance (sinews, muscles) together to make a person mobile (30.8–9). According

---

72. Roots, "The *De Opificio Dei*," 476.

to Gregory, "the brain contributes to life in a special degree" as the cause of life; "for if the tissue surrounding it receives any wound or lesion, death immediately follows the injury" (30.10). The membrane surrounding the brain is a "foundation for the senses," and the brain itself is a (Platonic/Galenic) charioteer, steering the body through the muscles and nerves (12).[73] The heart is the source of heat in the body, as the nature of heat is motion and the heart is in perpetual motion (30.11). The lungs bring air into the body for consumption by the heart to produce heat; since some matter must be consumed in order to generate more heat, it is the stomach's role to bring in more material (30.15–20). Thus, the fact that the openings for bringing air and food into the body are near to one another and are in close proximity to the heart where they are consumed provides Gregory even more evidence of the hand of the Divine in anatomy. "Under the influence of Galen, Gregory saw the functioning of the body in its natural state as the idealized form of the body with which the intelligible mind had the innate affinity to interact."[74]

John Chrysostom in his *Homilies on the Statutes* 11 also responds to those who would find fault with the apparently weak and corruptible nature and general structure of the human body. To the familiar question of human frailty in comparison to other animals having their weapons in their own body (such as horns and claws) (10), John replies that God made the weapons of human beings extraneous to them in order to show that a human being is a "gentle animal," that we might be free and unfettered in our body, and that we surpass the animals not only with our rationality but also with our superior body, a body designed for gentleness (11). Again, human physical weakness is a necessary part of God's providential design of humanity fit for reason. God made the human body to reflect the soul's nobility and execute its commands: "Nor did he [the Creator] create the body in such a way for its own sake, but so that it might participate in the rational soul; for if it were not [created] in such a way, the activities of soul would be certainly impeded; and this is clear from diseases. For, if the complexion

---

73. Cf. Galen, *Caus. resp.* 4.469.10–14K: "Choice is like a charioteer who moves the reins and the horses; the nerves are like the reins, and the muscles are like the horses."

74. Susan Wessel, "The Reception of Greek Science in Gregory of Nyssa's '*De hominis opificio,*'" *VC* 63, no. 1 (2009): 40; cf. Lactantius, *Opif.* 14.2. Although I do not deal with the relationship of the body and the soul in Gregory here, on that topic see J. P. Cavarnos, "Relation of the Body and Soul in the Thought of Gregory of Nyssa," in *Gregor von Nyssa und die Philosophie: Zweites Internationales Kolloquium über Gregor von Nyssa*, ed. H. Dörrie et al. (Leiden: Brill, 1976), 61–78.

of the flesh deviates from its accustomed constitution in any minor way, many activities of the soul are indeed impeded—the sort of thing I mean is if the brain becomes hotter or colder" (11).[75]

As we can see, the common ground for these Christian authors is that the human body is specifically attuned for reason.[76] Any change or adjustment would result in diseases of the soul and subvert the Creator's providential design geared toward "a useful purpose" (11). Moreover, as Jessica Wright notes, "the precision necessary for the design of a rational being itself comprises a bodily weakness, since the very 'complexion of the flesh' is held in tenuous balance."[77] We can also see the medico-philosophical significance of the brain as a crucial factor for psychic health through his example of the brain as the instrument of the rational soul whose imbalance impedes the activities of soul.[78] As Gregory noted, John also notices the brain's soft structure for receiving sensation and its protection with a middle membrane (*De statuis hom.* 8)[79] and the heart's "preeminence over all the members of our body" (9). Finally, in this homily we see the combined argument of utility and beauty again. John marvels at the beauty and visual power of an eye—how "it can at once survey the high aerial expanse, and by the aid of a small pupil embrace the mountains, forests, hills, the ocean, yea, the heaven, by so small a thing" (6)! God the Creator's manifold wisdom can also be seen in the design of the hairs of the eyelids, which "keep dust and light substances at a distance from the eyes," and the design of the eyebrows, which receive "perspiration as it descends from the forehead, [not] permitting it to annoy the eyes" (7). These not only provide the necessary protection of the eyes but also contribute to their beauty (7).

We now turn to Nemesius, whose work *On the Nature of Man* is the most scientific of all the Christian writings of this time, although his medical knowledge "has been judged to be that of an educated layman, not of a former practitioner."[80] As Nemesius articulates a teleological account

---

75. This translation is by Jessica Wright, "John Chrysostom and the Rhetoric of Cerebral Vulnerability," *Studia Patristica* 81 (2017): 120.

76. Cf. Wright, "John Chrysostom," 120.

77. Wright, "John Chrysostom," 120; see also Jessica Wright, "Brain and Soul in Late Antiquity" (PhD diss., Princeton University, 2016), 210–11.

78. Cf. Wright, "John Chrysostom," 121.

79. See Galen, *UP* 1.479 (8.6); 2.3–4 (9.1).

80. R. W. Sharples and P. J. van der Eijk, introduction to *Nemesius: On the Nature of Man*, translated with an introduction and notes by R. W. Sharples and P. J. van der Eijk (Liverpool: Liverpool University Press, 2008), 3.

of the structure and workings of the human throughout the treatise, his thesis is clear: "Man is eminently constructed of an intellectual soul and a body, indeed so well that he could not have come to be, nor be composed, well in any other way" (1.2).[81] Here Nemesius agrees with other Christian leaders that the body is an instrument of the soul; the body's suitable constitution means that it is united with the soul and works with, not against, the soul, which is incorporeal, unique to each individual, and created at the beginning of the world (2.24, 25). The human as the "best of possible worlds" is manifest because "the foresight of providence is revealed in every part of the body, which the industrious can catalogue from various works" (42.122). Providence for Nemesius is: (1) God's care of the existing things and (2) God's purpose whereby all existing things receive their most favorable way of life (42.125). Thus Nemesius, following Galen, elaborates in great detail cognitive and psychological capacities such as thinking, memory, perception, emotions, and voluntary movement, together with other bodily functions and processes like pulsation, nutrition, digestion, and reproduction.[82] As it relates to the internal constitution of the body's parts, yellow bile, for example, "contributes to digestion, stimulates excretion, and thus in a way becomes one part of the nutritive powers," but "it also provides a certain heat to the body" (28.92). Kidneys purify the blood and are responsible for the appetite for sexual intercourse (28.92). As it relates to the external makeup of the body, skin is "both of the soft flesh and of all the other internal parts"; bones are "a support of the whole body, especially the spine"; and nails are "not only for scratching and tearing the hardness of the skin but also for picking up small objects" (28.93). In human beings, the internal and external parts of the body work together perfectly, "and in such a way that it could not have been well otherwise" (4.46.16–17). All of these illustrate God's detailed care and purpose for the best outcomes for the human body.

Augustine, in the *City of God* (*De civitate Dei*) in the early fifth century, also sees that the beautiful, purposeful structure of the body is a manifestation of God's goodness and providence, designed for the service of a rational soul: "Moreover, even in the body, which is something we have in

81. See also 4.46: "Not every animal possesses all the parts of the body, . . . but man has them all and perfect, and in such a way that it could not have been well otherwise."

82. Cf. Philip J. van der Eijk, "Galen and Early Christians on the Role of the Divine in the Causation and Treatment of Health and Disease," *Early Christianity* 5 (2014): 351.

common with the brute creation—which is in fact weaker than the bodies of any of the lower animals—even here what evidence we find of the goodness of God, of the providence of the mighty Creator! Are not the sense organs and the other parts of that body so arranged, and the form and shape and size of the whole body so designed as to show that it was created as the servant to the rational soul?" (22.24). Augustine continues with examples of the erect posture of the human, facing toward the sky, and admonishes "him to fix his thoughts on heavenly things," and the wonderful mobility of the tongue and hands, so adapted for speaking and writing as well as for a multitude of arts and crafts (22.24). This is sufficient proof that the body was designed to be joined to the immaterial soul, and it also shows the excellence of the soul it serves (22.24), since the soul is created in the image of God.[83] Like Lactantius and John Chrysostom, Augustine argues that "the harmony of the body means that no part has been created for utility which does not also contribute to its beauty."[84] That means "a sublime creative *final causality* runs through every part of such an instrument";[85] and "this above all is a motive for the praise of the Creator" (22.24).

Finally, Theodoret of Cyrrhus, in his *On Divine Providence* (*De providentia*) (c. 437), seeks to demonstrate providence from the composition of the human body and also from the composition of human hands and crafts, among other things. Theodoret likens those who deny divine providence to those afflicted with disease and proposes providence as their doctrinal remedy (4.1–2). Theodoret is adamant about how providence manifests in human organs, inside and outside (such as the heart, the liver, veins, arteries, the spine, the cranium, the neck, the eyes, the nose, and the ears), and in human faculties (such as speech, sight, smell, and hearing), affirming the status of human beings as rational creatures stamped with the image of their Creator (3.10, 35). Among these, he highlights the heart as "the most important part of a human being in its continuous motion" (for arteries) (3.11) and the head "as the acropolis of the body, raised aloft," protecting the brain with the membranes and the pia mater (3.27–28). Like Galen, Lactantius, Gregory of Nyssa, and Augustine, Theodoret points out the uniqueness of the upright stature of the human. Rather than pointing out

83. *Gen. litt.* 3.20.30–3.2.33; 6.7.12; *Trin.* 14.4.6. On the soul in Augustine, see John M. Rist, *Augustine: Ancient Thought Baptized* (Cambridge: Cambridge University Press, 1994), especially chap. 4.

84. Kevin Corrigan, "The Soul-Body Relation in and before Augustine," *Studia Patristica* 63 (2006): 67; cf. 22.24.

85. Corrigan, "The Soul-Body Relation," 67.

our upright stature as a way to extol the virtues of human rationality or human dignity, however, Theodoret focuses on the structure of the leg's three sections; the leg is jointed at hips, knees, and ankles, each joint bound with sinews and muscles capable of voluntary movement, making it possible for us to walk or run, sit or stand with ease (3.20). Theodoret especially sees human hands and arms as the instruments of reason and therefore signs of God's providential care. The Creator made them in three parts, jointed at the shoulders, elbows, and wrists, which link the forearms with the metacarpal bones, which terminate in five fingers on each side (4.14). The fingers, divided into three sections separated by joints, are enclosed in muscles to facilitate movement, and the skin enveloping the fingers is soft in order not to impede their movements (4.14). The nails protect the soft edges of the fingers and thus are different in function from the claws of an animal, which are, for animals, natural weapons (4.15, 16). Rather, human hands are for holding and gripping, made possible by a thumb and the rest of the fingers (4.33). Like Galen, Lactantius, Gregory, and John Chrysostom, Theodoret avers that because of the hands, which are designed for a rational animal and work according to reason, humanity is not given animals' natural weapons like paws and claws (4.16, 18); and with their hands under the power of reason, humans cultivate agriculture and crafts and discover many sciences and techniques, including medicine, and the ability to communicate in writing (4.16, 18, 22, 38, 32).

Understanding nature and natural phenomena as God's work, and not claiming a divine character separate from the will and divinity of the Creator, early Christian leaders incorporated Hippocratic and Galenic naturalism and teleological anatomy into their own ideas about health.[86] Greco-Roman notions of naturalism also acknowledged the harmony between the soul and the body in a human being, a harmony in which the body is subordinate and instrumental to the soul; and this was considered to be according to nature (*kata physin*). Lastly, then, what is created is what is natural, and what is natural is God-designed. This provided a certain normative notion of health, especially in terms of bodily constitution and functions.

---

86. Owsei Temkin, *Hippocrates in a World of Pagans and Christians* (Baltimore: Johns Hopkins University Press, 1991), 190. Cf. Nemesius, *De nat. hom.* 43.128: "not all that is providential is natural, even if natural events occur providentially. For many works of nature, as was shown in the case of murder. For nature is part of providence, not providence itself."

While this understanding of health was presumably pervasive in early Christianity, another notion of health was portrayed in the emerging ascetic and hagiographic literature, especially from the fourth century onward. As some scholars have already noted, Athanasius of Alexandria, in his *Life of Antony* (c. 356–360), portrays Antony's health via rigorous ascetic discipline (*askēsis*) as a restoration of prelapsarian health.[87] His *askēsis*, far from injuring his body, in fact benefits it. After he spent two decades in a deserted fortress "pursuing the ascetic life by himself," Antony comes out completely unaged—"neither fat from lack of exercise, nor emaciated from fasting or combat with demons, but was just as they had known him prior to his withdrawal" (14). To the bewilderment of all who observed him, Antony's body maintained its former condition and his soul was pure, not affected by grief, pleasure, laughter, or dejection (14). Antony's twenty years of *askēsis* then allowed him to maintain "utter equilibrium, like one guided by reason and steadfast in that which accords with nature" (14). His regimen comprised only one meal of bread and salt per day (sometimes only every second and even fourth day) and water to drink (7). He regularly engaged in physical struggles with the devil overnight that left him in severe pain (8); slept on the bare ground or occasionally a rush mat; disapproved of anointing his skin with oil; and refused to wash, even his feet, for the rest of his life (7, 93). Andrew Crislip characterizes Antony's health of psychic and somatic unity in asceticism as more protological than medical;[88] Owsei Temkin also notes Antony's breach of all the rules of Hippocratic hygiene.[89] In this unity, however, soul and body are locked in dialectical influence in both directions, and it is "the body's signification of the monk's ascetic merit (rather than the soul's signification of bodily purity) that occupies the central role in Athanasius's *Life of Antony*."[90] Hence, through his fervent *askēsis*, Antony maintains this protological health throughout his life until his death—he retained his eyesight, all his teeth, health in his feet and hands, and "generally he seemed brighter and of more energetic strength than those who make use of baths and a variety

---

87. Such as Andrew T. Crislip, *Thorns in the Flesh: Illness and Sanctity in Late Ancient Christianity* (Philadelphia: University of Pennsylvania Press, 2012), 60–66; Robert H. von Thaden Jr., "Glorify God in Your Body: The Redemptive Role of the Body in Early Christian Ascetic Literature," *Cistercian Studies Quarterly* 38, no. 2 (2003): 201–3.

88. Crislip, *Thorns in the Flesh*, 63.

89. Temkin, *Hippocrates*, 154.

90. Crislip, *Thorns in the Flesh*, 64.

of food and clothing" (93). Antony's *askēsis* becomes the means by which he regains—in his body—the lost Edenic paradise.

Disease and Illness in Early Christianity

In their notions of disease, early Christian authors also followed the general medico-philosophical understandings of ill-balance (*dyskrasia*) of the elementary qualities and loosening of bodily coherence, where medical remedies were thought appropriate (Nemesius, *De nat. hom.* 1.9; Athanasius, *C. gent.* 27.54–62).[91] Tertullian in the early third century saw disease as a harmful excess or a harmful deficit contrary to nature, as most physicians did (*An.* 43.8).[92] As the Cappadocian fathers in general were knowledgeable about Galenic medical theories, they accepted medical theories of causation as well. For example, Basil of Caesarea, who himself was familiar with various illnesses, mentions a medical theory of disease that understands the body as "vulnerable to various kinds of harm," and that "it is distressed either by excess or deficiency" in diet and labor (*LR* 55.1). According to Basil's younger brother Gregory of Nyssa, bodily diseases are spread by corrupt humors (*Cat. magn.* 29). John Chrysostom, who also struggled with various ailments, likewise regarded an imbalance of the four humors as the cause of disease, especially whenever it involved an excess of bile and phlegm (*De statuis hom.* 9.9; 10.4). As Augustine regards health unity as a balanced order among the parts of the body, he identifies the essence of sickness (*aegritudo*), whether physical or spiritual, as an absence of that unity.[93]

Christians share the earthly lot of the human race with non-Christians: famine, invasion, drought, shipwreck, "and the diseases of the eyes, and the attack of fevers, and the feebleness of all the limbs is common to us with others," says Cyprian of Carthage (*Mort.* 8). For Lactantius, sickness and affliction in this life are God's design for humanity: "Since then, man was to be so formed by God that he would die at some time, the matter itself demanded that he be made of an earthly and weak body. . . . Thus it will be a consequence that man is subject also to diseases, for nature does not allow infirmity to be absent from that body which is at some time to be dissolved" (*Opif.* 4).

---

91. Cf. Galen, *Meth. med.* 13.1 (874K).

92. For a natural cause of disease, see also *Marc.* 1.24.7; *Idol.* 12.5.

93. Donald X. Burt, "Health, Sickness," in *Augustine through the Centuries*, ed. Allan D. Fitzgerald (Grand Rapids: Eerdmans, 1999), 417.

For specific examples of descriptions of disease, Clement of Alexandria describes indigestion as a result of heat caused by bodily movement; when heat is lacking due to insufficient movement, it leaves the food cold and brings about indigestion (*Paed.* 3.11). For Origen, food in the wrong quantity and of conflicting kinds develops fevers in the body (*Princ.* 2.10.4). John Chrysostom elaborates a list of diseases coming from excess: "As for instance, by a superabundance of bile fever is produced; and should this proceed beyond a certain measure, it effects a rapid dissolution. Again, when there is an excess of the cold element, paralyses, agues, apoplexies, and an infinite number of other maladies are generated. And every form of disease is the effect of an excess of these elements" (*De statuis hom.* 10.4). Then Gregory of Nyssa explains a fevered condition as follows: the heart's motion heats the body and works the lung incessantly and therefore the temperature rises unnaturally and breathing becomes more rapid and violent (*Hom. opif.* 30.17). Galen understood "leprosy" (*lepra*) (what he regarded as elephantiasis) as arising from an imbalance of choleric humors, particularly black bile (*Prorrheticum* 2.43), and Gregory of Nyssa "reiterates the prevailing medical theory that the putrid humors are the antecedent causes of health and malady and considers the disease's etiology in terms of humoral disorder, attributing the origin of the affliction to the oozing rotten humors of the blood that trigger the flow of the black bile."[94]

What about the various diseases of the soul? Just as the Hippocratics and Galen provided physiological explanations for psychic diseases, early Christian writers also engaged in similar somatic explanations for sickness of the soul. Origen was aware of the physiology of lunacy, in that "the moist humors which are in the head are moved by a certain sympathy which they have with the light of the moon, which has a moist nature" (*Comm. Matt.* 13.6). In one of his many homilies, John Chrysostom states that madness,[95] dementia, and phrenitis all have a humoral pathology. The person who is mad (*hō mainomenos*) is easily excused by onlookers because that illness is not from choice but from nature only; it is an imbalance of humors that leads to the following psychological diseases: "Whence have diseases their

94. Lampros Alexopoulos, "Medicine, Rhetoric and Philanthropy in Gregory of Nyssa's Second Sermon on the Love of the Poor," *Theologia* 3 (2015): 66. Alexopoulos reasons that Gregory follows Galen's causation etiology through Oribasius's encyclopedia.

95. On this topic, see Wendy Mayer, "Madness in the Works of John Chrysostom: A Snapshot from Late Antiquity," in *Concept of Madness from Homer to Byzantium: Manifestations and Aspects of Mental Illness and Disorder*, ed. Hélène Perdicoyianni-Paléologou, Byzantinische Forschungen 32 (Amsterdam: Adolf M. Hakkert, 2016), 349–73.

evil nature? Whence is frenzy (*phrenitis*)? Whence is lethargy? Is it not from carelessness? If physical disorders have their origin in choice, much more those which are voluntary. Whence is drunkenness? Is it not from intemperance (*akrasia*) of soul? Is not frenzy (*phrenitis*) from excess of fever? And is not fever from the elements too abundant in us? And is not this superabundance of elements from our carelessness?" (*Hom. 1 Thess.* 9). Moreover, he speaks of the physiology of aging and its impact on the brain. As he considers moderation (*sōphrosynē*) as mental health, he says that the elderly are slow, timid, forgetful, insensible, and irritable and that "there are among the old, some who rave (*luttōntes*) and are beside themselves (*paraphrones*), some from wine, and some from sorrow." He explains his claim physiologically by comparing womanhood to old age, in that the elderly, like women, are especially susceptible to wine (alcohol) because "the fumes mount more easily from beneath, and the membranes (of the brain) receive the mischief from their being impaired by age, and this especially causes intoxication" (*Hom. Tit.* 4).

Gregory of Nyssa also construes mental illness as the result of physical pathology. According to him, "insanity comes about not from heaviness of head alone, but also from the membranes underlying the ribs being in a pathological condition, just as the experts in medicine explain the sickness of the rational principle, calling the affection 'phrenitis,' since *phrenes* is the name of those membranes" (*Hom. op.* 12.4).[96] For Gregory, "mental disorder is caused . . . not by damage to any single part of the body, but injury to the body *per se*."[97] He further explains psychic disease's effect on the body, that grief, for example, causes a contraction and compression of many organs, which press on the lungs and give rise to violent drawing of breath in the effort "to widen what has been contracted, so as to open out the compressed passages; and such breathing we consider a symptom of grief and call it a groan or a shriek" (12.4); this bodily contraction caused by grief then makes bile pour on to the entrance of the stomach, and that is why those in grief become sallow and jaundiced (12.4).

Nemesius, who follows Galen's explanation of phrenitis, describes it as a symptom of bodily disease as well. As Galen localized phrenitis through

96. Translation by Jessica Wright in "Brain and Soul," 269. Wright discusses the contested localization of phrenitis in late antiquity between the brain (locus of *phronēsis*, "intelligence") presented by Galen and the diaphragm (*phrēn* or *phrenes* in older Greek) put forth by Gregory of Nyssa at 265–68. See also Jessica Wright, "Preaching Phrenitis: Augustine's Medicalization of Religious Difference," *JECS* 28, no. 4 (2020): 533–34.

97. Wright, "Brain and Soul," 161.

hot and dry swelling around the brain, unlike Gregory of Nyssa, he thought that people with phrenitis suffered damage in their intelligence (*phronēsis*). Nemesius then shares Galen's case story of a man with phrenitis suffering from disturbed cognitive function (*De nat. hom.* 13.70–71).[98] Moreover, for Nemesius, the gall bladder and the spleen generate bile, and the spleen pours out the excess of black bile into the stomach (28.92). Then "the boiling of the blood in the region of the heart arising from an evaporation of [the disturbed] bile" is none other than anger (20.81). He further notes that one of the reasons for "bad affections of the soul" is "a bad state of the body" (*kakhexia*); "for those with bitter bile are irascible, and those heated and moist in their bodily mixture are prone to sexual activity" (17.75). The remedy for the bad state of the body is to change it into "a mean bodily mixture by a suitable mode of life, by exercise, and by drugs if we need them as well" (17.76).

However, early Christians also understood many causes of diseases to be "much more than simply cases of physical affliction resulting from 'natural' causes" and conferred moral and symbolic meanings and religious significance on them in constructing their illness narratives.[99] Thus, following Old Testament, they constructed a divine etiology in relation to the sin-stricken fallen world. First, drawing on the Hebrew Scripture, and like their non-Christian neighbors, early Christian writers understood illness as a form of divine punishment or a sign of divine displeasure for human sin. Basil understands well that "not all illnesses come about naturally, or come from faults of lifestyle or from any other bodily origin," for which medical art is sometimes a benefit (*LR* 55.4; cf. *SR* 314).[100] Rather, illnesses are often "scourges for sins, sent for our conversion" (*LR* 55.4; cf. *SR* 314). In such cases, what we need is not medical art but acknowledgment of our transgressions, and we ought to "endure whatever comes our way" and "show our amendment by bringing forth fruits worthy of repentance" (*LR* 55.4; cf. *SR* 314). For Augustine, illness manifests a division of the body against itself, in which the illness of the body is the sign of a more fundamental ailment; in general, bodily ailments are the sign and the consequence of the first sin, that is, spiritual death. In other words, mortality, pain, and illness are understood as the punishment for sin.[101]

---

98. Cf. Galen, *Loc. affect.* 4.1–2.

99. Merideth, "Illness and Healing," 33.

100. *SR* 314 is an earlier version of *LR* 55, which treats the topic of medical remedy and illness in the life of Christian (monastic) life more fully and positively.

101. Isabelle Bochet, "Maladie de L'Âme et Thérapeutique Scripturaire selon Au-

Cyprian enumerates a number of physical and mental illnesses as a result of the sin of apostasy on the part of the lapsed during the Decian persecution in the mid-third century. A man who had denied Christ was struck dumb (*Laps.* 24). A woman, who went to the baths after her denial of Christ, became possessed by an unclean spirit and bit off her own tongue; later, "in the throes of internal pangs she expired" (24). Even a baby girl, who had been taken to the pagan magistrates "before the idol" by her nurse and had eaten bread dipped in wine, when later taken to the Eucharist by her mother, resented the prayer, refused the cup, and cried convulsively and vomited when she was made to drink it (25). Another mature girl who had received the Eucharist secretly experienced spasms and "collapsed in tremors and convulsions" (26).[102] Another lapsed woman who had tried secretly to receive the Eucharist found the eucharistic bread burnt; and countless many apostates were "driven out of their senses in a frenzy of fury and madness" as a result of their sin (26).

For Lactantius, God intervened to punish the persecutors of Christians by striking them with terrible diseases.[103] In his apologetic work *On the Death of the Persecutors* (*De mortibus persecutorum*), a couple of passages stand out in this respect. Furious at the outspokenness of the Roman people, Diocletian, his spirit sick with anger, left Rome at the height of winter, and the intense cold and rain made him contract a not very serious but chronic illness; it became increasingly grievous and oppressive, and he grew so languid and feeble that he would die a year later in a state of disorder and insanity (17.3). The most graphic and vivid passage is Lactantius's description of Galerius's disease and death, which also displays his physiological knowledge (33). "God struck him with an incurable plague"—a malignant ulcer at the bottom of his genitals (33.1). Despite the physicians' incision, the sore broke out again with such hemorrhaging that he was placed in mortal danger. Surgeons had to redo the operation, and at last cicatrized the wound. Nonetheless, any slight movement of the body reopened the wound, so that it released even more blood than before. Galerius turned livid and grew feeble and weak, and only then did the flow of blood stop. From then on the wound no longer responded to treatment; gangrene seized all the body parts close to the wound, and the more they

---

gustin," in *Les Pères de L'Église Face à la Science Médicale de Leur Temps*, ed. Véronique Boudon-Millot and Bernard Pouderon (Paris: Beauchesne, 2005), 385.

102. On another eucharistic punishment, see the Acts of Thomas 51.

103. This is a familiar *topos* of early Christian apologists. See Tertullian, *Scap.* 3.4.

treated the gangrene, the more it grew. Famous doctors were brought in from all over the world, but Lactantius emphasizes that "no human means had any success" (33.4). Deities were sought—Apollo and Asclepius were prayed to. Apollo prescribed a remedy, but the illness only spread and grew worse; the distemper had already taken hold of the whole lower part of Galerius's body. Despite their failures, the doctors continued to treat without hope of victory, and resisting all of their remedies, the illness came back, this time with worms. Its stench not only spread through the palace but invaded the whole city. Galerius was being devoured by worms, and his body was dissolving into rottenness. His suffering was intolerable (33.8). Through a complication of distempers, the different parts of his body had lost their natural shape. All of this lasted a whole year. At last, overcome by calamities, "he was obliged to acknowledge God, and he cried aloud, . . . that he would re-edify the Church, which he had demolished, and make atonement for his misdeeds; and when he was near his end, he published an edict" of toleration on April 30, 311 (33.11).

As Michel Jean-Louis Perrin notes, this vermicular disease with its horrible details (including the stench) recalls the dreadful death of Antiochus IV Epiphanius in Second Maccabees.[104] One can follow in detail a parallel story, which goes as far as the repentance of the persecutor, repentance that is useless; he dies in horrendous suffering, and the doctors and even the gods are powerless to fight against a disease that is the manifestation of the Christian God's vengeance, and thus of his power. In this sense, this passage is a rereading of Maccabees, adapted to the end of the reign of Galerius.[105] On the other hand, Eusebius's account resonates with Lactantius's in its generality (but with different details): Galerius died of a gangrenous ulcer, and in the midst of excruciating pain. This may indicate that the two stories were independent of each other and that Lactantius did not invent his account. The important point and demonstration for Lactantius here is his theological and apologetic interpretation of Galerius's illness.[106]

Another example of illness as divine punishment for sin comes from Christian apocalyptic literature in which blindness is among the eternal punishments in hell and Hades. Eye diseases and blindness occurred rather

104. Michel Jean-Louis Perrin, "Médecine, Maladie et Théologie chez Lactance," in Boudon-Millot and Pouderon, *Les Pères de L'Église Face à la Science Médicale de Leur Temps*, 349.

105. Perrin, "Médecine, Maladie et Théologie chez Lactance," 349.

106. Perrin, "Médecine, Maladie et Théologie chez Lactance," 350.

commonly in the ancient world. In the Apocalypse of Peter, Peter sees the punishment of those who are pretentious about their righteousness: "And nearby this place of torment shall be men and women who are dumb and blind and whose raiment is white. They shall crowd one upon another, and fall upon coals of unquenchable fire. These are they who give alms and say, 'We are righteous before God,' whereas they have not sought after righteousness" (12).

Meghan Henning provides a detailed study on blindness in Christian apocalypses, but here it suffices to say that dumbness and blindness are a consequence of the sinner's moral failure, particularly in neglecting the call to pursue the righteousness preached by Jesus in the Sermon on the Mount (Matt. 5:6).[107] In the Apocalypse of Zephaniah, Zephaniah witnesses the souls in torment in Hades: "And I saw some blind ones crying out, And I was amazed when I saw all these works of God. I said, 'Who are these?' He said to me, 'These are the catechumens who heard the word of God, but they were not perfected in the work which they heard.' And I said to him, 'Then do they not have repentance here?' He said, 'Yes,' I said 'How long?' He said to me, 'Until the day when the Lord will judge'" (10). Here blindness seems to be a consequence of the failure to understand the word of God, what Henning calls "an intellectual sin," with a window of opportunity to repent.[108] Finally, in the Apocalypse of Paul, Paul sees the following: "And I observed and saw men and women clothed in bright garments, but with their eyes blind, and they were placed in a pit, and I asked, 'Sir, who are these?' And he said to me, 'These are heathen who gave alms, and knew not the Lord God, for which reason they unceasingly pay the proper penalties'" (40).

Here we see a combination of the same sin-blindness trope in the Apocalypse of Peter and the Apocalypse of Zephaniah: external almsgiving but internal cognitive failure to know God ("intellectual sin"). These three texts as a whole highlight "the ancient *topos* of blindness as a metaphor for a person's ethical or philosophical obtuseness";[109] in this case, blindness is not merely a metaphor but is an actual physical punishment for the persons' moral and intellectual failure.

The connection between sin and illness is further seen in Irenaeus of Lyon's *Against Heresies* (*Haer.*). In book 5 chapter 17, where Irenaeus com-

---

107. Meghan Henning, "Metaphorical, Punitive, and Pedagogical Blindness in Hell," *Studia Patristica* 81 (2017): 139–51, here 145.

108. Henning, "Metaphorical, Punitive, and Pedagogical Blindness in Hell," 143.

109. Henning, "Metaphorical, Punitive, and Pedagogical Blindness in Hell," 139.

CHAPTER 2

ments on Jesus's healing of the paralyzed man from the Synoptics, he is concerned about affirming Jesus's identity and nature in and through his ministry in relation to the Father Creator. While a connection between the man's paralysis and sin is implicit in the Matthean version of the story (Matt. 9:2–7), Irenaeus explicitly draws the connection: "By this work of his [Jesus's healing] he confounded the unbelievers, and showed that he is himself the voice of God, by which man received commandments, which he [the paralyzed man] broke, and became a sinner; for the paralysis followed as a consequence of sins" (*ex peccatis autem paralysis subsecuta est*). After affirming the close connection between sin and illness,[110] Irenaeus goes on to affirm a corresponding relationship between forgiveness and healing: "Therefore, by remitting sins, he did indeed heal man" (*Haer.* 5.17.2, 3). Furthermore, Basil's *Longer Rules* for his monastic community features some cases that also make a clear connection between sin and illness. While Basil acknowledges natural causes for illness and the necessity of medical remedies, his larger concern is how physical sickness and remedies provide "a pattern for the care of souls" (*LR* 55.2, 3). A case in point is chronic sickness, "such that we seek healing over a long period of time and use remedies both painful and varied[;] this shows us that we ought to amend the sins of the soul by sustained prayer and prolonged repentance and a more laborious struggle than reason would suggest to us is sufficient for our healing" (55.3). When a Christian suffers illness but does not get well even after a doctor's treatment, that physical affliction reflects internal sin.[111] Finally, Augustine also links sin and sickness when he states that diseases may be caused by intemperate living (*Serm.* 156.2) and the excesses and perverse practices of even married life (*Bon. conj.* 20.20, 31). Augustine also refers to an *archiater* (chief city physician) named Dioscurus, an outstanding physician but a non-Christian; he made promises to the church out of devotion for his dying daughter but did not keep them. However, after divine punishment featuring blindness and paralysis, he became a Christian (*Ep.* 227).

Examples of these sorts are abundant, especially in ascetic literature.[112] For example, Theodoret of Cyrrhus's own mother was afflicted with a

110. See also *Pseudo-Clementine Homilies* 19.22.
111. Merideth, "Illness and Healing," 41.
112. An unusual ascetic denunciation of the sin-illness connection comes from the instructions of Theodore, the successor to Pachomius in the Pachomian community: "Let no one among us say, 'no doubt it is because he is a very wicked man that these tribulations [i.e., illnesses] have fallen on him!' He who will say that to himself does not perhaps yet deserve remedy himself" (*Instructions of Theodore* 15).

medically incurable disease in one of her eyes and sought to be cured by a holy man, Peter the Galatian.[113] However, when the man of God saw her with earrings, necklaces, and other golden jewelry, and in an elaborate dress woven from silk thread, he pointed out her sin of love of adornment and told her to bid farewell to doctors and medicines and accept God's medicine by allowing Peter to lay his hand on her eye (*Phil. hist.* 9.6–7). Only after she repented by rejecting all makeup, external ornament, and elaborate dress did she experience "a double cure: in [the] quest of healing for the body, she obtained in addition the health of the soul" (9.8). In Palladius's *Lausiac History*, he approached a holy man, Macarius, and begged him to cure a priest of the village whose head was terribly eaten away by "cancer," but Macarius, learning, through his clairvoyance, of this priest's sin of lust, did not receive him. Only after the priest repented by confessing his sin, refraining from saying Mass or performing other priestly functions, and amending his ways did Macarius lay his hand on him, and he was healed (18.19–21).[114] In another episode, in the *History of Monks in Egypt*, a child suffering from rabies was bound with a chain, and "his suffering was so unbearable that his whole body was convulsed by it" (22.3). The holy man Amoun exposed a crime that the child's parents had committed. They had killed a widow's ox surreptitiously, and Amoun told them to replace it. When they did that, Amoun prayed, and the child instantly recovered.

Second, while not seeing God as the direct cause of diseases/illness as divine punishment, early Christians still regarded illness as a part of divine pedagogy and providence; that is, God allowed it to occur to encourage the ill person's spiritual progress and the edification of others (cf. Basil, *SR* 314; *Quod Deus non est autor malorum* 3). Referring to the apostle Paul's own infirmity, Irenaeus speaks of the "beneficial effect" of an "infirmity of the flesh" (*Haer.* 3.1–2). According to Tertullian, "all the plagues of the world" (*omnes saeculi plagae*) come upon Christians for admonition and non-Christians for chastisement from God (*Apol.* 41.4). Clement of Alexandria states, "penury and disease, and such trials, are often sent for admonition, for the correction of the past, and for care for the future"; and a remedy in this case is prayer (*Strom.* 7.13; cf. 7.61). Origen also thought that bodily affliction was of spiritual value (*Princ.* 2.10.6) and that God used suffering to purify and reform people (*Comm. Matt.* 11.4). Gregory of Nazianzus

---

113. Theodoret, *History of the Monks of Syria* 9.5.
114. See also *HL* 17.4.

offered a helpful counsel to an ill friend that he should be "philosophical" about his suffering and show himself superior to the cause of his affliction, "beholding in the illness a superior way towards what is ultimately good for you" (*Ep.* 31.3). In addition, Augustine urged his listeners to bear illness and adversity as a forbearing patient as they were corrections from the hand of God (*Serm.* 9.9). According to John Chrysostom, "God does not permit illness to debase us but because he wanted to make us better, more wise, and more submissive to his will, which is the basis of our salvation. . . . It is for our good that we are victims of illnesses . . . , since the pride stirred up within us by relaxed attention finds a cure in this weakness and in these afflictions" (*Homilies on Annas* 1.2). In fact, John raises the question why God permitted saints (like Timothy) to fall into sickness, and he provides the following eight different but related answers (*De statuis hom.* 1.13–14): First, that they may not easily grow presumptuous due to their excellent good works and miracles. Second, that others may not have a greater opinion of them than appropriate for human beings, and take them to be more than mortals.[115] Third, that the power of God may be made manifest, "in prevailing, and overcoming, and advancing the word preached," through human weakness. Fourth, that the patience of the saints may become more striking, demonstrating that they serve God not for a reward (in this life). Fifth, that their examples lead us to ponder the resurrection and the last judgment, in which they will receive their recompense for their present labors. Sixth, that "all who fall into adversity may have a sufficient consolation and alleviation" by looking at such saints who suffered much. Seventh, that we understand that they are of the same nature as we are and thus we can imitate their virtues and surpassing good works. Eighth, that we know what consists of true happiness and unhappiness. Therefore, illnesses are profitable to God's people—God permits them for the end that they may exercise moderation and humility and that they may not be puffed up (1.15). An extreme example of valuation of illness as divine pedagogy would be a story of "an elder who was continually sick and unwell," but then he was not sick for one year. "He was terribly upset and wept, saying: 'God has abandoned me and not visited me.'"[116]

Third, another major etiology of illness for early Christians is that of demons, which was clearly seen in the Hellenistic Jewish literature and

115. On this, see also Basil of Caesarea, *LR* 55.4.

116. John Wortley, *The Anonymous Sayings of the Desert Fathers: A Select Edition and Complete English Translation* (Cambridge: Cambridge University Press, 2013), 149.

Christian Scripture (the New Testament), as part of the development of demonology in early Christianity.[117] Gary Ferngren argues against any prevalence of demonic etiology in earliest Christianity (first to early second centuries) and argues that "Christians typically accepted a natural causality of disease" in early Christianity in general.[118] He then attributes the rise of demonic etiology in the mid-second century to third century to the rise of Christian apologetic literature against pagan deities,[119] which was popularized by "the *Christus Victor*" theory of atonement; and the apologists' references to demons are rather generic and unspecified. Furthermore, the alleged "daimonization of religion" in the third century has to do with the popularity of exorcism as the established rite before baptism, which had little to do with demonic etiology and healing. Finally, the fourth-century resurgence of demonic etiology and healing has to do with the rise of asceticism, an increase in the practice of magic and relics, and the rise of holy men.[120] As we have seen, however, demonic etiology for Christians was not a new phenomenon in the second and third centuries. While parts of Ferngren's arguments are helpful, his study does not take into account the continuity of demonic etiology in Christian literature. As we examine Christian literature unmentioned by Ferngren, we will see that demonic etiology remained a relatively consistent part of how early Christians made sense of illness.

Evidenced by earlier and later versions with numerous translations, and also by widespread patristic references (despite repeated condemnation by the church), the apocryphal Acts of apostles of the late second and early third centuries were enormously popular throughout the church.[121] While the apologists of the late second and early third centuries presented

117. On early Christian demonology of the fourth and fifth centuries, see Dennis P. Quinn, "In the Names of God and His Christ: Evil Daemons, Exorcism, and Conversion in Firmicus Maternus," *Studia Patristica* 69 (2013): 3–14; Kathryn Hager, "John Cassian: The Devil in the Details," *Studia Patristica* 64 (2013): 59–64; Samantha L. Miller, *Chrysostom's Devil: Demons, the Will, and Virtue in Patristic Soteriology* (Downers Grove, IL: IVP Academic, 2020), especially chap. 2; R. J. S. Barrett-Lennard, *Christian Healing after the New Testament: Some Approaches to Illness in the Second, Third, and Fourth Centuries* (Lanham, MD: University Press of America, 1994), chap. 7 (on demon possession and exorcism in Athanasius).

118. Ferngren, *Medicine and Health Care*, 43.

119. For example, Tertullian, *Apol.* 23.1–16; 27.4–7; 37.9. However, there are more demonic references of specific nature: *Apol.* 22.4, 6, 11; *Bapt.* 5.3–4; *Spect.* 26; *An.* 57.4. See also Tatian, *Or. Graec.* 16.3; Minucius Felix, *Oct.* 27.6–7.

120. Ferngren, *Medicine and Health Care*, 51–57.

121. On the apocryphal acts' popularity and readership and their relationship with

Christianity mainly as true philosophy that surpassed the best of Greco-Roman philosophies, their contemporaneous apocryphal Acts presented Christianity as true power over the devil and its minions.[122] In the thought world of the apocryphal Acts, where the world is bewitched and dominated by hostile demonic forces, demons are often portrayed as causing physical illness; and the processes of exorcism, physical healing, and spiritual healing are all intertwined and interconnected. In the famous *Epitome of the Acts of Andrew* (*Laudatio*) by Gregory of Tours, where it recounts only the miracles, when Andrew sees a man with a wife and son, all blind, he declares, "Truly this is the work of the devil, for he has blinded them in mind and body" (32).[123] As Andrew restores their physical eyes, he also calls on the name of Jesus Christ to "unlock the darkness of [their] minds," which results in their conversion and the conversion of many witnesses. In the Acts of John, John's healing of the old women (30–36) is particularly illustrative of a triangular relationship of demonic power in illness, physical healing, and conversion. John commands all women over sixty years of age in Ephesus to be cared for. When he finds out that all but four women are sick, he regards it as the wicked mockery of the Ephesians by the devil, and in response plans a public spectacle of healing in the theater specifically to convert people to Christ, the God of whom John preaches. On the next day, as the whole city gathers in the theater, John first speaks to the whole city and to Andronicus a *strategos*; John's message is an encratic indictment of the Ephesians' unbelief and slavish bondage to the treasures, pleasures, and lusts of this world and is also a call to repentance and conversion. The text then abruptly states, "Having thus spoken, John healed all their diseases by the power of God" (36). Here the demonic work in illness is taken seriously, and the purpose of physical healing is directly linked to the "care of the souls" (34); the manifestation of the power of God is for the sake of spiritual healing—the exclusive worship of the true God of the apostle.

Moreover, Christoph Markschies examines a Christian magical amulet on papyrus presumably from Arsinoe, Egypt, from the fourth century, which was used against "a whole range of maladies."[124] Regarding a fever,

---

the apologists, see Helen Rhee, *Early Christian Literature: Christ and Culture in the Second and Third Centuries* (London: Routledge, 2005), introduction and chap. 1.

122. See Rhee, *Early Christian Literature*, chap. 2 in particular.

123. See also *Epitome* 2 and 5 for further examples of the healings of diseases of demonic cause. On the demonic cause of sickness, see further Pseudo-Clementine Homilies 8.9–11.

124. Christoph Markschies, "Demons and Disease," *Studia Patristica* 81 (2017): 18.

the text mentions a demon "which has the feet of a wolf, but the head of a frog." As Markschies notes, frogs were described as impure spirits coming from the mouth of the devil in the book of Revelation (16:13–14); and in antiquity, the wolf was seen as a "greedy hunter, bloodthirsty and sexually deviant." Then, the obvious connection was made that the frog-headed and wolf-footed demon was responsible for prolonged fever. The text then orders the fever to leave the body in the name of the "four gospels of the son" and the "God of Israel," and the demon leaves the body at the same time. According to Markschies, "the fact that first the fever and then the demon are ordered to leave the body in the text of this amulet shows that the relationship between the fever and the demon is perceived to be so close that the actual order in which they were invoked did not particularly matter."[125] Another amulet from the fifth-century Egyptian city of Oxyrhynchus calls upon Christ as a pursuer of demons to "chase away and banish from her [Joannia] every fever and ever kind of chill—daily, tertian, quartian—and every evil" through "the intercession of our mistress the Theotokos."[126] David Frankfurter notes that the post-Constantinian ecclesiastic authorities' varying efforts to define and root out magic in Christendom focused rather on secretive ritual acts or subversive circumstance. In the meantime, however, the preparation of amulets "became among the dominant functions of Christian shrines, holy men, and ecclesiastical scribes throughout the Roman Empire."[127] These examples in the apocryphal Acts and in Christian magical amulets show significant grassroots-level belief concerning demonic etiology, marking it as a definite undercurrent in popular Christianity, one that circulated with beliefs about divine etiology; they also show that the line between official, orthodox Christianity and popular, heterodox Christianity was, in practice, blurred.[128]

Indeed, in the fourth- and fifth-century (ascetic) texts, as Ferngren acknowledges and regrets, references to demonic etiology of illness (differ-

125. Marschies, "Demons and Disease," 18.

126. David Frankfurter, "Amuletic Invocations of Christ for Health and Fortune," in *Religions of Late Antiquity in Practice*, ed. R. Valantasis (Princeton: Princeton University Press, 2000), 342.

127. Frankfurter, "Amuletic Invocations," 340. On Christian healing by ascetic holy men and at the Christian shrines, see chap. 5.

128. On Christian use of amulets for illnesses and exorcism as a curative method in general, see Anastasia D. Vakaloudi, "Illnesses, Curative Methods and Supernatural Forces in the Early Byzantine Empire (4th–7th C. A. D.)," *Byzantion* 73, no. 1 (2003): 172–200, especially 197–99.

ent from exorcism) are readily available. Basil acknowledges that illness sometimes comes about "at the petition of the evil one" (*LR* 55.4); Augustine considers demon possession a proper cause of illness (*Civ.* 22.8) and attributes diseases to demons as follows: "all diseases of Christians are to be ascribed to demons; chiefly do they torment fresh-baptized Christians, yea, even the guiltless newborn infants" (*Div.* 1.2). Gregory of Nyssa maintains that the actual cause of the plague in Neocaesarea and in the Pontic countryside in the third century was a demon (*Vita Greg. Thauma.* 97–98). Both Athanasius and John Chrysostom, in their respective texts on the Gospel of Matthew, regard "falling sickness" (epilepsy) as caused by the devil/demon. For Athanasius, it is caused by "the scheming and deceitful devil"; for John, it is not the moon that is responsible but rather a demon (*daimōn*).[129] According to Amma Theodora, the devil "attacks your body through sickness, debility, weakening of the knees and all the members. It dissipates the strength of body and soul" (*Sayings*, Theodora, 3). Amma Syncletica also says, "the ambushes of the devil are many. . . . He was worsted by health? He makes the body sick" (7). Amma Syncletica was a fourth-century desert mother from Alexandria, and her ascetic prowess and heroic bearing of atrocious illness and pain are described in graphic detail in *The Life and Regimen of Blessed and Holy Teacher Syncletica*.[130] Syncletica held fast to her strict ascetic regimen, fasting in particular, and if ever she deviated from her *askēsis*, she became ill: "Her face was pale, and the weight of her body fell" (*Vita* 19). Like her saying, her last illness that led her to spiritual victory *and* eventual death is clearly understood by the author as coming from the devil; the author compares the illness to the trials of Job. We will come back to her illness and asceticism a bit later. As in Syncletica's case, the sickness comes from the devil/demon when the ascetic loosens one's *askēsis*, succumbs to the temptation of the devil, or commits sin. In another case, a man, wanting to cure her, brings his daughter, "afflicted with a serious illness from the demon," to Apa Pachomius (*The Bohairic Life of Pachomius* 43). When it is found out that she does not keep

129. Quoted in Nadine Metzger, "Not a *Daimōn*, but a Severe Illness: Oribasius, Posidonius and Later Ancient Perspective on Superhuman Agents Causing Disease," in *Mental Illness in Ancient Medicine: From Celsus to Aegina*, ed. C. Thumiger and P. N. Singer (Leiden: Brill, 2018). 90n42. Athanasius, *Fragmenta in Matthaeum* (*De lunaticis*); John Chrysostom, *Hom. Matt.* 57.3.

130. Unless noted otherwise, translation of this text comes from Elizabeth Bryson Bongie, *The Life & Regimen of the Blessed & Holy Syncletica by Pseudo-Athanasius* (Toronto: Peregrina, 1999).

monastic purity, she confesses her sin and promises to keep guard over herself. Only then does Pachomius pray over some oil and send it to her for her healing, which takes place when she anoints herself with that oil, in faith (43). As one can see, demonic etiology of illness is not necessarily confined to the ascetic literature (although it is most prominent there); it "gives expression to the tension that Christians felt in a world which was not, despite Constantine's conversion, 'converted' to Christianity";[131] it certainly coexisted with natural and divine etiologies as one of the antique and late antique paradigms of understanding and interpreting illness.

Finally, we now move to the topic of ascetic causes of illness and the relationship between illness and asceticism. Earlier we discussed Antony's prelapsarian health through his *askēsis*, which became a theological and symbolic ideal for the monks to follow. However, the reality was that rigorous *askēsis* often led ascetics to ill health rather than good health, and the ascetic had to deal with that reality head-on when faced with the formidable and influential model of Antony's ascetic health. Andrew Crislip's masterful analysis of ascetics' painstaking attempts to make meaning and use of illness within ascetic communities starts with a full acknowledgment of the puzzle (or scandal) of horrible illness suffered by an ascetic at "the height of ascetic perfection."[132] A famed healer, Benjamin, is the perfect example of this phenomenon. Benjamin contracted dropsy eight months before his death; his body was so swollen "that another person's fingers could not reach around one of his fingers." When he died, "the door and jambs were pulled down so that his body could be carried out of the house, so great was the swelling" (*HL* 12.3). According to Papias in the second century, Crislip notes, dropsy was the very grotesque disease that befell Judas Iscariot, the worst of sinners; and traditions of Judas swelling and bursting were well-known in antiquity. According to Crislip, to the ancient readers, Benjamin's illness and death could evoke accounts of Judas's well-deserved punishment, which might explain Palladius's concluding caveat: "I felt that I must tell about this sickness so that we might not be too perplexed (*xenizometha*) when some accident befalls just men (*andrasi dikaiois*)" (*HL* 12.3). Palladius anticipates his audience's puzzlement over "[the] disjunction between sanctity and the spectacle of illness."[133] Recognizing the deep ambiguity in the meaning and function of illness among

---

131. Merideth, "Illness and Healing," 39–40.
132. Palladius, *HL* 12; Crislip, *Thorns in the Flesh*, 16–17.
133. Crislip, *Thorns in the Flesh*, 16.

ascetics, Crislip finds a clue about the potential value of illness to ascetics in John Cassian's *Conferences* as it draws on the Stoic triad of good, bad, and indifferent (6.3.1).[134] Illness, along with health, wealth, and poverty, etc., is indifferent and "can have good or bad consequences according to the character and desire of the user" (6.3.2). In fact, illness "can function as a mode of sanctification and self-improvement,"[135] as poor Lazarus in the gospel "very patiently bore deprivation and bodily sickness, and for this he deserved to possess Abraham's bosom as his blessed destiny" (6.3.5). Then Crislip writes: "For Cassian, thus, the illness or injury of the ascetic is potentially scandalous: a challenge to meaningfulness, especially since the symbolic world of Christianity offers ambiguous and contradictory models for making sense of illness. Yet illness itself may be of special use as a type of asceticism, a 'spiritual exercise' or a 'technology of the self,' in the terminology of Pierre Hadot or Michel Foucault. Proper behavior in the face of illness facilitates the process of paying attention to oneself to transform the self."[136]

Taking this as our cue in light of Arthur Kleinman's notion that expectations about how to behave when ill can change through negotiations in different social situations,[137] we can trace this transformation without repeating Crislip's work. On the one hand, because of the inherent ambiguity of ascetic illness, the concern for illness due to *askēsis* led ascetics such as Jerome to exhort Eustochium (384 CE), "the first virgin of noble birth in Rome," not to engage in excess ascetic regimen, which only leads to vainglory and boastfulness, which are "tokens of the devil" (*Epist.* 22.27, 28). He repeats this caution in his letter to Demetrius (414 CE), a noble Roman lady who had recently embraced the ascetic lifestyle, not to engage in "extreme fasting or abnormal abstinence from food" (130.11). Such practices destroy one's bodily constitutions and cause bodily sickness before one has laid the foundations of a holy life. Virtues are means, whereas "all extremes are of nature of vice." While curbing the desires of the flesh, she must have sufficient strength to read Scripture, sing psalms, and observe vigils (130.11). According to Abba Theonas, if an ascetic refuses to relax a rigorous fasting when "fleshly weakness and frailty demand" it, he should be considered "the cruel murderer of his own body rather the procurer of

---

134. Crislip, *Thorns in the Flesh*, 21.
135. Crislip, *Thorns in the Flesh*, 21.
136. Crislip, *Thorns in the Flesh*, 22.
137. See introduction, pp. 6–7.

his salvation."[138] Amma Syncletica, despite her own strict adherence to fasting, also mitigated the severity of her asceticism so that "the parts of her body might not become totally debilitated" (*Vita* 19); she thought "physical collapse was a sign of defeat [by the enemy]." In contrast to those who inflict self-mortification in fasting without moderation or discernment, she "would take care of her body when her ship, so to speak, was in calm waters" (19). Accordingly, her exhortation to her female disciples is one of moderation: "Do not expend all your defensive weapons at one time. . . . Our armour is the body and our soul is the soldier; take care of them both against the time of need. . . . Fast reasonably and yet scrupulously" (100). In these cases, illness is clearly seen as a threat to the proper goal and function of *askēsis*, and it is an ascetic's responsibility to take care of the body even within the context of a scrupulous ascetic regimen.

On the other hand, when even the scrupulous ascetics get ill, it becomes necessary to develop an ascetic discourse that embraces illness or incorporates it into ascetic life. Here the sick ascetics' dispositions and behaviors are critical in determining positive valuation and the utility of illness for them and their community.[139] For Basil, like Chrysostom, one of the reasons the apostle Paul got sick was so that "he might not seem to exceed the bounds of human nature" and that no one should think that his capacity was beyond human nature—"illness was his lot continually, in order to make plain his own human nature (cf. 2 Cor. 12.7)" (*LR* 55.4). Theodoret of Cyrrhus also relates the story of a holy man, Julian Saba, who was expected to heal numerous people, just like the apostle Paul, but, "so that everyone should learn that he was a human being, he had a violent attack of fever" in front of everyone gathered; when it was clearly known that he was only human through his fever and he prayed for healing, he was cured so that he could cure others gathered for him.[140] In these cases, the illnesses of the holy men ground them as human and are, as Anne Merideth notes, "not a punishment for sin, but a means by which to avoid sin"—that of pride on the part of the holy men and that of presumption on the part of the witnesses.[141]

Another way to incorporate illness into ascetic discourse is to acknowledge *askēsis* as both a threat and a remedy for illness. Apa Pachomius's

---

138. Cassian, *Conferences* 21.14.4.

139. Note the positive valuation of enduring illness implying the athletic discipline like *askēsis* already in Ignatius's letter to Polycarp: "Bear the diseases of all, as a perfect athlete" (*pantōn tas vosous bastaze, ōs teleios athlētēs*) (Ignatius, *Pol.* 1.3).

140. Theodoret, *Phil. hist.* 2.18.

141. Merideth, "Illness and Healing," 53.

mentor Apa Palamon became sick in his spleen on account of his "manifold *ascesis* and above all because he took no respite in his old age from his exercises" (*The Bohairic Life of Pachomius* 16). His ascetic neighbors brought to him a famous doctor, who told him to take a little appropriate food and he would get well. However, when Palamon followed that instruction and ate some of the dishes given to the sick, he did not get well. Telling his brothers that healing and strength come through Christ, Palamon went back "once more to his *ascesis* with great mortification until the Lord saw the constancy of his courage . . . and cured him of his illness" (16). Thus in this case, Palamon's *askēsis* was a means by which both Palamon's illness and cure were acquired.

The carefully constructed stories of Apa Pachomius's chronic illness put forth a vision of enduring illness by an ascetic as a marker of sanctity especially in contrast to the vision of ascetic health in the *Life of Antony*.[142] Described as being "not sturdy" even when he joined the army as a young man (7), Pachomius had a series of serious illnesses throughout his life due to his rigorous *askēsis* (47, 61, 94, 114, 117, 121). A consistent theme in his repeated illnesses is his refusal of a special dietary and physical treatment reserved for sick monks and his refusal to alter or modify his *askēsis* in illness even while he mandated the lessening or suspension of the ascetic rules for the sick and those caring for the sick (48, 61, 117, 118, 120). Similar to his mentor Palamon's case, Pachomius acquired both his illnesses and his cures through his *askēsis*. While his mundane, faithful endurance of his illnesses did not result in the recovery of prelapsarian health, it elevated his sanctity to that of "a sick saint."[143]

Indeed, the most positive way to incorporate illness is to regard the ascetic's illness as a form of involuntary asceticism. However, the illness of the ascetic is not self-induced through one's *askēsis* here; in fact, Amma Syncletica is critical of self-inflicted illness like that of Palamon.[144] Syncletica exhorts her community not to be saddened if they are unable to carry on their *askēsis* due to the frailty and afflictions of their bodies; those ailments have been *given* to them for the "destruction (*kathairesin*) of [their] desires," and *askēsis* has the same purpose (*Vita* 99). Then, if sickness has

---

142. Crislip, *Thorns in the Flesh*, 109, 137.
143. Crislip, *Thorns in the Flesh*, 131, 136.
144. Cf. Andrew T. Crislip, "'I Have Chosen Sickness': The Controversial Function of Sickness in Early Christian Ascetic Practice," in *Asceticism and Its Critics: Historical Accounts and Comparative Perspective*, ed. Oliver Freiberger, AARCC (New York: Oxford University Press, 2006), 196.

"dulled these pleasures, austerity is superfluous"; for "potentially fatal lapses have been checked by illness as if by some strong and powerful medication"; therefore, this is the great *askēsis*: "to remain strong in illness and to keep sending up hymns of thanksgiving to the Almighty" (99). She faced her illness from the devil in her final days just as she exhorted her community. For three and a half years she battled the devil, and through the torment of a high fever and infected lungs, her suffering, according to her hagiographer, surpassed that of the noblest martyrs (106). Then, even in her unspeakable suffering, and amid the putrefaction of her gums and the inhuman stench of her body,[145] the hagiographer says that Syncletica's body "was supported over all by divine power" until her death (112). Here is an ascetic's *askēsis* (i.e., endurance of illness) that is transformed into ascetic holiness, which is honored by divine power.

Illness serves to display ascetic sanctity in Gregory of Nyssa's *Life of Macrina* as well. Having had an ominous vision about Macrina, Gregory visited her in her severe illness but found her refusing to take any physical comfort or care for what ailed her. She was on the ground rather than on a couch or bed, supported by a board covered with a coarse cloth; another board supported her head and neck. Gregory compares the suffering of Macrina to that of Job and describes her saintly endurance of illness as follows:

> Just as we hear in the story of Job, that when the man was wasting away and his whole body was covered with erupting and putrefying sores, he did not direct attention to his pain but kept the pain inside his body, neither blessing his own activity nor cutting off the conversation when it embarked upon higher matters. Such a thing as this I was seeing in the case of this Superior [Macrina] also; although the fever was burning up all her energy and leading her to death, she was refreshing her body as if by a kind of a dew, she kept her mind free in the contemplation of higher things and unimpeded by the disease. (*Vita* 175–176)

Macrina faithfully endures her disease by focusing on "her discourse on the soul," and in "all of this, she went on as if inspired by the power of the Holy Spirit," according to Gregory (176). Thus her endurance of illness becomes a marker of her spirituality and sanctity even as she was preparing herself for death with hope for greater good (177).

---

145. On the significance of stench in an ascetic suffering, see Susan A. Harvey, "On Holy Stench: When the Odor of Sanctity Sickens," *Studia Patristica* 35 (2001): 96–97.

Lastly, Theodoret highlights the remarkable endurance of Limnaeus's illness as a marker of his ascetic holiness. A healer, exorcist, and miracle worker, he was "assailed by the disease of the gripes" (*Phil. hist.* 22.4), which caused within the afflicted an agonizing pain, one that led them to roll about like "lunatics," turning over on one side to the other, and "at the same time stretching out and then bending back their legs"; there was no position of repose whatsoever. Wrestling with this disease and enduring this kind of and magnitude of pain, Limnaeus "could not endure a bed, and got no relief from medicine or food"; rather, "seated on a plank lying on the ground," he lulled his pains by prayer, the sign of the cross, and the spell of the divine name (22.4). On another occasion, he was bitten by a viper more than ten times and left with bitter pains all over. Rather than accepting medical help, he relied on the same sign of the cross, prayer, and invocation of God to endure his pain. Comparing Limnaeus to Job, Theodoret confidently concludes: "I am therefore of the opinion that the God of the universe allowed the beast to rage against his sacred body in order to reveal undisguised the endurance of this godly soul. . . . How else would we know the courage of the one and the endurance of the other, if the adversary of piety had not been given room to shoot all kinds of arrows against them? Therefore this is sufficient to teach the endurance of the man" (22.6). Limnaeus's single-minded bearing of his illness and pains only by means of faith, that is, his *askēsis,* merits Theodoret's praise of his ascetic valor in victory over the devil.

Thus, in these ascetic discourses of illness, the valiant endurance of illness affords the ascetics paradoxical sanctity and empowerment; but it also offers them a point of connection with those who come to them for healing of their own illnesses and afflictions. It is then through the experience of illness, the universal dimension of human existence, that the holy men and women are, on the one hand, brought low and become relatable in their very human experiences; on the other hand, it is through the experience of illness that the ordinary faithful are lifted up and able to approximate the extraordinary sacrifices of the saints and, thereby, the sanctity of the saints as well.[146]

Conclusion

This chapter has examined notions of health and disease/illness in the Hebrew Bible, Christian Bible, and early Christianity. In the Hebrew Scripture, Yahweh, as the Creator of life and punisher of human sin, is the direct source of health and disease, which establishes the sin-illness paradigm,

---

146. Merideth, "Illness and Healing," 179.

for which God is the only healer. In the Hellenistic Jewish texts, we see the emerging role of demons as the inflictors of disease; and in the Christian Gospels, we see, in addition to an etiology of natural causation for the majority of cases, stories that, in qualified ways, identify a demon-illness connection as well as a sin-illness connection as the reason for disease. As the Gospels portray Jesus, through his healing ministry, as God's agent of healing par excellence, Jesus's healings function as foretastes of God's eschatological *shalom*. While in Acts the apostles carry on Jesus's healing ministry, Paul's own "thorn in the flesh" in 2 Corinthians lays a meaningful foundation that not healing, but God's power in weakness, illness, and hardships, is a source of his sanctity. Early Christians both continued and further developed their notions of health and disease in close contact with established ideas in Greek medicine and philosophy. In terms of the concept of health, they shared the Greek medico-philosophical notion of health as a constitutional balance, and developed a Christianized Galenic functional teleological anatomy in light of God's creation and providence. In terms of the concepts of disease and illness, they not only shared with Greco-Roman writers natural or medical etiologies but also constructed their own comprehensive illness narratives with their own supernatural etiologies. With regard to supernatural etiology, illnesses were believed to come from God either as God's punishment for sin or God's pedagogy, or from the devil/demons. These beliefs constituted significant meaning-making frameworks for illness, especially in ascetic literature, where the ambiguity of illness was eventually overcome with the idea that the endurance of illness constituted a new *askēsis*, which was the marker of ascetics' holiness and strength. This paradoxical relationship leads to a positive valuation of illness, one that opens up the possibility of a formation of a community of sufferers. This "plurality of etiologies" would run parallel to "a plurality of therapies" for both non-Christians and Christians. It is to these therapies that we will turn in chapters 4 and 5.[147] In the meantime, we will deal with the related but unique topic of pain in Greco-Roman and early Christian works, where we will develop the theme of a community of cosufferers in mimetic identification with Christ.

---

147. Merideth, "Illness and Healing," 63, quoting David Westerlund.

3

PAIN IN GRECO-ROMAN CULTURE
AND EARLY CHRISTIANITY

In the previous chapter, we examined not only how early Christians represented and reconstructed existing Greco-Roman notions and experiences of health, disease, and illness, but also how they did so both in light of their developing theology and in close interaction with Greco-Roman medicine and philosophy. This chapter will examine their similar endeavors regarding pain, which is related to illness as its symptom or sign, but distinct from it, especially when caused by violence, in particular. Paying special attention to pain's relationship with language and (in)expressibility, I will first show that pain was understood and experienced in Greco-Roman medicine mainly as a physiological phenomenon, but with affective and social implications. I will then examine the construction of pain narratives and pain pedagogies in Greco-Roman literature and philosophy as they intersect with not only rational medicine but also early Christian narratives. It is in these literary and philosophical texts that we detect the broadening construction and display of pain not only as it manifests in illness but also as it manifests in violence; these texts illustrate some of the specific ways in which people navigated and endured pain in the public arena. Drawing on this wider cultural milieu of pain construction and drawing on the fundamental pain narrative of Jesus's passion, early Christians, across their literature, framed pain as an operative meaning-making category in imitation of Christ's passion. In constructing their own pain narratives and pedagogies, early Christians shaped their communal identity as the community of cosufferers in mimetic identification with Christ and thus participated in and contributed to larger cultural discourses on and representations of pain.

Pain in Greco-Roman Medicine

Recent scholarship on pain, whether in the Greco-Roman world or the contemporary Western world, has expanded significantly and reflects two broad approaches—the history of science or medicine especially in bio-medicine and neurology, and the history of culture especially in anthropology, literature, history, and philosophy.[1] The former primarily understands pain in physicalist terms, such as "a neurobiological sensory function of the brain and spinal cord,"[2] which is universal in all cultures and historical ages; the latter challenges the medical and reductionist perspective and tends to emphasize specific cultural-historical contexts and interpretive frameworks of pain, and people's perceptions, expressions, and experiences of pain. How did people make meaning of their experiences of pain? In what ways did they assign social significance to it? One way to integrate both approaches is shown by Sarah Coakley: "It seems that our particular sensitivity to pain, and the anxiety that attends it, is not simply a matter of genetics, physiology, and circumstance, vitally important as these are, but also one of learned 'hermeneutics': the way we interpret our pain is all important for the mode of our suffering it."[3] We can take a cue from this attempt but would need to integrate "scientific" and "cultural" views in a more nuanced and contextualized way since a physical body in Greco-Roman and Roman imperial contexts is not, as Coakley seems to suggest, a fixed a priori reality

1. For the approach of the history of science or medicine, see Ronald Schleifer, *Pain and Suffering*, Routledge Series Integrating Science and Culture (London: Routledge, 2014); Patrick Wall, *Pain: The Science of Suffering* (New York: Columbia University Press, 2000); for the history-of-culture approach, see Daniel King, *Experiencing Pain in Imperial Greek Culture*, OCM (Oxford: Oxford University Press, 2018); Keith Wailoo, *Pain: A Political History* (Baltimore: Johns Hopkins University Press, 2014); Javier Moscoso, *Pain: A Cultural History* (New York: Palgrave Macmillan, 2012); Roselyne Rey, *The History of Pain*, trans. Louise E. Wallace, J. A. Cadden, and S. W. Cadden (Cambridge, MA: Harvard University Press, 1995); Judith Perkins, *The Suffering Self: Pain and Narrative Representation in the Early Christian Era* (London: Routledge, 2009); David B. Morris, *The Culture of Pain* (Berkeley and Los Angeles: University of California Press, 1991); Elaine Scarry, *The Body in Pain* (Oxford: Oxford University Press, 1987).

2. Clifford Woolf, "Deconstructing Pain: A Deterministic Dissection of the Molecular Basis of Pain," in *Pain and Its Transformations: The Interface of Biology and Culture*, ed. Sarah Coakley and Kay Kaufman Shelemay (Cambridge, MA: Harvard University Press, 2007), 27.

3. Sarah Coakley, introduction to Coakley and Shelemay, *Pain and Its Transformations*, 1–2.

to which the sufferer can simply apply a personal interpretation. Rational medicine's understanding of the body and of sense perception was already culturally embedded, and Greco-Roman medicine, literature, and philosophy were, as we have seen in chapter 1, closely interrelated.

Starting from the massive Hippocratic Corpus, ancient medical texts are consistent in viewing pain as a critical element in diagnosis. Usually listed first in a taxonomy of symptoms, the precise location of a patient's pain was regarded as an important indicator of various diseases. According to *Diseases I*, "in most cases, [a collection of pus in the abdomen] is to be recognized by where the pain (*odynē*) is felt" (17).[4] Also, *Aphorisms IV* 33 states: "if previous to an illness a part be in a state of pain (*ponesei*), the disease settles in that part." The semantics of pain in the Hippocratic treatises reveals several of the most pervasive terms of pain that will persist (also in literature and philosophy) throughout the Roman imperial and Byzantine period: the family of *algos* and *algēma* appears over 400 times; the members of the *ponos* (*poneō*) family almost 700 times; *odynē* exactly 772 times; *pathos* some 150 times; and finally, *lypē* 59 times.[5] While Peregrine Horden sees these terms used rather synonymously and interchangeably without any specific resonances,[6] Helen King and Roselyne Rey distinguish the meanings of words like *ponos* and *odynē*. King recognizes that *ponos* was often used for long-lasting (chronic) pain, or dull pain, while *odynē* was used to describe sharp, piercing pain; similarly, Rey sees the verb *poneō* to indicate a general state of suffering or illness, and *odynē* to describe pain that was qualified or localized.[7] In fact, King further points out the wider meaning of *ponos* as "primarily pain with a goal, a means to an end."[8] For instance, in *Epidemics V* 2, Timocrates of Elis took, for his

4. Also in *Aphorisms II* 47.

5. Peregrine Horden, "Pain in Hippocratic Medicine," In *Religion, Health, and Suffering*, ed. J. R. Hinnells and R. Porter (London: Kegan Paul International, 1999), 300; Rey, *The History of Pain*, 18–19. See also E. Astyrakaki, A. Papaioannou, and H. Askitopoulou, "References to Anesthesia, Pain, and Analgesia in the Hippocratic Collection," *International Anesthesia Research Society* 110, no. 1 (2010): 190–92.

6. Horden, "Pain in Hippocratic Medicine," 301.

7. Helen King, "Chronic Pain and the Creation of Narrative," in *Construction of the Classical Body*, ed. James I. Porter (Ann Arbor: University of Michigan Press, 1999), 275; Helen King, "The Early Anodynes: Pain in the Ancient World," in *The Management of Pain: The Historical Perspective*, ed. Ronald D. Mann (Carnforth, UK: Parthenon, 1988), 58; Rey, *The History of Pain*, 19. Cf. *Hippocrates*, translated and introduced by W. H. S. Jones, vol. 1 (Cambridge, MA: Harvard University Press, 1957), lx.

8. Helen King, "Chronic Pain," 276.

madness, the necessary purgative *pharmaka* to cure his excess of black bile; he had "much pain (*ponos*) in the purging" but was cured. While this pain was part of the healing process that led to a cure, King contrasts it with the fatal case of Eupolemos of Oineiadae, whose condition began with great *odynē* in the right hip and groin. After he drank a purgative, his condition first improved, but "the pains (*odynai*) did not leave him" (*Epidemics V* 7) and grew worse, eventually leading to his death. In the former, the pains are *ponoi* and imply a "normal and necessary" part of the (healing) process and "thus something to be endured";[9] in the latter, they are the *odynai* that are excessive and therefore should be treated immediately (but in this case remain unchanged by the treatment).[10] This philological examination of pain moves our understanding of pain from a purely physiological sensation to its broader cultural meanings and cultural framework, a framework that shapes the language of pain. We will come back to this theme later.

In terms of the etiology of pain, the Hippocratic Corpus is not uniform. A fundamental notion is that feeling pain is a part of the human condition. According to the *Nature of Man*, we feel pain because we are not in unity— if we were made up of only a single element,[11] there would be no cause of pain (2).[12] Hence, *Places in Man* 42 discusses the way in which pain (*odynē*) arises from cold and heat, excess and want. Pain is brought on by excessive heat in cold people, by excessive cold in hot people, by moisture in those with a dry constitution, and by dryness in those with a moist constitution (*physis*). Since pains are produced every time there is a change that corrupts the natural constitution (*physis*), cures are proposed by the principle of opposites (allopathy),[13] that is, "treatment by agents producing effects contrary to those of the disease" as the fundamental principle of treat-

---

9. Helen King, "Chronic Pain," 276.

10. Helen King, "Chronic Pain," 276; *Epidemics V* 7.

11. Here it is in the sense of "one in form and power" (Galen, *Hipp. Elem.* 1.416) rather than numerically one. See R. J. Hankinson, "Philosophy of Nature," in *The Cambridge Companion to Galen*, ed. R. J. Hankinson (Cambridge: Cambridge University Press, 2008), 212.

12. Agreed by Galen in *The Constitution of the Art of Medicine* 247–250K. See also *Nature of Man* 4: "Pain (*algei*) is felt when one of these elements is in defect or excess, or is isolated in the body without being compounded with all the others. For when an element is isolated and stands by itself, not only must the place which it left become diseased, but the place where it stands in a flood must, because of the excess, cause pain and distress (*odynēn te kai povon*)."

13. For the principle of the cure by opposites, see also *Breaths* 1: "Opposites are cures for opposites." Cf. *Aphorisms II* 22: "In general contraries are cured by contraries."

ment.[14] For example, people with hot constitutions made sick by the cold need a remedy that heats the body, and people with dry constitutions made sick by moisture need a remedy that dries the body. This etiology differs from the one proposed in *On Ancient Medicine*, which traces the origin of suffering to the (unbalanced) strength and quality of the humors; what brings on pain is the predominance of qualities like bitterness, mildness, acidity, and so on, rather than hot or cold.[15] What is important is that both the etiology and the treatment of pain should be understood within the Hippocratics' explanatory framework, that is, ancient humoral systems. As Helen King explains, borrowing the "master metaphor" of cooking from anthropologist Byron Good, "normal bodily processes involve a series of stages of cooking that gradually transforms food into blood, tissue, and humors, while abnormal symptoms appear when rotting occurs due to too much or too little heat being applied, leading to the generation of morbid humors."[16] As mentioned in previous chapters, humoral medicine is holistic in that it demands changes in one's entire way of life; diet, exercise, and sexuality were all thought to influence the internal cooking of humors.[17] Within a basic framework in which life is heat and death is cold, both the causes and the cures of disease and pain are constructed toward the goal of restoring a right balance and improving the quality of humors.

These various and specific etiologies of pain reflect various and specific pain treatments. As mentioned in *On Drugs,* the authors of the Hippocratic Corpus knew and used narcotic plants as anodynes. Three types of henbane (*hyoskyamos*, black, yellow, and white), belladonna (*strychnos manikos*), black nightshade (*strychnos*), mandrake (*mandragoras*), and opium, among others, are widely mentioned[18] and were in fact readily available by other kinds of healers. In humoral systems opium is cooling and wild lettuce is a little cooling, for instance, so too can much opium lead the patient to death.[19] A more typical treatment was fomentation. In treating

14. Jacques Jouanna, *Hippocrates*, trans. M. B. DeBevoise (Baltimore: Johns Hopkins University Press, 1999), 343.

15. Rey, *The History of Pain*, 21.

16. Helen King, "Chronic Pain," 273.

17. Helen King, "Chronic Pain," 273.

18. On pharmacologic analgesia, see Astyrakaki, Papaioannou, and Askitopoulou, "References to Anesthesia," 193; Helen King, "The Early Anodynes," 51–62. William V. Harris raises doubt on the Hippocratic knowledge of opium as an analgesic in "Pain and Medicine in the Classical World," in *Pain and Pleasure in Classical Times*, ed. William V. Harris (Leiden: Brill, 2018), 62–65.

19. J. M. Riddle, "High Medicine and Low Medicine in the Roman Empire," *ANRW* II.37.1 (1993): 106.

pain in the lung, *Diseases III* recommends the application of light moist fomentations to warm and moisten the pained area (7). For gastrointestinal pains, a common treatment was purging: "Pains (*odynēmata*) above the diaphragm indicate a need for upward purging; pains below indicate a need for downward purging" (*Aphorisms IV* 18). At times, surgeons made incisions; as in the case of a patient who, experiencing pain in the front of her head, might be treated with an incision made at the pained site, so that the site of pain might be healed by fresh air (*Diseases III* 3; cf. *Diseases II* 18). In the case of pleurisy with erratic and unyielding pain and pus, the physician might be persuaded to cauterize the affected area as close to the diaphragm as possible, discharging the pus using a tent of raw linen. This treatment is also recommended for pneumonia (*Diseases III* 16). Addressing laypeople (*idiōtai*), the author of *Affections* provides rather comprehensive pain remedies from head to foot. For example, for headaches, the patient should wash, sneeze, and clean out phlegm and mucus and take a regimen of gruel and water; if the pain persists in intensity, there should be an incision of the patient's head and cauterization of the veins (2). For earaches, the patient should wash with hot water and administer a steam bath to the ears; if necessary, the patient should drink a medication (*phamakon*) to draw phlegm upward (4). For sciatica, the patient should soften the affected area with fomentations and a steam bath to evacuate the cavity downward; the patient should also take a medication and drink "boiled ass's milk" (29). *Epidemics VI* 6 sums up the pain treatment in the following way: "In pains (*Odyneōn*): purge the nearest part of the intestine. Cautery, excision, heating, cooling, sneezing, the juices of plants for things they affect and *kykeon*" (a traditional therapeutic drink made of barley mush, cheese, wine, etc.). Finally, three sections of the Hippocratic Corpus propose the principle of likes (homeopathy) for curing pain instead of opposites. *Epidemics V* advises: "Match like with like, for example pain calms pain";[20] the same *Places in Man* 42 where it proposes the principle of opposites also suggests the principle of likes in other cases. In case the two pains (*ponoi*) occur simultaneously in different places, "the strongest obscures the other," weaker pain, according to *Aphorisms II* 46. These examples reveal the therapeutic necessity of medically induced pain in treating pain. These multiple, contextual, and sometimes inconsistent etiologies and treatments of pain in the Hippocratic Corpus testify against the idea of a single anodyne applied to all situations, or a single universal method of treating pain; rather, they disclose doctors' careful observations and assessments of the patients' ill-

---

20. Quoted in Rey, *The History of Pain*, 22.

ness/pain based on their bodily constitutions (e.g., *Regimen in Acute Diseases (Appendix)* 31–32).[21]

Indeed, in the case studies of the Hippocratic Corpus, we can detect both the words of the patient and her unique relationship with the doctor who interprets her expressed pains and symptoms; but we should also recognize that we hear more of the doctor's interpretation than the patient's voice, since the doctor was in charge of the situation and the diagnosis.[22] Still, it is evident that the doctor's reliance on the patient's expression of her pain pushes the doctor to face the suffering individual as a whole person, and not just a diagnostic puzzle to solve.[23] In this doctor-patient relationship, the author of *Breaths* interestingly considers medicine as the painful arts that rather cause grief (*lypērai*) *to the doctors* but are useful (*onēistoi*) for their patients: "For the medical man sees terrible sights, touches unpleasant things, and the misfortunes of others bring a harvest of griefs that are peculiarly his; but the sick by means of the art rid themselves of the worst of evils, disease, suffering, pain and death. For medicine proves for all these evils a manifest cure" (1.2). This indicates the physician's affective engagement with the patient's pain, which we will see in Aretaeus of Cappadocia.

By the time of the early Roman imperial period, there was a greater concern about rational medicine's need to integrate patients' reports (and narratives) about their felt sensations of pain with more developed anatomical theories.[24] In that process, a new view of pain emerged, one that emphasized the combination of physiology, language, subjectivity, affective aspects, and interaction within the doctor-patient relationship.[25] In *On Medicine* by Aulus Cornelius Celsus, the Roman medical encyclopedist in the first century (CE), the word "pain" (*dolor*) appears over two hundred times and is at the core of every moment of medical practice from diagnosis to therapeutics and prognosis.[26] In diagnosis, as in the Hippocratic

21. Cf. Daniel King, *Experiencing Pain*, 274.

22. On patients in Hippocratic case histories, see Susan P. Mattern, *Galen and the Rhetoric of Healing* (Baltimore: Johns Hopkins University Press, 2008), 28–31.

23. On patients' role in Greco-Roman medicine and relationship to medical professionals, see Georgia Petridou and Chiara Thumger, eds., *Homo Patiens: Approaches to Patients in the Ancient World* (Leiden: Brill, 2015).

24. Daniel King, *Experiencing Pain*, 36.

25. Daniel King, *Experiencing Pain*, 37.

26. This section draws on Aurélien Gautherie, "Physical Pain in Celsus' *On Medicine*," in *"Greek" and "Roman" in Latin Medical Texts: Studies in Cultural Change and Exchange in Ancient Medicine*, ed. Brigitte Maire (Leiden: Brill, 2014), 137–54.

treatises, locating pain as a sign of an imminent illness is the first step toward potential recovery. Among the body parts, the head is most prone to pain, especially because of headaches associated with fevers (45 times), followed by internal organs (28 times), eyes (22 times), and hips (14 times). A patient's pain is also a main factor in distinguishing one illness from another, similar one: "But those called myrmecia are less prominent and harder than thymion; their roots are more deeply fixed and they are more painful" (*Med.* 5.28.14c). In the context of therapeutics, the word *dolor* appears nearly half as often, and in two-thirds of those cases, the doctor or medical practitioner (*medicus*) chooses the best treatment or performs the proper surgical act based on the presence of pain.[27] In other cases, the presence or absence of *dolor* is a contraindication to a treatment (1.5.2; 7.20.4); depending on whether there is an outbreak or the persistence of *dolor*, treatment is to be stopped or modified (5.28.1b; 4.29.2); and if *dolor* ends, the existing treatment should be stopped or a new remedial step begun (6.6.9c; 7.20.3); hence, how a patient's pain is experienced and expressed is carefully attended to in every step of the therapeutic process. In prognosis, pain can help predict the result of a disease by its presence (4.13.1) or absence (5.26.27b). For example: "If there is a great swelling without pain, and dryness, there is no danger; if there is dryness, accompanied by pain, there is general ulceration, and at times the result is that the eyelid sticks to the eyeball" (6.6.1c). Moreover, Celsus specifies and classifies the nature or degree of *dolor* according to its size, severity, and chronology with twenty-two adjectives, which would be perfected by Galen a bit later (cf. 4.29.1–2; 7.7.8d; 7.20.3).[28]

In granting pain such importance in the medical arts, Celsus recognizes two significant points: first, Celsus acknowledges that for all his efforts to classify the various pains, pain is inherently a subjective experience and depends on the sufferer. Each pain is specific to the person, and one person's "moderate pain" (*dolor mediocris*) could be another person's "intolerable pain" (*dolor intolerabilis*), and vice versa. Furthermore, there is an unbridgeable gap between the pain perceived and the pain expressed, and the language of pain is inevitably ambiguous and always approximate between the patient and the doctor. Therefore, second, the context of a "pain dialogue" requires a mutual trust (*fides*) between patient and doctor as a way to minimize the linguistic incommunicability and ambiguity of pain. Hence it is no surprise that Celsus, in his preface, promotes medicine

27. Gautherie, "Physical Pain," 140.
28. Gautherie, "Physical Pain," 143.

between friends (*amicis*) and equals rather than between strangers (73).[29] Even in that setting, the doctor's task is challenging—one must, with compassion (*misericordia*), try to comprehend the subjective perceptions and feelings of the patient, perceptions and feelings that can never be translated and transmitted accurately.

In the work of Celsus's younger contemporary, Roman physician Aretaeus of Cappadocia, pain emerges as a "psychosomatic and interpersonal experience."[30] He writes, in *On the Causes and Signs of Chronic Disease*:

> Either the diseases are not driven out entirely, or they return at a slight error. . . . If there is also pain (*ponos*) from an extremely painful remedy—from thirst, hunger, bitter and painful medicines, surgery, or cauterization—of the sort which is sometimes necessary in difficult diseases, the patients run off, since they really prefer death itself. In these circumstances, the virtue of the medical man is revealed, from his great-spiritedness, his variety [in treatment], his indulgence of pleasant things which are not damaging, and his encouragement. But the patient also needs to be courageous, and to stand with the physician against the disease. For, when it takes a firm hold, it quickly wastes and corrodes not only the body, and frequently disorders the sense perception as well, and even deranges the soul by the disorder of the body. (3.1.1.2–2.7)

Here pain emerges as part of the extended therapeutic process—the patient and doctor jointly and courageously manage the sufferer's capacity to cope. Thus pain and disease move from a biological phenomenon to an illness experience as a result of the patient's and doctor's combined negotiation of a biological reality. In this context, Aretaeus, like the Hippocratic physicians, pays close attention to the emotional responses of the pained that might shape their perception of their pain. Bringing up blood from the lung indicates a sense of mortality and fear in the patient; "for pain, even if it is slight, induces fear of death, but in the majority of instances it is more fearful than evil" (2.2.18.6–7). This sense of fear through which pain is seen reinforces the doctor's role as an encourager of the patient in "his great-spiritedness." Indeed, Aretaeus presents "pain as something [that] must be

---

29. "Presuming their state to be equal, it is more useful to have in the practitioner a friend (*medicum esse amicum*) rather than a stranger (*extraneum*)" ("Prooemium," 73).

30. Daniel King, *Experiencing Pain*, 59. On Aretaeus's construction of pain, I draw on King, 61–65.

seen," as Daniel King points out.[31] In discussing tetanus, Aretaeus says its condition is an "inhuman calamity" (*exanthrōpos hē symphorē*): "the sight is unpleasant, a painful spectacle even for the one who looks on. . . . With those, then, who are overcome by the condition, no longer applying his craft, [the physician] only feels distress in common (*synaxthetai*) with the patient" (1.6.8.8–9.7). As the patient experiences his "inhuman calamity," the viewers experience that "unpleasant sight" that is a "painful spectacle," which in turn draws them and the doctor into a common feeling of distress with the patient. Here pain is not just a physical sensation but becomes a felt perception and a public spectacle that binds the pained individual, doctor, and viewers together in emotional responses and social relationships.[32] In addition, Aretaeus's designation of tetanus as a calamity compares pain to a type of moral tragedy in the traditional Greek tragedy.

The most famous and influential imperial physician, Galen, further develops a complex understanding of pain embedded in a consistent natural physiology, perceived and expressed by the patient, and in close relationship with the patient. Largely following the Hippocratic Corpus (and Plato), Galen presents a much more elaborate theory of the generation of pain in which pain arises when an overwhelming or violent and contrary-to-nature experience takes place in our body and its natural state (*Caus. symp.* 7.115K).[33] "In regard to the patient's symptoms, pain is either indicative of the composition of humors (*diathesis*) or it reveals the affected organ."[34] This localization of pain corresponds to his (general) structural and functional understanding of disease, which is "any condition (*diathesis*) contrary to nature by which function is harmed, primarily" (*Symp. diff.* 7.43K). Galen's etiology of pain presupposes first an organ to receive outside impressions, then a connecting passageway, and an organizational center to transform the sensation into a conscious perception.[35]

31. Daniel King, *Experiencing Pain*, 65.
32. Daniel King, *Experiencing Pain*, 65.
33. Cf. Daniel King, *Experiencing Pain*, 71.
34. Quoted in Rudolph E. Siegel, *Galen on Sense Perception: His Doctrines, Observations, and Experiments on Vision, Hearing, Smell, Taste, Touch, and Pain, and Their Historical Sources* (Basel: S. Karger, 1970), 185. See *Loc. affect.* 2.1 (8.70K): "Pain is symptomatic of a certain condition or location, cough of others, and in this manner [we diagnose the source of] vomiting, bleeding, loose stools, cramps. chills, shivering and delirium." See also *The Art of Medicine* 20 (356–357K): "Pain (*algēma*) strongly established in a particular place indicates dissolution of continuity or a sudden complete change."
35. Rey, *The History of Pain*, 32.

Using the terms *algēma* (*algos*) and *odynē* most often,[36] Galen places pain under the sense of touch, which is distinguished from other senses in the ways in which sensory organs interact with external and internal stimuli.[37] Although sight, sound, smell, and taste are all affected by external sensibilities, pain in the sense of touch arises from specific external stimuli such as things that warm, cool, cut, or bruise the sensory organ; and from internal ones such as humoral changes (*Caus. symp.* 7.116K; *Symp diff.* 7.57K).[38] These two causes of pain then reveal pain's evident utility in Galen's usual teleological understanding: "pain served to warn and to protect every living being, and these functions accounted for the arrangement of the nerves and membranes which enveloped the brain, the pia mater and dura mater in particular."[39]

Drawing on an Aristotelian system of classification, Galen then classified the different kinds of pain according to their character, the location of the diseased organ, and a general humoral imbalance, which had been used until modern times: sharp, tension pain; punctuating, dull, or numb pain; pulsating (throbbing) pain from inflammations; pain of ulcerating character; and heavy or oppressive pain with a feeling of weightiness in organs like the kidneys, liver, or lungs, and so on (*Loc. affect.* 2.2–4 [8.70–79K]).[40] Furthermore, he "differentiated a periosteal pain caused by inflammation of the membrane lining the bone, or pain resulting from external injury."[41] He even observed sympathetic pain (i.e., referred pain) spreading from the stomach to the heart, often leading to the individual's collapse, reasoning that the communication of painful irritation was through the nervous system.[42] As one can see, the rational system set up by Galen almost establishes a "universal law" concerning the human body's susceptibleness to pain: a direct link exists between each type of pain and each part afflicted, characterized by its own specific nature (*physis*).[43] With a correct analy-

36. Nicole Wilson, "The Semantics of Pain in Greco-Roman Antiquity," *Journal of the History of the Neurosciences* 22 (2013): 133.

37. Cf. Siegel, *Galen on Sense Perception*, 184–93.

38. Cf. Daniel King, *Experiencing Pain*, 74. See *Symp. diff.* 7.57K: "In the case of touch, pain comes not only from what is external, but also far more from conditions in the body itself, and often in fact so strongly that some who are overcome by suffering may kill themselves. . . . The most severe pains happen to the sense of touch."

39. Rey, *The History of Pain*, 33. See Galen, *UP* 5.9.

40. Siegel, *Galen on Sense Perception*, 190.

41. Siegel, *Galen on Sense Perception*, 190.

42. Siegel, *Galen on Sense Perception*, 187–88.

43. Rey, *The History of Pain*, 35.

sis, "the symptom made it possible to identify the specific reality of the disorder." While the Hippocratic treatises also drew indications from the regional localization of pain, the actual center of the disorder had a much more precise anatomical significance in Galen. Hence Roselyne Rey sums up the significance of Galen's contribution in this way: "The qualitative description of the different pains was closely linked to their presumed cause, and their meanings were part of a perfectly elaborated rational system of thought based on humours and qualities."[44]

Again, this rational system presupposes both a trusting relationship between the patient and the doctor, and the patient's ability to perceive and communicate pain in a diagnostic setting.[45] Recognizing the limits of scientific language to describe only the common properties of pain, to obtain a more precise and intimate sense of the pain symptoms and disease of the sufferer, the doctor has to rely on the metaphorical language used by his patients to explain what troubled them (*Loc. affect.* 2.9 [8.116K]). When Galen asked his patients to describe clearly the different symptoms of pain and examine the changes in the affected parts, the patients described pain as if "being pierced by a needle; at another time again they thought they were being perforated by a trepan, bruised, torn apart, drawn or pulled up or down, or they had the feeling of an oppressive weight, or as if something were hanging or lying on them, or as if they were pressed against things surrounding them" (2.9; 8.116K). At the same time in that setting, the doctor had a superior authority in interpreting and analyzing the patient's shared symptoms, as in the Hippocratic Corpus. Discussing the importance of clear terminology in diagnosis and medical knowledge, Galen states:

> This is difficult to evaluate, since we have to rely on many other persons: either on those who suffer but do not understand clearly (*saphōs*) their experiences because their minds [souls] are weak, or on those who understand but are unfit to communicate clearly (*saphōs*), being totally unable to formulate their suffering in words [or rational discourse—*logos*], since it requires a considerable effort or it is inexpressible (*tōi mēo eivai rēton auto*). Consequently a person who wants to describe each type of

44. Rey, *The History of Pain*, 35.
45. On this aspect of the patient-physician relationship in Galen and his "patient-centeredness," which Galen himself challenges at times (see the paragraphs following), see Mattern, *Galen*, 124–25, 138–58, and Courtney Roby, "Galen on the Patient's Role in Pain Diagnosis: Sensation, Consensus, and Metaphor," in Petridou and Thumger, *Homo Patiens*, 304–22.

pain (*algēmatōn*) should have experienced it personally, should also be a physician and able to speak clearly (*hermēneuein*) to others, and should observe it with understanding while suffering, and with his mental powers [souls] intact. (*Loc. affect.* 2.7 [8.88–89K], trans Siegal with modification)

This passage is telling for several reasons. Whereas Aretaeus emphasized a sense of rapport between the patient and the doctor, for all its emphasis on a patient's voice and expression, Galen reveals a substantial suspicion of patients' ability to express and articulate their perceptions of their pain using rational discourse; this is in contrast to his confidence concerning a doctor's rational capacity, thus suggesting a hierarchical relationship between patient and doctor. Pain resists easy explanation either because its perception and communication require a high level of ability imagined to be rare among the patients, or because the experience of pain is inexpressible in words.[46] When the patient is unable to express the pain, the doctor, since he rationally understands the patient's pain and possesses the linguistic ability to express it clearly, can solve the problem of inexpressibility by filling in the gaps of a patient's testimony. Thus, there still can develop a sense of "common experience" between patient and doctor as long as it is managed and controlled by the doctor (Galen himself) as the latter shapes the patient's narrative of pain.[47] In constructing a "common experience," Galen, like Aretaeus, also paid careful attention to a patient's emotional responses to pain. As Daniel King notes, Galen, in general, commonly uses emotional experiences like distress (*lypē*) and anxiety (*ania*) in close association with pain and disease.[48] Distress and other emotions (*pathē*) such as anger, fear, worry, and desire are "both antecedent causes of diseases and caused by them," so much so that "the boundary between the physiological aspects of pain (or disease in general) and its emotional or psychological elements was always fluid in Galen's work."[49] As we will see, ideas about the emotional or psychic repercus-

46. See also Galen, *Loc. affect.* 2.6–9 (8.86–87K).

47. Cf. Daniel King, *Experiencing Pain*, 80; Mattern, *Galen*, 124–25.

48. Daniel King, *Experiencing Pain*, 89. On emotion in Galen's case histories, see Mattern, *Galen*, 132–36; Susan P. Mattern, *The Prince of Medicine: Galen in the Roman Empire* (Oxford: Oxford University Press, 2013), 249–56. On distress (*lypē*) and anxiety (*ania*) in Galen in particular, see Clare K. Rothschild and Trevor W. Thompson, eds., *Galen's De indolentia: Essays on a Newly Discovered Letter*, STAC 88 (Tübingen: Mohr Siebeck, 2014).

49. Daniel King, *Experiencing Pain*, 89; Galen, *Caus. symp.* 1.6. Also, Mattern, *The Prince of Medicine*, 255.

sions of physical pain *and* the physicality of emotional states are pervasive in Galen's contemporary literary and philosophical works. Nonetheless, unlike Aretaeus, Galen was reticent about engaging with the patient's pain in a more compassionate way.

In the aforementioned quote, Galen, moreover, recognizes the inexpressibility of pain due to the nature or quality of pain itself. He often describes painful sensations as "unspeakable" (*arrētos*) throughout his corpus[50] and acknowledges that no adequate terms exist for describing the various personal impressions of pain. This leads him to note pain's irreducible subjectivity, because no doctor has experienced all types of pain (*Loc. affect.* 2.7 [8.89K]):[51] "It is evidently impossible to transmit the impression of pain by teaching, since it is only known to those who have experienced it. Moreover, we are ignorant of each type of pain before we have felt it" (2.9 [8.117K]). Here Galen's stance is to discredit his rival Archigenes of Apamea's approach (also Aretaeus's to an extent) to pain language, as the latter wrote about the differences in all types of pain "as if he has experienced all forms of pain" (2.7, 9 [8.89–90, 117K]).[52] According to Galen, Archigenes's vocabulary used for pain is impossible to understand or use. Descriptors like "harsh," "soft," "sweet," "salty," and "astringent" are adjectives more appropriate for descriptions of humors recognized by the tongue; "moist," "dry," "hot," and "cold" are appropriate for humors recognized by touch, and so forth (2.6, 9 [8.87–88, 117K]). "Therefore one cannot know more about the pain which Archigenes calls harsh or rough than if one should call it blue, red or some other color" (2.9 [8.114K]).

Still, beyond his polemic, Galen might have found agreeable Elaine Scarry's argument that pain is inherently unshareable because it essentially shatters language; (physical) pain has no referential content: "It is not *of* or *for* anything. It is precisely because it takes no object that it, more than any other phenomenon, resists objectification in language."[53] This inaccessibility to language with its fundamental subjectivity challenges the possibility of understanding and empathizing with the pain of others. According to Scarry: "For the person in pain, so incontestably and unnegotiably present is it that 'having pain' may come to be thought of as the most vibrant ex-

---

50. E.g., Galen, *Loc. affect.* 8.117K; 8.339K.

51. Horden, "Pain in Hippocratic Medicine," 308.

52. On Galen's detailed criticism of Archigenes's pain language, see Wilson, "The Semantics of Pain," 134–37, and Roby, "Galen on the Patient's Role," 307–12.

53. Scarry, *The Body in Pain,* 5.

ample of what it means to 'have certainty,' while for the other person it is so elusive that 'hearing about pain' may exist as the primary model of what it is 'to have doubt.' Thus pain comes unsharably into our midst as at once that which cannot be denied and that which cannot be confirmed."[54] This understanding of pain's inexpressibility, subjectivity, and inshareability, while seemingly negating the possibility and expectation of its opposite (the expressibility, objectivity, and shareability of pain), eventually compels writers (ancient and modern) to pursue a means by which pain might be clearly communicated and shared. This counterintuitive impulse should be viewed in the larger context of the early Roman imperial culture of representation, to which we now turn.

Pain in Greco-Roman Literature and Philosophy

In these medical texts, we have seen pain's centrality in diagnosis (and therapeutics). We have also seen the affective and social implications of successfully communicating one's experience of pain via language (expressibility), and failing to communicate one's experience of pain via language (inaccessibility). All these factors complicate the inherent complexity of the patient-doctor relationship. In this section we discuss pain in literary and philosophical texts of the early imperial period (the first two centuries CE). As pain and its experience arise at the intersection of body, mind, and culture,[55] shifts in the way pain is expressed, represented, and valued can be significant indicators of larger social and cultural changes.[56] In turn, such changes influence the way individuals perceive, experience, and endure pain. According to Judith Perkins's *The Suffering Self*, as mentioned in chapter 1, one such momentous shift occurred in the first two centuries of the Roman Empire.[57] During this period a new kind of cultural discourse appeared in all different locations, "projecting a particular representation of the human self as a body liable to pain and suffering," in contrast to the discourses in the classical and (early) Hellenistic period that represented the mind as exercising control over the body, and downplayed the expe-

54. Scarry, *The Body in Pain*, 4.
55. Morris, *The Culture of Pain*, 1, 18–19, 28–29.
56. Susanna Elm, "Roman Pain and the Rise of Christianity," in *Quo Vadis Medical Healing: Past Concepts and New Approaches*, ed. Susanna Elm and Stefan N. Willich (New York: Springer, 2009), 43.
57. Perkins, *The Suffering Self*, 7, 193.

rience of pain.[58] In this very transition of "discursive struggle over these representations," Perkins locates the "triumph" of Christianity; it is around a type of represented subject, "the suffering self, that Christianity as a social and political unity would form and ultimately achieve its institutional power."[59] We will address her focus on Christianity in the next section. For now it is important to recognize and proceed with her point of how the experience of pain and suffering is embedded in cultural discourses, which extend beyond, and intersect with, medicine, literature, and philosophy.[60] How is pain constructed in these texts, and how does pain emerge as part of a coherent cultural discourse?

Perkins uses the terms "pain" and "suffering" interchangeably, as though the former and the latter are equivalent to each other without any distinction or explanation. As we have seen, pain in medical texts was primarily a physiological sensation and perception but included emotional and even social dimensions in the right contexts; and in diagnostic settings, the pained individual was also referred to as the suffering patient. It is in literary and philosophical texts that pain moves beyond its anatomical basis and structure and is treated as a (long-term) "hardship or trial to be endured and navigated" with one's perseverance and moral strength and also in connection with disaster (*symphora*) or some evil (*kakon*).[61] As Daniel King notes, this is a "moment when pain perception and symptoms move close to what some scholars have designated as 'suffering.'"[62] Pain indeed constitutes "suffering when it is overwhelming, uncontrolled, unexplained, or in some other way associated with a dire *meaning* that calls into question the continued, integrated existence of the personal self"[63] or community. In this sense, the interchangeable use of "pain" and "suffering" has a more nuanced legitimacy in these texts, unless the texts imply otherwise or distinguish them.[64] In this context, then, it is appropriate to speak of a "pedagogy of pain,"[65] a way

---

58. Perkins, *The Suffering Self*, 3.

59. Perkins, *The Suffering Self*, 3.

60. Cf. Daniel King, *Experiencing Pain*, 9.

61. Daniel King, *Experiencing Pain*, 109.

62. Daniel King, *Experiencing Pain*, 110.

63. Howard Brody, *Stories of Sickness*, 2nd ed. (Oxford: Oxford University Press, 2003), 49.

64. Note that Galen equates pain (*odynēn*) and suffering (*ponon*) in *De placitis Hippocratis et Platonis* (7.6.33–34), although he distinguishes between a painful condition (*ponos*) of the body and the perception of pain (*odynē*) attached to this condition.

65. This phrase itself comes from Antigone Samellas, "Public Aspects of Pain in Late Antiquity: The Testimony of Chrysostom and the Cappadocians in Their Graeco-

to navigate, endure, overcome, structure, or deal with pain with a certain moral agency and even with rituals "to school the human mind in extreme endurance, whether aided by human or divine example."[66]

In literature the social and cultural construction of pain is mainly expressed through "pain narratives," which address the aforementioned (in)expressibility, subjectivity, and (in)shareability of pain (and suffering) through a thematic story of a main character or main characters. In fact, they have already been embedded in Greek tragedies, which continue to be read and referred to and exert influence during the imperial period; one of them was Sophocles's influential tragedy *Philoctetes*. In this play the center stage is set on a hero in pain, Philoctetes ("swell-foot"), who, on his way to Troy, was inflicted with a serpent's bite on his foot by the nymph Chryse on account of his inadvertent transgression of her sacred precinct. Subsequently, the wound festered ceaselessly, tormenting him with chronic yet sudden bouts of overwhelming, severe pain. As David Morris observes, Philoctetes, after nine years of relentless suffering, in some sense, becomes his wounds; his character not only has become inseparable from his pain but also has been changed by his pain.[67] Now Odysseus and Neoptolemus (Achilles's young son) arrive on the deserted island of Lemnos, where Philoctetes has been living entirely alone as a reluctant exile. They come to bring him back to join the Greek attack on Troy because they heard from the Trojan seer Helenus that the Greeks could take Troy only if Philoctetes rejoined their ranks, equipped with the famous bow he has received from the dying Heracles. However, Philoctetes does not trust Odysseus and the Greeks because it was they who had deserted him on Lemnos nine years earlier (on their way to Troy). Their reason for deserting Philoctetes was his festering wound—because of its foul stench, and because his "savage (*agrios*) and ill-omened cries, shouting, and screaming" disrupted their sacrifices and made it impossible for their community to function (*Phil.* 8–11). Indeed, his continuous pain is language shattering and prevents him from rational speech or discourse (*logos*) but permits him only to project a

Roman Context," *ZAC* 19, no. 2 (2015): 279. Talal Asad suggests pain not merely as a passive state but also as action itself and intentionally collapses (or at least reduces) the distinction between physical pain and psychological or social suffering in *Formations of the Secular: Christianity, Islam, Modernity* (Stanford, CA: Stanford University Press, 2003), 79–80, 81–85.

66. Elm, "Roman Pain," 51.

67. Morris, *The Culture of Pain*, 249. Cf. Nancy Worman, "Infection in the Sentence: The Discourse of Disease in Sophocles' *Philoctetes*," *Arethusa* 33 (2000): 6.

voice or a scream. Even before his actual entrance to the scene, the chorus sings of "the far-off grievous cry of a man in distress" (208), "his cry can be heard from afar. . . . His bitter cry" goes before him (215–219). Only then does Philoctetes appear, dragging his wounded leg.

The significance of his incommunicative pain continues in the subsequent scenes between Philoctetes and Neoptolemus. Odysseus had devised a plan that Neoptolemus should befriend and beguile the deserted hero with a false story of his own betrayal by the Greeks, and thereby lure Philoctetes to board his ship; this was under the pretense of offering Philoctetes a safe journey to his home in Malis. As they are making for the ship, Neoptolemus urges Philoctetes, "Why are you silent like this, . . . and stand as though struck dumb?" (730–731). Philoctetes's pain prevents him from communicating with his people "and dissolves his humanity" in a series of unintelligible cries, "Ah, ah, ah, ah!" (732, 739).[68] He cannot explain his pain and can only manifest it in his scream and body. In the absence of any reasonable expression and explanation of his pain, his cry and body are what allow us to "grasp the undissembled truth of human affliction" and suffering.[69] Neoptolemus then asks with frustration, "What is the matter with you? Will you not tell me, but remain silent as you are?" (740–741). Only then Philoctetes utters words in stumbling, broken phrases, interrupted by a long, frightful cry: "[my pain] (*kakon*) goes through me, it goes through me (*dierchetai*)! O misery unhappy as I am! . . . I am devoured (*brychomai*), my son! A-a-a-a-a-h!" (743–746). His words can only express his pain, which is indeed grievous and unspeakable (756).

Yet, as Neoptolemus stands witness to this "spectacle" of Philoctetes's incommunicative pain, he finally begins to change in momentous ways that transform him to show pity to Philoctetes's suffering *and* to act against the deceptive and pragmatic Odysseus, fighting for his noble reputation. After Philoctetes gives Neoptolemus his bow in the midst of his bouts of pain and he regains his consciousness from his pain, Neoptolemos confesses, "I have been in pain long since, lamenting for your woes" (*algō palai dē papai soi stenōn kaka*; 805–806). After Neoptolemus finally tells Philoctetes the truth about sailing to Troy instead and Philoctetes, now betrayed by Neoptolemos, unleashes his tirade against him, Neoptolemus tells the chorus, "a strange pity (*oiktos*) for this man has fallen upon me" (965). This sense

---

68. Anthony J. Podlecki, "The Power of the Word in Sophocles' *Philoctetes*," *GRBS* 7 (1966): 235.

69. Morris, *The Culture of Pain*, 251.

of pity/mercy seems pivotal in the direction of the play—Philoctetes has repeatedly appealed to pity (501, 759a, 870), and Neoptolemos gets it at last; and his "new compassion brings him from simulated friendship to genuine concern, from speech to embodied deeds."[70] He eventually decides to return the bow to Philoctetes. Then, has Philoctetes's pain finally become communicative and thus shareable—with Neoptolemos? With the audience? Certainly not, at least from Philoctetes's perspective—he is betrayed again by his fellow Greek, and Neoptolemus's lies have destroyed any hope of reestablishing the old bond of communication; it has all become futile, and Philoctetes's attempt to communicate his pain has now imbued him with a greater bitterness, and a more hardened heart. Thus, says David Morris, "the sternest wisdom of Greek tragedy may be that suffering cannot be shared: only witnessed."[71] Furthermore, Odysseus's witness of Philoctetes's pain leaves him only with scorn for the latter. Therefore, "pain—compounded by the treachery of Odysseus and the hardness of Philoctetes—has created an impasse that neither persuasion nor compassion can unblock."[72] Philoctetes still remains bitter, suspicious, friendless, and alone through his unshareable pain. Only a god, Heracles, finally resolves that impasse by his divine command at the end (1409–1444).

Hence, Philoctetes secures a place in the Greco-Roman literary memory as an icon of inexplicable prolonged pain and unendurable suffering. In *That Epicurus Actually Makes a Pleasant Life Impossible* (*Non posse suaviter vivi secundum Epicurum*), Plutarch, speaking on long-term pain and its gnawing effect, refers to Philoctetes via a version of his story written by Aeschylus: "Once it had struck, the snake did not release its hold, but lodged in me [Philoctetes] its fangs of tempered steel, that grip my foot" (1087F). He goes on to say, "For there is nothing smooth and gliding in pain (*algēdōn*), nor does its scratching and tickling propagate an answering smoothness in the body. . . . So pain (*ponos*) broadcasts its hooks and roots and entangles itself in the flesh, lasting not only for the space of days and nights, but in some person for whole seasons and Olympiads, and is barely got rid of when new pains (*ponōn*) thrust it out, like nails more strongly driven" (1088A). By referencing Philoctetes, Plutarch highlights, in vivid imagery, pain's permanence, its capacity to take root in every part of the

---

70. Morris, *The Culture of Pain*, 253.
71. Morris, *The Culture of Pain*, 253.
72. Morris, *The Culture of Pain*, 254.

body, and its lasting implications for the individual.[73] For Plutarch, this aspect of pain is particularly poignant as he contrasts it with the temporal nature of pleasure, and thus critiques the Epicurean position that pleasure is the highest good. It does not make sense that they base their good in a body, easily susceptible to both pleasure and pain since a body "receives pleasures in few of its parts, but pain (*algēdonas*) in all" (1088D). In his *Sacred Tales*, Aristides also alludes to Philoctetes, focusing more on pain's subjectivity and inexplicability. According to Aristides, speaking about Smyrna is challenging, for "just as, they say, the one who was bitten by the serpent did not wish to speak to another, but only that man who has experienced it, so also having seen the beauty of the city [I wish] to make it common only to one who has seen it" (*Or.* 17.18). Although the connection between Philoctetes and Smyrna is not direct here, Aristides, as he styles himself as a kind of heroic sufferer of prolonged incommunicative pain in the *Tales*, emphasizes the limits of language to share a certain experience and hence pain's inshareability.[74]

We indeed come to the six orations of the *Sacred Tales*, a record and representation of Aelius Aristides's pained existence that we encountered in chapter 1. Deeply concerned about the perception of pain, Aristides represents his pain symptoms and experience and his reactions to the doctors' treatment and regimen in ways akin to contemporary medical texts—coughs, vomiting, swellings, bloodletting, fever, headache, shivering, and intermittent inability to eat, breathe, or walk properly (*Or.* 48.37–45). He was very ill in 144 CE, and "the doctors were wholly at a loss not only as to how to help, but even how to recognize what the whole thing was" (48.5; cf. 48.69). When his intestines swelled and his breathing was blocked, the doctors first purged him with elaterium (squirting cucumber) and made an incision from chest to bladder and used cupping instruments, at which his breathing left him (48.62–63). "And a pain, numbing and impossible to bear (*odynē narkodes kai aporos pherein*), passed through me, and everything was smeared with blood and I felt extreme pain (*hyperalgeinos*), and I perceived that my intestines were cold and hanging out, and the difficulty in my respiration was intensified" (48.63). Here Aristides is hyperaware of what is being done to his body deep inside, and with his men-

---

73. Daniel King, *Experiencing Pain*, 119.

74. Later Gregory of Nazianzus would also style himself as a type of Christian Philoctetes. See the section "Gregory of Nazianzus: Ontological Pain and Community of Cosufferers" (pp. 182–85).

tal acuity and diagnostic terminology, fully capable of understanding and expressing what he perceives and suffers—the very opposite of Galen's patient who lacks "a high level of ability" to perceive and communicate her own pain.[75]

Ironically, it is his linguistic ability and mental acuity to communicate his pain that enable him to construct his narrative in a manner that highlights the insurmountable enormity and incomprehensibility of his pain experience as both the sufferer and observer of his own pain. His narrative first involves the doctors' failure to comprehend and treat his suffering in contrast to the treatment by Asclepius. In addition to the aforementioned episode, earlier at Smyrna, he says, doctors and gymnastic trainers were unable to help him, failing to recognize "the complexity of his disease" (48.69; cf. 48.5). On another occasion, when he developed a painful swelling in the groin, doctors urged surgery or cauterization by drugs with a prognosis of death if not followed (47.62). In contrast, Asclepius commanded Aristides to "endure and foster the growth"; and for Aristides there was no choice between obeying the doctors or the god (47.63). However, the growth worsened, and others marveled at his endurance or criticized his credulity in dream and his cowardice in the face of the knife. The god's explanation of the condition was that it was dropsy diverted downward, so it was a safe and appropriate swelling. Asclepius then gave him a dramatic treatment: run a race barefoot in winter, ride horses, sail in stormy weather after eating honey and acorns, and then vomit (47.65). Lastly, Asclepius the "Savior" told his foster father Zosimus in a dream to apply a certain drug containing salt to the growth, which led to its disappearance. While the doctors stopped their criticism, they still thought surgery was necessary to treat the loose skin left after the disappearance of the growth. The god then told Aristides to smear on an egg, and it cured his disease to the extent that no one could even tell on which thigh the tumor had been (47.68). Asclepius's diagnosis of diverted dropsy could have been given by doctors, and Aristides continues to use doctors elsewhere in the *Tales*. However, as Helen King notes, Asclepius offers something that the doctors do not or cannot: "an explanation of suffering acceptable to the patient" and a "holistic" approach to the patient's condition as seen in

---

75. See also Aristides, *Or.* 48.39: "Doctors gave up and finally despaired entirely, and it was announced that I would die immediately. However, even here, you could use the Homeric phrase, 'his mind was firm.' Thus I was conscious of myself as if I were another person, and I perceived my body ever slipping away, until I was near death."

this episode.[76] Going back to his illness in 144 CE, when his breathing was blocked, Asclepius not only recognized that Aristides's main concern was his breathing, but he also understood his patient's anxieties about oratory. Thus, he not only gives Aristides medicine but also eases his inability to breathe, convincing him that once he starts to speak, his problems will diminish (cf. 48.7–10; 71–73; 50.13).[77] Understanding that oratory is central to his patient's life, Asclepius in fact exhorts him back to the oratory he had, in hopelessness and despondency, abandoned (50.14). Just as the god told him, when Aristides starts to speak, even with a shortness of breath, he finds that he is soon able to breathe well and eventually resumes his oratory (50.22).

The second aspect of the incomprehensibility of his pain experience involves his rhetorical emphasis on his incapacity to narrate his pain experiences and thus his choice to keep silent. Modern studies of trauma reveal that testimonies of extreme trauma are often characterized by the vacillation between the sufferers' compulsion to speak and their inability to capture all of their experiences.[78] Aristides would have identified with that vacillation, as his detailed narrative conveyed the overwhelming power of pain that undermined his capacity for narrative description. "In the beginning it did not occur to me to write about any of these [events and experiences], because I did not believe that I would survive. Next my body was also in such a state that it did not give me leisure for these things. . . . So I thought that it was better to *keep completely silent* than to spoil such great deeds. And for these reasons I made many excuses both to the god and to my friends, who from time to time asked me to speak and write about these things" (48.1 [emphasis added]).

In light of his central concern on oratory, it is important for us to recognize that the relationship between his overwhelming bodily pain and his choice to remain silent is closely linked with his incapacity to engage in oratory (48.6; 50.14). Thus, he describes his pain as "unspeakable" (*arrētos*) in the same language used by Galen (48.23), and even inconceivable (49.17). Aristides's pain is language shattering, silencing, and thus inexpressible in utter subjectivity similar to Scarry's argument; and the doctors' inability to recognize "the complexity of his disease" further heightens this incom-

---

76. Helen King, "Chronic Pain," 279, 280.

77. Cf. Helen King, "Chronic Pain," 280.

78. Dominick LaCapra, *History and Its Limits: Human, Animal, Violence* (Ithaca, NY: Cornell University Press, 2009), 60–62, in Daniel King, *Experiencing Pain*, 143.

municability and incomprehensibility of pain.[79] Paradoxically, however, "in Aristides' hands, the ineffability that undermined medical accounts of pain becomes a way of communicating the nature of his experiences."[80]

Indeed, compelled by Asclepius, Aristides chooses to excavate his past and narrate his experiences (48.2). Suffering the ineffable pain and facing the loss of his public career, he comes to believe that it is the god, not doctors, who offered him an explanatory framework for his suffering, understood his anxieties about oratory, and helped him speak, breathe, and eventually recover his oratory. As he submitted himself to the god's orders to endure painful experiences contrary to the demands of doctors and even, at times, common sense, the result was not pain but well-being and union with the divine (48.19–23).[81] Therefore, Aristides comes to understand that his suffering, a "seemingly incoherent mass of symptoms"[82] and "tempests of [his] body" (47.1; cf. 42.7), has meaning: "If someone should take these things into account and consider with how many and what sort of sufferings and with what necessary result for these he bore me to the sea and rivers and wells, and commanded me to contend with the winter, he will say that all is truly beyond miracles, and he will see more clearly the power and providence of the god, and will rejoice with me for the honor which I had, and would not be more grieved because of my sickness" (48.59).

Aristides narrates the full extent of his suffering because for him it is teleological and redemptive in that his readers and audience will not grieve his suffering but "see more clearly the power and providence of the god" which he experienced for himself and also "the honor which [he] had." Thus his pain

---

79. See Morris, *The Culture of Pain*, 72–73: "The normal failure of language under the assault of acute pain, . . . is a common but not devastating experience. . . . [But a chronic pain] constitutes a radical assault on language and on human communication. There is simply nothing that can be said."

80. Daniel King, *Experiencing Pain*, 144.

81. See *Or.* 48.21–23: "When we reached the river, there was no need for anyone to encourage us. But being still full of warmth from the vision of the god, I cast off my clothes, and not wanting a massage, flung myself where the river was deepest. . . . When I came out, all my skin had a rosy hue and there was a lightness throughout my body. There was also much shouting from those present and those coming up, shouting that celebrated phrase, 'Great is Asclepius!' . . . There was a certain inexplicable contentment, which regarded everything as less than the present moment, so that even when I saw other things, I seemed not to see them. Thus I was wholly with the god."

82. C. A. Behr, *Aelius Aristides and* The Sacred Tales (Amsterdam: Adolf M. Hakkert, 1968), 162.

becomes not only expressible but also sharable through his narrative; and his pain narrative becomes his pedagogy of pain through which he endures and makes sense of his pain in his intimate relationship with Asclepius. In the words of Helen King: "The body, susceptible to pain, disease, old age, and death, seems to be a sign of distance from the divine world, but [Aristides's] creation of a story from the minute details of its physicality paradoxically seeks to transcend its materiality and make it into a sign of divine favor."[83]

However, Aristides's narrative and pedagogy of pain stand in intriguing relationship to Stoic teachings on pain. Stoicism, one of the most dominant schools of thought in the Roman Empire, provided a "philosophic *koine*" as a common currency of moral philosophy and imperial ideology, especially in the first three centuries of the empire.[84] Along with (Middle) Platonism, it was the most influential philosophical school for emerging Christian theology and ethics, particularly in its teaching on virtue in relation to enduring pain. In some sense, Stoic discussions on (pleasure and) pain may be viewed as a response to the agenda set by Epicureanism, which was roughly contemporary with Stoicism.[85] For Epicureans, the highest good and *telos* toward which each individual must strive was pleasure (*hēdonē*), not in a popular "hedonistic" sense but as a state of ultimate tranquility (*ataraxia*), free from any distress or pain.[86] Since "every pleasure must be enjoyed and every pain must be rejected," for them, "certain pleasures must be rejected in order to savor greater pleasures later, while some pains must be patiently endured to escape more severe ones" (Cicero, *Fin.* 2.4.14). This doctrine demands a rather rigorous lifestyle, but regards pain (any injury, discomfort, sickness, etc.) as the intrinsic and ultimate evil or adversity to be avoided at all costs. As they were materialists, although pleasure and pain were not limited to the physical senses, nothing worthwhile was possible without them.[87] In this ideology virtue is subordinate to pleasure—the Epicurean sage pursues it not for its sake, but so that it can lead him to pleasure, the intrinsic and ultimate good.

---

83. Helen King, "Chronic Pain," 282.

84. James A. Francis, *Subversive Virtue: Asceticism and Authority in the Second-Century Pagan World* (University Park: Pennsylvania State University Press, 1995), 1.

85. Catharine Edwards, "The Suffering Body: Philosophy and Pain in Seneca's Letters," in *Constructions of the Classical Body*, ed. J. I. Porter (Ann Arbor: University of Michigan Press, 2007), 254. See Cicero, *Tusc.* 2.7.17–2.13.33, for Cicero's engagement with Epicurean-Stoic discussions on pain.

86. Cf. Cicero, *Fin.* 2.6.18–2.7.22.

87. Rey, *The History of Pain*, 38–39. Cf. Cicero, *Fin.* 2.33.107–108.

For Stoics, the highest good is virtue (*virtus*), which is living life accord-ing to nature, and this should bring about a life without passions (*apatheia*), the Stoic *telos*. Living life according to nature means bringing one's life into conformity with the actual course of events by exercising control over what is in one's power, including one's opinions, attitudes, and emotions/passions (Epictetus, *Ench.* 8; 1).[88] Stoic emphasis on self-mastery and self-discipline as the way to attain virtue focuses on moral choice (*proairesis*) based upon one's ability to distinguish between what is under one's control and what is not (*diairesis*). Stoics also emphasize the need for inner forti-tude, and for freedom from outside constraints such as changes in fortune, health, wealth, slavery, sickness, or social status or upheavals (Epictetus, *Diss.* 1.22.10; 2.6.24). Since virtue alone is good and vice alone is evil, every-thing else in life belongs to the *indifferentia* or *adiaphora*—external things that do not affect one's ability to live a good life. As mentioned in chapter 1, for Stoics, every emotion (*pathos*; *affectus*; Cicero, *perturbationes*) is "both a compound false judgment concerning indifferent values and a violent movement of the soul inseparable from this judgment," and as such, in-compatible with happiness and wisdom.[89] Pleasure (*hēdonē*) and distress (*lypē*; *aegritudo*),[90] along with desire and fear, belong to the four classes of emotion, from which a Stoic sage must detach himself.[91] Yet Stoics also acknowledge good emotions caused by true judgment (*eupatheiai*; Cicero, *constantiae*).[92] Pain (*ponos/algos*; *dolor*), however, along with other con-ditions like sickness, poverty, exile, and death, belongs to the category of *indifferentia*. While Stoic writings in general were hostile to pleasure (part of *pathē*), Roman Stoicism in particular was much more concerned with turning pain into something useful.[93]

88. Cf. Perkins, *The Suffering Self*, 79.

89. Jan Edward Garrett, "Is the Sage Free from Pain?" *Volga Journal of Philosophy and Social Sciences*, no. 6 (1999): 2.

90. In common English translation, *lypē* is translated as "pain," which is not incor-rect in a philological sense, as it is used as one of the terms for pain in the Hippocratic Corpus and pain also has an emotional aspect in Galen. However, Jan Edward Garrett notes that in the Stoic texts, especially by Diogenes Laertius, Epictetus, and Marcus Aurelius, the distinction between *lypē* (as an emotion of distress) and *ponos* (as pain) is fundamental, especially with Latin equivalents starting from Cicero, and I follow his argument in this section: Garrett, "Is the Sage Free from Pain?," 2.

91. Cicero, *Tusc.* 3.11.24–25.

92. Cicero, *Tusc.* 3.11.24–25; 4.6.11–4.7.14.

93. Edwards, "The Suffering Body," 254. For a nuanced discussion on the problems

Epictetus, an ex-slave and early-second-century Stoic, seems to affirm a person's full control of her reactions to external stimuli such as pain at different times. The sage does not worry about things he cannot change, and we "must make the best of what is under our control" and take what is not (*Diss.* 1.1.17). Whether I am executed is not up to me, but groaning (*stenonta*) while being executed is under my control; whether I am fettered is not up to me, but wailing while being fettered is under my control (1.1.22). However, Arrian also records the exchange between a student and the Stoic sage: "'I have a head-ache (*kephalēn algō*).' Well, do not say 'Alas!' I have an ear-ache (*ōtion algō*).' Do not say 'Alas!' And I am not saying that it is not permissible to groan, only do not groan in the center of your being" (1.18.19; cf. *algō tēn kephalēn*, 1.18.16). Here Epictetus indicates that pain in our head ("headache") *as well as* groaning is not up to us; there are an acceptable groaning (i.e., involuntary reaction) and an unacceptable groaning (i.e., groaning coming from "within"—"the center of our being"). As for the (physical) pain, Epictetus confirms that it is indeed outside our control and suggests that the sage is not insensitive to pain with his self-control: "Is anything disgraceful to you which is not your own doing, for which you are not responsible, which has befallen you accidentally, as a headache (*kephalalgia*) or a fever?" (3.26.8). Rather, a sage is free from (emotional) distress or grief (*lypē, alypia*), not from pain (*algos, odynē*) throughout the *Discourse* (3.18.11; 3.22.48; 4.3.7; 4.6.8; 4.6.19; *Frag.* 3),[94] although he is to endure pain without complaint as self-mastery (*Ench.* 10.5). Exemplified in Epictetus and earlier classical (Greek) Stoics and in contrast to Aristides's self-portrayal, pain then is (for the most part) given "no place within the arena of moral good or evil and, in this sense, it [has] no value as far as the wise person [is] concerned and [has] no place among his preoccupations."[95]

While in chapter 1 we encountered Marcus Aurelius's indulgent illness narrative, a narrative that contrasts sharply with his Stoicism, it might be productive to turn to another unique representative of Roman Stoicism, Seneca, Emperor Nero's tutor and apostle Paul's contemporary. In his writ-

---

of pleasure and pain in Old (Greek) Stoa, see John M. Rist, *Stoic Philosophy* (Cambridge: Cambridge University Press, 1969), 37–53.

94. In all these passages, Oldfather (LCL) mistakenly translates *lypē* and its cognates as "pain" rather than "distress" or "grief" and obscures the philosopher's point.

95. Rey, *The History of Pain*, 40.

ings, Seneca, in his paradoxical preoccupation with pain, sees it as potentially useful. Verily, pain belongs to the *indifferentia*, but virtue can only be displayed in its dealings with the *indifferentia*, as he emphasizes:[96] "none of these things (*indifferentia*) is in itself glorious; but nothing can be glorious without them" (*Ep.* 82.11).[97] Indifferent hardships such as bodily pain and infirmity (*dolor corporis et debilitas*), the loss of friends or children, or the country's ruin do "buffet" the sage but do not overthrow him: "I do not deny that the wise man feels (*sentire*) these things; for we do not claim for him the hardness of stone or of steel" (*Const.* 10.4). Virtue is always self-conscious, and this self-consciousness of virtue is impossible unless there are felt hardships.[98]

In his *Epistle* 78 Seneca claims the power of the mind in overcoming pain and suffering. Writing to someone who has been suffering from chronic catarrh, Seneca recounts his own painful experience of illness in early years, and shares how he consoled himself during that period with a philosophical remedy and became well (78.3–4). Detailing diseases accompanied by great pain (*cruciatus*) such as that experienced with gout and pain in the teeth, eyes, and ears, Seneca offers several strategies to deal with them—his pedagogy of pain. First, a high-minded and sensible man is to divorce "soul from body" and to dwell "much with the better or divine part" as far as he must with the affected part (78.13). The sensible man, if he chooses to think of his pains as slight, will make them so since "it is according to opinion we suffer" (78.13). This, based on the mind-body dualism, fits the classic Stoic approach because one's opinion is under one's control.

Second, as the letter progresses, however, Seneca shows a greater engagement with physical pain by fighting against it:

> What blows do athletes (*athletae*) receive on their faces and all over their bodies! Nevertheless, through their desire for fame they endure every torture, and they undergo (*patiuntur*) these things not only because they are fighting but in order to be able to fight. Their very training means torture (*exercitatio ipsa tormentum est*). So let us also win the way to victory in all our struggles—for the reward is not a garland or a palm or a trumpeter

96. Epictetus similarly sees *adiaphora* as potentially helpful or advantageous in exercising one's virtue, mentioning disease (*nosos*), death, poverty, abuse (*loidoria*), and trial on capital charges in *Diss.* 3.20 (3.20.12).

97. Cf. Edwards, "The Suffering Body," 254.

98. *Const.* 10.4; see also 16.2; *Ep.* 71.27; 85.29.

who calls for silence at the proclamation of our names, but rather virtue (*virtus*), steadfastness of soul, and a peace that is won for all time, if fortune has once been utterly vanquished in any combat. (78.16)

It is revealing that Seneca compares an aspiring sage, struggling to conquer pain, with the athlete; just as the athlete's training is torture, the pained sick person's torture may be training. Elsewhere Seneca uses a similar comparison of external hardship serving as training for the body and especially for the soul: "All his adversities [the good man] counts as mere training. . . . Without an adversary, prowess (*virtus*) shrivels" (*Prov.* 2.2–4). Coming back to this passage, Seneca acknowledges an immensity of challenge to battle bodily pain—the language of (painful) training and struggle notes endurance as the quintessential Stoic virtue (cf. Epictetus), which is also a masculine virtue in the long philosophical and military tradition of noble death;[99] and this language also implies a strict discipline required of the soul (and the body) throughout the battle. The metaphor of athletics is further significant because an athletic game is a public spectacle.[100] Similar to the image used by Aretaeus, the public spectacle presupposes an audience observing the suffering of the sick person struggling for the victory of virtue over the worldly glories offered to the victorious athlete. The struggle for virtue, tranquility, and peace is a deeply subjective experience, yet Seneca frames it as an ostensibly visible and even shareable phenomenon in the popular Roman context.

Furthermore, Seneca sets physical pain in the context of another type of familiar public spectacle in *Epistle* 14:

The other kind of evil comes, so to speak, in the form of a huge parade (*pompa*). Surrounding it is a retinue of swords and fire and chains and a mob of beasts to be let loose upon the disemboweled entrails of men. Picture to yourself under this head (*cogita hoc loco*) the prison, the cross, the rack, the hook, and the stake which they drive straight through a man

---

99. On endurance as Stoic virtue in the ancient construction of masculinity, see, for example, L. Stephanie Cobb, *Divine Deliverance: Pain and Painlessness in Early Christian Martyr Texts* (Oakland: University of California Press, 2017), 136, 204nn100–102; Nicola Denzey, "Facing the Beast: Justin, Christian Martyrdom, and Freedom of the Will," in *Stoicism in Early Christianity*, ed. Tuomas Rasimus, Troels Engberg-Pedersen, and Ismo Dunderberg (Grand Rapids: Baker Academic, 2010), 182–83.

100. Catharine Edwards stresses how the idea of spectacle in the Roman context is central to Seneca's approach to bodily suffering in "The Suffering Body," 258.

until it protrudes from his throat. Think of human limbs torn apart by chariots driven in opposite directions, of the terrible shirt smeared and interwoven with inflammable materials, and of all the other contrivances devised by cruelty, in addition to those which I have mentioned. It is not surprising, then, if our greatest terror is of such a fate; for it comes in many shapes and its paraphernalia are terrifying. For just as the torturer accomplishes more in proportion to the number of instruments which he displays (*instrumenta doloris exposuit*),—indeed, the spectacle (*species*) overcomes those who would have patiently withstood (*patientia*) the suffering,—similarly, of all the agencies which coerce and master our minds, the most effective are those which can make a display. Those other troubles are of course not less serious; I mean hunger, thirst, ulcers of the stomach, and fever that parches our very bowels. They are, however, secret; they have no bluster and no heralding; but these, like huge arrays of war, prevail by virtue of their display and their equipment (*aspectu apparatuque*). (14.4–6)

Here Seneca asks his correspondent to imagine graphic and specific details of bodily evil (pain) *experienced and endured* as a literal torture, deliberately inflicted by an external agent, the torturer. As we will see, these kinds of descriptions and images of pain and gruesome public torture have a close parallel with Christian martyr texts. Seneca presupposes the endurance of torture here, which he elsewhere closely associates with other masculine virtues such as bravery, honor, and courage (*Ep.* 67.4; cf. 78.19–21). While he places his readers as imaginary observers of this public spectacle of torture, he invites and expects them to align themselves with the victim and imagine the unspeakable pain of the victim; the visual effect of torture is expected to create a certain bond between the reader and the imagined victim of this terrifying judicial torture enacted by the Roman power.[101] Ironically, Seneca was part of that terrorizing power structure and at the same time an eventual victim of that power personified in Nero. Then he pays attention to the "private" yet no less serious pain of illness, which the reader is also expected to endure. Through endurance, there is "room for virtue even on a sickbed" (78.20). There is a tension in Seneca given his personal and political contexts: while a wise man should not try to be preoccupied with the pains of the body, he himself—almost in

101. The Christian martyr texts create and employ similar methods and effects on their audience and readers.

Aristides's fashion—obsessively describes the extremes of bodily suffering (through violence), both here and elsewhere.[102]

Finally, he encourages his reader to think of all the brave men who have conquered pain as examples (*exempla*) of how to depart from pain (*Ep.* 78.19–21; cf. 24.5). The descriptions of those brave men echo the afore-mentioned victims of torture as they conquered pain at the stake and on the rack and the red-hot plates with their smiles and silence: "If pain can be conquered by a smile, will it not be conquered by reason (*non vincetur dolor ratione*)? . . . There have been men who have not uttered a moan amid these tortures (*Inter haec tamen aliquis non gemuit*). 'More yet!' says the torturer; but the victim has not begged for release. . . . Can you not bring yourself, after an example like this, to make a mock at pain (*dolorem deridere*)?" (78.19). In the progression of Seneca's exhortations, we see a distinction between the mastery of passions (*apathēs*) while one is subjected to pain and the mastery of passions while one is insensitive to pain (*anaisthēsia*). His narrative construction of pain in relation to virtue provides a concrete pedagogy of pain in the imperial context that will profoundly influence emerging Christian narratives and pedagogies of pain.

Christ's Passion Narratives and Imitatio Christi

In this rich cultural milieu of the representations and pedagogies of pain, we see that central to the Christian narrative and theology is the pained body of Jesus, hanging in excruciating torture and torment on a Roman cross, his painful death endured for the salvation of the world. With this we enter into the realm of institutionalized violence and its attendant pain and suffering as a public spectacle—the primary context of the early Christian pain narrative. The canonical gospel narratives present not only physical pain but also the emotional pain and social suffering experienced by Jesus. In Luke, at Gethsemane (the Mount of Olives), Jesus was so deeply distressed and troubled at the prospect of his impending death that even his sweat became like drops of blood falling to the ground.[103] He was tried by the Roman governor (and the Sanhedrin and Herod); mocked with a crown of thorns on his head; stripped, spat on, and struck on the face and head repeatedly by the Roman soldiers; betrayed and deserted by his dis-

---

102. Also in *Ep.* 13.4. Edwards, "The Suffering Body," 255, reminds the readers that Seneca was equally obsessed with describing pleasure and the excesses of luxury.

103. Mark 14:32–34; Matt. 26:36–38, 67; Luke 22:39–44.

ciples; forced to carry his own cross outside of Jerusalem, the cross upon which he would be hung; derided by the crowd; and crucified.[104] While he kept his silence during the trial, he died with a cry of dereliction (see Ps. 22:1), and upon his death his side was pierced with a spear.[105] Jesus's passion, narrated in the Gospels, involved the all-encompassing pain experienced by a whole person, and as such was clearly communicated to the observers. Jesus's death also wrought salvation in ways that manifested visible, material effects such as the tearing the curtain of the temple in two; the shaking of the earth; the splitting of the rocks; the opening of the tombs; and a Roman centurion (an inflictor of violence) confessing Jesus's identity.[106] Jesus's pain was agentive and active.[107] Jesus's painful suffering and death are the necessary culmination and fulfillment of his becoming human,[108] which had already been interpreted by Paul as God's condescending, self-humiliating, and self-impoverishing *act* for humanity (Phil. 2:7-8; 2 Cor. 8:9). The divine suffering through the incarnation, along with Jesus's passion, would occupy later theological development and debates.

In the earliest and subsequent Christian texts, particularly in the martyr texts, Christ's all-consuming passion and death occupy a pivotal theological and practical significance, providing an essential narrative of pain and suffering as the explanatory framework for Christians' suffering.[109] Paul is the first one who links the suffering of Christ and its sharing by Christians in imitation of Christ: "I want to know Christ and the power of his resur-

✠ rection and the sharing of his sufferings by becoming like him in his death" (*tēn koinōnian tōn pathēmatōn autou, symmorphizomenos tōi thanatōi autou*) (Phil. 3:10). In fact, God granted grace to the believers not only to believe in Christ but also to suffer for him (*hyper autou paschein*) (Phil. 1:29). A Pauline author also ties Paul's apostolic suffering to that of Christ with a claim

✠ that Paul's suffering completed what was lacking in the sufferings of Christ

---

104. See Mark 14:44–50, 65–71; 15:1–32; Matt. 26:47–56, 69–74; 27:11–44; Luke 22:47–23:37; John 18:18, 25–40; 19:1–27.

105. See Mark 15:3–4, 34; Matt. 27:12–14, 46; John 19:34; cf. Mark 14:60–61; Matt. 26:61–63; Luke 23:9.

106. Mark 15:38–39; Matt. 27:51–54; Luke 23:45, 47.

107. See n. 65 above. According to Talal Asad, pain is active (not merely passive) and creates the conditions of action and experience; see *Formations of the Secular*, 79, 85.

108. Cf. John 1:1–5, 10–11, 29; Mark 10:45; Matt. 20:28.

109. This and the next paragraphs are largely drawn from Helen Rhee, *Early Christian Literature: Christ and Culture in the Second and Third Centuries* (London: Routledge, 2005), 91–92.

(*thlipseōn tou Christou*) for the sake of the church (Col. 1:24). The author of 1 Peter exhorts his readers to "rejoice insofar as [they] are sharing Christ's suffering" (*katho koinōneite tois tou Christou pathēmasin chairete*) (4:13) and give glory to God when they suffer as Christians (*hōs Christianos*) (4:16); for they are suffering according to God's will (4:19). Indeed, nascent churches' experiences during the period of persecution sharpened the church's understanding of what it meant to be Christian by interpreting the meaning of Jesus's death anew in the face of Christians' violent deaths.[110] Ignatius of Antioch (c. 110 CE) was famously (or notoriously) eager to experience his martyrdom by the wild beasts in Rome so as to imitate Christ's passion by suffering a violent death in the arena, thus showing himself to be a real disciple, and thereby to "attain to God" (*theou epitychein*) (*Rom.* 4.1; 5.3; 6.3): "If I suffer (*pathō*), I will be a freedman (*apeleutheros*) of Jesus Christ, and rise up free in him [4.3].... Fire and cross and battles with wild beasts, mutilation, mangling, wrenching of bones, the hacking of limbs, the crushing of my whole body, cruel tortures of the devil—let these come upon me, only that I may attain to Jesus Christ (*hina Iēsou Christou epitychō*) [5.3]!... Allow me to be an imitator of the suffering of my God (*mimētēn einai tou pathous tou theou mou*)" (6.3). "By his passion he [Jesus Christ] destroyed the roots of sin," declares the martyr Apollonius (*Mart. Apol.* 36); he "suffered for the redemption of those who are saved in the entire world," writes the author of the Martyrdom of Polycarp (Mart. Pol. 17.2). Christ's passion effected the atonement and redemption of the people, and therefore, his death was seen as an atoning sacrifice.[111] Christ is not only the "eternal and celestial high priest" (14.3) but also a victim, offering himself to God as the perfect sacrifice. Thus, he is the first "martyr" (a witness by death): "the true and faithful witness, the first-born of the dead, and the prince of God's life" (cf. Heb. 8:3; 9:12–14).[112] It is precisely through his death that he overcame his adversary, the devil, and thus received glory (*Mart. Lyons* 2.3). Christ the Sufferer and the Victim is Christ the Victor and "the mighty and invincible Athlete."[113] *– Ignatius quote*

In this way, Christ modeled martyrdom "in accordance with God's will" so that the saints might follow and imitate (Mart. Pol. 2.1; 1.2). Moreover,

---

110. Cf. Karen King, "Rethinking the Diversity of Ancient Christianity: Responding to Suffering and Persecution," in *Beyond the Gnostic Gospels: Studies Building on the Work of Elaine Pagels*, ed. Eduard Incinschil et al., STAC 82 (Tübingen: Mohr Siebeck, 2013), 76.

111. Cf. *Mart. Carp.* 5 (Greek).

112. *Mart. Lyons* 2.3.

113. Cf. *Mart. Lyons* 1.27, 44; Mart. Pol. 17.1.

it is Christ who strengthens and empowers his martyrs in their contests with the Adversary and who manifests his glory in and through their sufferings.[114] The "glorious Christ . . . represented as actively present in the life of the Church" and the individual martyrs, is "thought of as a heavenly source of life, strength, dynamism, and consolation" to his martyrs, as he is united with them in their struggles and sufferings.[115] This interpretation of Christ's passion and death in the martyr texts carried practical implications for late-second-century Christians faced with occasional persecutions and sufferings, whether real or perceived. It set the theological (christological) basis for and vindication of Christian suffering and martyrdom; since "Christ suffered and died in the same [physical] way that we do ourselves, can our suffering and death imitate his."[116]

And yet, not all texts that affirmed the reality of the Lord's suffering and death interpreted the significance of pain and suffering in the same way. The second-century apocryphal Gospel of Peter reworked the Jesus tradition and presented his passion differently than did the canonical Gospels. While it broadly follows the canonical order, such as the mocking of Jesus with a purple robe, a crown of thorns, and the spitting and scourging that followed the judgment (6–9), it places Jesus's silence at the crucifixion rather than during the trial with a phrase, "But he was silent as if he felt no pain" (*hos mēdena ponon echōn*) (10).[117] This phrase has long been considered a docetic element of this gospel. However, the rest of the gospel clearly records the Lord's death with no denial of his actual suffering and his resurrection (11–20, 37, 40, 42, 50).[118] Nevertheless, what does "as if he felt no pain" mean? Did Jesus still suffer while not feeling pain? Or did Jesus suffer while mastering pain (while still feeling pain)? How about the relationship between silence and immunity to pain? Although this text is not docetic in Christology, it opens a possibility of Jesus's insensitivity to pain while hanging on the cross. In fact, as we will see, later orthodox theologians minimize Jesus's pain while affirming his suffering, and some

114. *Mart. Lyons* 1.28, 41; cf. *Mart. Pol.* 2.2; *Mart. Apol.* 47; *Mart. Perp.* 15.6.

115. Denis Farksafalvy, "Christological Content and Its Biblical Basis in the Letter of the Martyrs of Gaul," *SecCent* 9 (1992): 23.

116. Elaine Pagels, "Gnostic and Christian Views of Christ's Passion," in *The Rediscovery of Gnosticism*, ed. B. Layton (Leiden: Brill, 1980), 1:266.

117. For detailed discussion on the Christology of this gospel, see P. M. Head, "On the Christology of the Gospel of Peter," *VC* 46, no. 3 (1992): 209–24.

118. It is not without miraculous elements such as a walking and talking cross, however.

of the martyr texts during the same and later periods present martyrs being subjected to all kinds of heinous tortures, who were impervious to pain in various ways. This interpretation of Jesus's passion and the valuation of pain and suffering provides an alternative approach to pain.

Some Nag Hammadi texts show a similar but further developed tendency. For example, in the Apocryphon of James, which purports to be an account of revealed teachings of Jesus 550 days after his resurrection, yet before his ascension, the risen Savior secretly explains to James and Peter the true meaning of his suffering and death. He affirms his real suffering and death against Peter's docetic Christology and insists that his disciples must (or at least must be willing to) suffer and die believing in his cross if they wish to be saved (5:7–9, 26–29). The Savior reproves them to stop loving the flesh and not to fear suffering, for they are transient (5:33–6:1). However, their willingness or readiness to suffer and face martyrdom in the face of temptation, oppression, and persecution from Satan means that "one has turned away from the flesh to the spirit, and it is this spiritual orientation that marks salvation and makes Peter and James beloved of God" (4:31–5:6).[119] As Karen King notes, while this text upholds the material reality and theological significance of Jesus's cross and physical resurrection, "it seems that ultimately for this work eternal life is spiritual, not fleshly."[120] Other Nag Hammadi texts such as the Apocalypse of Peter certainly take a docetic view, rejecting the reality of the incarnation, passion, and bodily resurrection of Christ, and attacking enthusiasm for martyrdom; and thus these texts were widely denounced by heresiologists such as Irenaeus, Tertullian, Clement of Alexandria, and Origen.

These same first theologians and apologists, while strongly affirming the incarnation, passion, and bodily resurrection of Christ, did so in light of contemporary philosophical language and concepts in an attempt to present Christianity as a true philosophy on par with or superior to the best of Greco-Roman philosophical schools (e.g., Middle Platonism and Stoicism)—against the anthropomorphic gods of Greco-Roman religions. They explicated Christian monotheism and divine transcendence as compatible with Middle Platonism, claiming that the former's superiority meant adopting the latter's philosophical monotheism, including divine

---

119. Cf. Karen King, "Rethinking the Diversity," 71.

120. Karen King, "Rethinking the Diversity," 72. King adds that the body is not necessarily evil, but its moral condition depends on the condition of the soul in this work in a footnote, 72n56.

immutability and impassibility (at least partially) in delicate (and uneasy) balance with the gospel accounts. Their twofold problem was first the relationship between the transcendent God and the Son of God, the agent of the divine activity in the world, and then the relationship between the divine Logos-Son and Jesus Christ. In addressing the first problem, they adopted the Middle Platonic concept of the preexistent Logos as the mind of God sharing essential unity with God in substance, yet distinct in name, function, and personality.[121] For the second problem, they all insisted that Jesus Christ was the historical embodiment of the eternal Logos-Son but emphasized the unmixable, distinct natures functioning in ways only appropriate to each nature. For example, replying to pagan critic Celsus's argument that God cannot suffer and be crucified, Origen states: "The person and the essence of the divine being in Jesus is quite a different matter from that of his human aspect" (*Cels.* 7.16). "The Word remains Word in essence. He suffers nothing of the experience of the body or the soul," but (human) Jesus suffers and feels pain appropriate to that nature (4.15; 2.23). And for Clement of Alexandria, divine *apatheia* is the goal of a true Christian gnostic ("perfected Christian") in imitation of God and Christ (e.g., *Strom.* 5.11.67; 6.9.7). On the other hand, Tertullian ascribes passions to the Son of God in the paradox of Christian faith: "The Son of God is crucified. Because it is so shameful, it does not shame me. The Son of God dies. It is credible because it is so unlikely. The buried One rises. It is certain because it is impossible" (*Carn. Chr.* 5.4).

In the midst of changing contexts, early Christians tried to relate various experiences of suffering to that of Christ, all of which shaped early Christian perspectives on and experiences of pain and suffering; their ideas on pain and suffering developed and were shaped by a number of interrelated and yet divergent interpretations of Christ's passion. In the wake of the newly established peace between the church and the empire in the early fourth century, the Arians, in their conceptualization of divine transcendence, posited that the passible Son was inferior to the impassible Father in that he was created and was subject to human sufferings. To safeguard absolute divine impassibility of the High God, which precluded any divine involvement in human history and the *pathos* of human existence, they relegated the Son to a *tertium quid* (third thing) between the High God and a mere human so as to imagine him as capable of change and suffering. Then, during the subsequent Trinitarian and christological contro-

---

121. See Rhee, *Early Christian Literature*, 56–59.

versies, against Arian and neo-Arian doctrines of mutability, passibility, and the subordination of the Son, some prominent Nicene theologians downplayed or minimized Jesus's pain (and suffering) and spoke of the Son's divinity as consistently aloof from Jesus's sufferings (as some of the earlier theologians did). For instance, Hilary of Poitiers (c. 315–367 CE) examined Jesus's suffering in confrontation with Arian Christology and undermined pain's significance in the passion. The divine Son assumed his human nature at the time of the incarnation. Since the divine by definition is impassible and immutable, the Son must have had a human condition that befitted his divinity and was therefore subject to no pain or sorrow or fear. In explaining the gospel passion narratives, to which the Arians appealed to prove the Son's inferiority to God, Hilary distinguished suffering (*pati*), the external infliction of blows, wounds, cutting, and such, from pain (*dolere*), the perception and inner experience of such suffering, which Jesus Christ could have never felt:

> When he was struck with blows or inflicted with wounds or lashed with whips or lifted upon the cross, he felt the force of suffering (*pati*), but without its pain (*dolere*). When a dart cuts through water or pierces a flame or slashes the air, it inflicts all the "sufferings" which belong to its nature—it cuts, pierces and slashes. Yet the "suffering" which it inflicts does not have an effect on the things that it strikes, for it is not in the nature of water to be cut or of flame to be pierced or of air to be slashed. . . . So our Lord Jesus Christ "suffered" (i.e., received) blows, hanging, crucifixion and death. But the suffering which assailed the Lord's body, though it was "suffered" (i.e., received), still did not convey the nature of our suffering (*naturam passionis*) . . . for his body did not have a nature susceptible of pain (*natura ad dolendum*). (*The Trinity* 10.23.206–209; cf. 10.47.246–249)

Hilary wants to separate the unique humanity of Jesus from any trace of human weakness, since Jesus's human nature is one from above and thus different from ours.[122] In the incarnation, Christ's body was "trans-

---

122. Carl L. Beckwith notes the indebtedness of Hilary's theological anthropology to a Stoic moral psychology in that he distinguished between a "weak soul" (*animae infirmis*) and a strong or complete soul. "A person feels the pain inflicted on the body because he possesses a soul weakened by sin. To assert that Jesus felt pain would be to attribute a 'weak soul' to him" (78). However, because Christ possesses a perfect soul (nature), he does not feel the pain of a weak human soul (nature). See Carl L.

figured" to such an extent that it could drive away fevers by its touch and walk on water; thus it was "peculiar and proper" to him that Jesus's body could not feel pain (*natura ad dolorendum*) (10.23.208). In a certain sense, this seems as though Hilary is fleshing out the Gospel of Peter's cryptic mention of Jesus's immunity to pain in his passion with a full-blown theological commentary.

However, Gregory of Nazianzus, one of the Cappadocian fathers, who were also champions of the Nicene orthodoxy, put forth a more sophisticated unitive Christology, which confessed Jesus Christ as "one and the same" Son of God. "God was conceived and born" that

> God came to an end as man, to honor me,
> So that by the very things he took on, he might restore,
> And destroy sin's accusation utterly,
> And, by dying, slaughter the slaughterer. (*Carm.* 1.1.10.22, 6–9)[123]

"For Nazianzen God suffers," notes Frederick Norris.[124] "Humanity is not inserted into the equation so that divinity will be kept from full involvement in incarnation."[125] Rather, Christ the Son of God is a "single whole . . . by combination (*synodos*)" through the "intermingling" (*synkrasis*) of the two natures of Christ (*Or.* 30.8, 6). For Gregory, the theological import of the incarnation is essentially soteriological, and the climax of Christ's saving work is his passion on the cross. As both Frederick Norris and Christopher Beeley note, Gregory frequently speaks of God—a single subject—suffering and dying on the cross for our salvation (134; 138).[126] When the Son cries out from the cross, "My God, my God, look upon me, why have you forsaken

---

Beckwith, "Suffering without Pain: The Scandal of Hilary of Poitiers' Christology," in *In the Shadow of the Incarnation: Essays on Jesus Christ in the Early Church in Honor of Brian E. Daley, S. J.,* ed. Peter W. Martens (Notre Dame: University of Notre Dame Press, 2008), 79–80, 83–84.

123. Cf. *Ep.* 101.7: "What is not assumed is not healed and what is united with God, that indeed is saved."

124. Frederick W. Norris, introduction to *Faith Gives Fullness to Reasoning: The Five Theological Orations of Gregory Nazianzen,* by Frederick W. Norris, trans. Lionel Wickham and Frederick Williams, Supplements to Vigiliae Christianae 13 (Leiden: Brill, 1991), 50.

125. Norris, introduction to *Faith Gives Fullness to Reasoning,* 50.

126. Norris, introduction to *Faith Gives Fullness to Reasoning,* 50; Christopher Beeley, *Gregory of Nazianzus on the Trinity and the Knowledge of God: In Your Light We Shall See the Light,* OSHT (Oxford: Oxford University Press, 2008), 134, 138.

me?" (Ps. 21:1 LXX; Matt. 27:46), it does not indicate, Gregory argues, that he was abandoned by the Father or his own divinity—as though he were afraid of suffering! Rather, the Son's ultimate cry of abandonment shows his authentic (and complete) assumption and representation (*typon*) of our (fallen) condition, making "our thoughtlessness and waywardness his own" (*Or.* 30.5). God's association with human existence without actually becoming human is not enough; the Son himself must assume and experience human suffering (including pain)[127] and death in order to purify "like by like" (*Ep.* 101.51). The Son thus takes on suffering for our sake, and we are saved by the suffering of the impassible one (*Or.* 30.5), whereby God includes our forsakenness "within his saving embrace and his healing presence in the midst of our desolation and death."[128] Therefore, "We needed an incarnate God, a God put to death, so that we might live, and we were put to death with him" (45.28). For Gregory then, Christ's *mimesis* of our condition in his incarnation and suffering grounds the possibility of our transformation in imitation of him in our suffering in and for him.[129]

[margin: IMITATION]

In his controversy with Nestorius in the early fifth century, Cyril of Alexandria further advanced unitive Christology in the incarnation and passion of the Word. As recent studies note, Nestorius, while affirming Nicene orthodoxy, shared the same concern of Arians in protecting absolute divine impassibility from being compromised by any involvement in human weakness and suffering; and thus his two-subjects account of the incarnation in the conjunction (*synapheia*) of divine and human subjects in Christ was a means to preserve an absolute divine impassibility.[130] Therefore, for Nestorius, the subject of *kenōsis* in Philippians 2:5–8 was not the impassible divine Word but a passible man indwelt by the Word;

---

127. Note that there is no distinction made between suffering and pain in Gregory or, later, Cyril of Alexandria.

128. Beeley, *Gregory of Nazianzus*, 138.

129. Cf. Ben Fulford, "Gregory of Nazianzus and Biblical Interpretation," in *Re-Reading Gregory of Nazianzus: Essays on History, Theology, and Culture*, ed. Christopher A. Beeley (Washington, DC: Catholic University of America Press, 2012), 40.

130. See John J. O'Keefe, "Impassible Suffering? Divine Passion and Fifth-Century Christology," *TS* 58 (1997): 39–42; Paul L. Gavrilyuk, *The Suffering of the Impassible God: The Dialectics of Patristic Thought* (Oxford: Oxford University Press, 2004), 141–44; Joseph M. Hallman, "The Seed of Fire: Divine Suffering in the Christology of Cyril of Alexandria and Nestorius of Constantinople," *JECS* 5, no. 3 (1997): 385–89. Cf. John A. McGuckin, *St. Cyril of Alexandria: The Christological Controversy; Its History, Theology, and Texts* (Leiden: Brill, 1994), 126–74; Thomas G. Weinandy, *Does God Suffer?* (Notre Dame: University of Notre Dame Press, 2000), 177–81.

[handwritten margin notes: "not agency in suffering", "something involves entering into a suffering", "union w/ God", "He has made suffering holy", "that Christ has pioneered this path"]

it was a man indwelt by God (cf. John 1:14b) who became poor, suffered, was emptied out of his human life, and died. In contrast, Cyril's starting point was the voluntary self-emptying of a single divine subject, God the Word, who "is the exact image of God the Father, the likeness, and visible expression of His person."[131] It is only he "who took the form of a slave, that is, became a man, and made Himself poor" by accepting full human nature with its limitations, emotions, and suffering,[132] "not for his own sake, but for ours who lack everything."[133] The emptying of a human being, however much indwelt by God, if not of the very Word who became flesh (cf. John 1:14a), was not an emptying at all. For Cyril, something unique and unparalleled happened in the incarnation in that, while the Word always remained in his divine glory, power, and the perfection of his nature, the Word-in-the-flesh feared, wept, suffered, and was crucified as a participant in human weaknesses.[134] Thus Cyril insists on the paradox of the impassible who voluntarily assumed the conditions of *pathos* in the following way:

> It would consequently be fitting also for the Word to fear death, to look upon danger with suspicion, to weep in temptations, and in addition to learn obedience by what he suffered when tempted. Nevertheless, I think it completely foolish either to think or say this, since the Word of God is all-powerful, stronger than death, beyond suffering, and completely without a share in fear suitable to man. But though he exists this way by nature, still he suffered for us. Therefore, neither is Christ a mere man nor is the Word without flesh. Rather, united with a humanity like ours, he suffered human things impassibly *(pathoi apathōs)* in his own flesh. Thus these events became an example *(hypotypōsin)* for us in a human fashion, as I said to begin with, so that we might follow in his steps *(tois ichnesin autou epakolouthēsomen)*.[135]

131. *Commentary on the Gospel of Luke*, 11.
132. *Commentary on the Gospel of Luke*, 11.
133. *Commentary on Isaiah*, 2.4.
134. Cf. Gavrilyuk, *The Suffering of the Impassible God*, 156; McGuckin, *Christological Controversy*, 191–92. See also *Second Letter to Succensus* 4: "When you insist that the Only-begotten Son of God did not personally experience bodily sufferings in his own nature, as he is seen to be and is God, but suffered in his earthly nature. Both points, indeed, must be maintained of the one true Son: The absence of divine suffering and the attribution to him of human suffering because his flesh did suffer (*kai to mē paschein theikōs kai to legesthai paschein anthrōpinōs hē autou gar peponthe sarx*)."
135. *On the Right Faith*, 163, trans. Rowan Greer, cited in Warren J. Smith, "Suffering Impassibly: Christ's Passion in Cyril of Alexandria's Soteriology," in *Suffering*

For Cyril, similar to Gregory, Christ's "impassible suffering"[136] is the necessary culmination of his incarnation and thus a necessary means of our salvation and sanctification. Christ suffers impassibly in that his voluntary suffering does not change the divinity of the Word but transforms human compulsion, self-preoccupation, and fear into "an incomparable courage."[137] Christ's impassible suffering then serves "as an archetype of revealing the character and power of human nature sanctified by the indwelling of Christ's Spirit and thereby giving an example for those who would take up his cross for themselves."[138]

Christian Pain Narratives and Pedagogies of Pain

We have thus far examined the "pain narrative" of Jesus's public death and the subsequent early Christian interpretations of that narrative within the context of Christians' experiences and developing Christology. We should remember here that emerging Christian pain narratives, including that of Jesus and its ensuing interpretations, were part of the larger literary and philosophical representations of the "suffering self," in which notions of pain and suffering typically merge and are understood interchangeably, and with a sense of moral agency, by the sufferer.[139] Christian texts that distinguish pain from suffering in Jesus's passion are usually explicit and clear for their particular theological purpose, as we have seen. In this section we delve into how early Christians constructed their own narratives and pedagogies of pain in relation to that of Jesus's, and in mimetic identification with Christ. We will see the same pattern of interchangeability of pain and suffering in those texts in general and explicit distinction between them in a few texts.

---

and Evil in Early Christian Thought, ed. Nonna Verna Harrison and David G. Hunter (Grand Rapids: Baker Academic, 2016), 191–92.

136. See also Cyril's *Third Letter to Nestorius*, 6: "We confess that the very Son begotten of God the Father, the Only-begotten God, impassible (*apathēs*) though he is in his own nature, has, as the Bible says, suffered in flesh (*sarki peponthen*) for our sake [1 Pet. 4:1], and that he was in the crucified body making the suffering of his flesh his own, in an impassible way (*ta tēs idias sarkos apathōs oikeioumenos pathē*)." Thus Gavrilyuk notes that the "Word was in a qualified sense passible to the degree to which he made the sufferings of humanity his very own," in *The Suffering of the Impassible God*, 171.

137. *Commentary on John*, 8; *Second Oration to the Royal Ladies* (PG 76:1393B).

138. Smith, "Suffering Impassibly," 205.

139. See pp. 140–42.

Martyr Texts and Texts concerning Martyrdom:
Mimetic Identification with Christ

An obvious starting point is the Christian martyr texts, which were written in the mid-second century through the fourth century with some consistent themes and familiar cultural symbolism.[140] The martyr texts describe and celebrate, sometimes with gruesome detail and arresting images, the undaunted faith, noble endurance, and matchless valor of Christians arrested, imprisoned, tried, tortured, and brutally executed by Roman authorities. Through their public suffering, endurance, and courage, they nullify or reverse the intended effects of tortures (Satan's instruments), such as: breaking the will of the tortured and having them recant; magni-

---

140. There is a plethora of modern studies on the early Christian martyr texts. Some of the major monographs include, most recently, Cobb, *Divine Deliverance*; Stephanie L. Cobb, *Dying to Be Men: Gender and Language in Early Christian Martyr Texts* (New York: Columbia University Press, 2008); Candida R. Moss, *Ancient Christian Martyrdom: Diverse Practices, Theologies, and Traditions* (New Haven: Yale University Press, 2012); Candida R. Moss, *The Other Christs: Imitating Jesus in Ancient Christian Ideologies of Martyrdom* (Oxford: Oxford University Press, 2010); Thomas J. Heffernan, *The Passion of Perpetua and Felicity* (New York: Oxford University Press, 2012); Pierre Maraval, *Actes et Passions des martyrs chrétiens des premiers siècles*, Sagesses Chrétiennes (Paris: Cerf, 2010); Gail Streete Corrington, *Redeemed Bodies: Women Martyrs in Early Christianity* (Louisville: Westminster John Knox, 2006); Johan Leemans, *More Than a Memory: The Discourse of Martyrdom and the Construction of Christian Identity* (Leuven: Peeters, 2006); Paul Middleton, *Radical Martyrdom and Cosmic Conflict in Early Christianity* (London: T&T Clark), 2006); Rhee, *Early Christian Literature*; Lucy Grig, *Making Martyrs in Late Antiquity* (London: Duckworth, 2004); Elizabeth A. Castelli, *Martyrdom and Memory: Early Christian Culture Making* (New York: Columbia University Press, 2004); Robin Darling Young, *In Procession before the World: Martyrdom as Public Liturgy in Early Christianity* (Milwaukee: Marquette University Press, 2001). Some of the recent major edited volumes include Éric Rebillard, ed., *Greek and Latin Narratives about the Ancient Martyrs* (Oxford: Oxford University Press, 2017); Hans Reinhard Seeliger and Wolfgang Wischmeyer, eds., *Märtyrerliteratur*, TUGAL 172 (Berlin: de Gruyter, 2015); Paul Hartog, ed., *Polycarp's Epistle to the Philippians and the Martyrdom of Polycarp: Introduction, Text, and Commentary*, Oxford Apostolic Fathers (Oxford: Oxford University Press, 2013); Jan Bremmer and Marco Formisano, eds., *Perpetua's Passions: Multidisciplinary Approaches to the* Passio Perpetuae et Felicitatis (Oxford: Oxford University Press, 2012). Important earlier monographs include Glen W. Bowersock, *Martyrdom and Rome* (Cambridge: Cambridge University Press, 1995); Daniel Boyarin, *Dying for God: Martyrdoms and the Making of Christianity and Judaism* (Stanford, CA: Stanford University Press, 1999); and Perkins, *The Suffering Self*.

fying and displaying the absolute might of the Roman power; and instilling fears in the spectators and deterring them from committing similar crimes. Just like Jesus's pain and suffering, theirs must be publicly seen and require their own "witnesses" (*martyres*)—audiences' attestation—beyond private, subjective experience, because their pain and suffering make a difference not only to the martyrs themselves but also to the world in which they lived, including their audiences and "enemies." As such, their openness to pain and suffering was "precisely part of the structure of their agency as Christians."[141] Therefore, they are the partakers of Christ in his sufferings and the imitators of Christ in his death, who "willed to have his fill of joy in suffering (*saginari patientiae uoluptate discessurus uolebat*)."[142] Similar to Christ, their crown "can only be earned in the *agōn* of the passion."[143] The martyrs are the heroes of the church whose deaths stand as witnesses to their unshakable allegiance to Christ. Just like Jesus, they are noble athletes (*athlētēs*) and gladiators for Christ, who are engaged in supernatural combat (*agōn*) against the devil, a battle that extends beyond earthly struggles against the crowd or governors. When Judith Perkins writes of the "triumph" of the Christian representation of "the suffering self" as Christians' self-definition, she has the martyr narratives in mind in particular.[144] We may be reticent about this kind of triumphalism, but we can still trace the indelible and unique impact of these pain narratives and pain pedagogies throughout the formative centuries of Christianity in both intratextual and extratextual ways.

Scholars have traditionally distinguished between two genres of martyr texts: *acta* (acts) and *passiones* (passions/martyrdoms). The *acta* consist of purported official records of the martyrs' court trials before the authorities (*commentarii*), and the *passiones* refer to the descriptive accounts of the martyrs' torture and death as conveyed in diverse literary forms such as letters, narratives, etc.[145] In relation to the more imaginative apocryphal Acts

---

141. Asad, *Formations of the Secular*, 85.

142. Tertullian, *Pat.* 3.9. Tertullian goes on to say: "For those to whom there has been granted the gift of faith they suffice to make it very clear, not only by the words our Lord used in His prophets, but also by the sufferings (*passionibus*) which He [Christ] endured, that patience is the very nature of God, the effect and manifestation of a certain connatural property [of His being]" (3.11).

143. Pseudo-Cyprian, *De centesima* 18, cited in Jan Den Boeft and Jan Bremmer, "Notiunculae Martyrologicae," *VC* 35 (1981): 44.

144. Perkins, *The Suffering Self*, 12–33, 114.

145. Rhee, *Early Christian Literature*, 39. With this distinction comes a thorny

in the second and third centuries, martyr texts recast the past as a part of the continuum of history.[146] Regardless of modern scholarly debate on the historical authenticity of martyr texts, for our purpose, what is important is that the martyr texts, rather than a complete, historical story, present a version of truth that was deemed necessary and significant to their authors, their audiences, and their readers. The martyr texts had multilayered intention and purpose: to defend the Christian faith and confer meaning on perceived or real experiences of persecution; to teach and encourage their audience (and readers, whether in local or distant or future Christian communities), especially in challenging and changing contexts; to elicit empathy and action from their audience/readers so that the latter would internalize the Christian principles exemplified by the martyrs;[147] and thus, to help the audience/readers cultivate particular kinds of selves—"the suffering selves"—and to shape Christian communal identities and boundaries through the constructed memories of martyrs' heroic imitation of Christ's suffering.[148] Their audience/readers (local, distant, or future) took that authorial intention and purpose for granted and believed that these stories were trustworthy and true. Our discussion of martyr texts proceeds from these starting points.

The martyr texts portray Romans' extraordinary "willingness to make extremes of pain a public spectacle" built into their penal system[149] and depict the martyrs' willing exploitation of it for the Christian purpose. In

---

issue of authenticity and historical reliability of the martyr narratives. While earlier scholarship tended to assume that they were eyewitness testimonies (e.g., Barnes), more recent (North American) scholarship largely regards the martyr texts as theological and rhetorical discourses and a means of identity formation (e.g., Elizabeth A. Clark, *History, Theory, Text: Historians and the Linguistic Turn* [Cambridge, MA: Harvard University Press, 2004]; Moss, *Ancient Christian Martyrdom*; Cobb, *Dying to Be Men*); still, among those who consider martyr texts as literary products such as Moss, there is a recognition that the pre-Decian martyr texts have a greater historical verisimilitude (or less miraculous elements); Cobb does not make that distinction. See also Rhee, *Early Christian Literature*, 40–41.

146. Cf. Rhee, *Early Christian Literature*, 33.

147. Nicole Kelley, "Philosophy as Training for Death: Reading the Ancient Christian Martyr Acts as Spiritual Exercises," *CH* 75, no. 4 (2006): 729.

148. Cf. Kelley, "Philosophy as Training for Death," 729; Rhee, *Early Christian Literature*, 47.

149. Gillian Clark, "Bodies and Blood: Late Antique Debate on Martyrdom, Virginity, and Resurrection," in *Changing Bodies, Changing Meanings: Studies on the Human Body in Antiquity*, ed. Dominic Montserrat (London: Routledge, 1998), 102.

that exploitation, by both adopting and inverting Stoic and Roman virtues, and using athletic and military imagery, martyr texts uniformly present martyrs displaying hypervirility (*virtus*) through their persistent endurance, courage, and martyrdom regardless of their biological sex or age. *Letter of the Churches of Lyons and Vienne* (c. 177 CE) immortalizes one of the most dramatic scenes of early Christian martyrdom through the death of Blandina, a slave woman. Blandina, along with her fellow confessors, has experienced "torments beyond all description" (*hyperanō pasēs eksēseos hypemenon kolaseis*) with her entire body broken and torn; but the torturers confessed that they were beaten instead, amazed that she was still breathing (1.16, 18). When she was brought to the amphitheater, she was hanged on a post to be exposed to the animals. She looked like one "hanging in the form of a cross, and by her fervent prayer, she aroused intense eagerness in those who were contending; for during their contest (*agōni*), with their physical eyes, they saw in the person of their sister the one who was crucified for them, in order that she[150] might persuade all who believe in him that all who suffer (*ho pathōn ekei*) for Christ's glory will have eternal communion (*koinōnia*) with the living God" (1.41).

The author further states: "tiny, weak, and insignificant as she was she would give inspiration to her brothers, for she had put on Christ, the mighty and invincible athlete, . . . and through her contest had won the crown of immortality" (1.42). Blandina, while hanging from the stake, appears in the eyes of fellow believers present as the crucified Christ. Here, the intimate connection between Blandina's suffering body and her complete identification with Christ crucified could not be clearer. Particularly, it is her suffering in a female slave's body—cheap, ugly, and contemptuous—that is deemed worthy of the glory of God by embodying a union with Christ (1.17). Blandina enters communion (*koinōnia*) with him and in the eyes of Christian observers; the two blend into one without negating each other. In her identification with Christ, Blandina "becomes the mediating Christ," the one who prays and comes to share actively in his power of persuasion.[151] By doing so, she demonstrates that indeed such

150. Here I follow the translation of Elizabeth Goodine and Matthew Mitchell, which I find grammatically and theologically not only possible but also convincing, giving more agency to Blandina: Elizabeth A. Goodine and Matthew W. Mitchell, "The Persuasiveness of a Woman: The Mistranslation and Misinterpretation of Eusebius; *HE* 5.1–41," *JECS* 13 (2005): 1–19. The prevalent translation has been "he," so it is Christ doing the work of persuasion and thus eclipsing Blandina at this point.

151. Goodine and Mitchell, "The Persuasiveness of a Woman," 10. Later her me-

communion and mediating power are a possibility for all who suffer for and with Christ; and Blandina's experience, as unique as it is, becomes paradigmatic for all who suffer for and with Christ. Truly her inspiration comes from putting on Christ, the mighty and invincible athlete, whom she is also called; and by doing that she endures and wins the contest and attains the crown of immortality even as a slave woman. This passage underscores the purpose of martyrdom (eternal communion with the living God); and suffering for Christ (through torture and pain in this context) in mimetic identification with Christ spurs that longing for union with Christ and is a means by which a Christian can attain that. At her death this is replayed, fulfilled, and confirmed: "After the scourges, the animals, and the hot griddle, she was at last tossed into a net and exposed to a bull. After being tossed a good deal by the animal, she no longer felt what was happening because of the hope and possession of all she believed in and because of her intimacy with Christ. Thus she too was offered in sacrifice (*etythē*), while the pagans themselves admitted that no woman had ever suffered so much in their experience" (1.56).

We see a similar dynamic between pain and union with Christ at work in the *Passions of Perpetua and Felicitas,* one of the most beloved and influential early martyr texts from early-third-century Carthage. Felicitas is eight months pregnant at her impending martyrdom on the birthday of the emperor Geta (203 CE), but the Roman law prohibits the execution of pregnant women. Afraid that her pregnancy might prevent her from dying with her fellow confessors, she and her fellow Christians pray earnestly that she will give birth before her contest. As she feels intense birth pangs (*dolelet*), one of the prison guards mocks her: "You are in such pain now (*Quae sic modo doles*)—what will you do when you are tossed to the beasts?" (15.5). She answers: "What I am suffering now, I suffer by myself, but then another inside me will suffer for me, since I will also be suffering for him" (*modo ego patior quod patior; illic autem allius erit in me qui patientur pro me, quia et eco pro illo passura sum*) (15.6). Felicitas's response establishes a reciprocal relationship of suffering with Christ; she and Christ will suffer for each other, and she implies that it will mitigate the pain that she will feel during her contest in the arena.[152] There is a deeper dynamic here, however. Hartmut Böhme calls attention to the different categories

---

diating role is highlighted and compared to the Maccabean mother. See *Mart. Lyons* 1.53–55.

152. Cf. Candida R. Moss, "Miraculous Events in Early Christian Stories about

of pain in Felicitas's "economy of suffering": labor pain is a natural pain felt by the embodied self; the other kind of pain is that "in which the spiritual self merges in a mutual representation with the other (Christ), identifying in a form of projection with the crucified Christ."[153] Here Böhme speaks of a "transformation" or even a "transubstantiation of pain" for Felicitas: "Felicitas succumbs to the 'natural force' of the pains of labour or the 'power' of torture. However, this surrender is turned into an active, intentional position, one in which the suffering can be transformed into a desire—longing for and bringing about the closeness to Christ."[154] Felicitas then delivers a premature baby girl—a clear sign of God's favor and intervention for her martyrdom—who is taken up by a Christian sister; thus in the arena, she is "glad that she had safely given birth so that now she could fight the beasts, going from one blood bath to another, from the midwife to the gladiator, ready to wash after childbirth in a second baptism" (18.3). Her transubstantiation of pain into longing for union with the suffering Christ is completed at the martyrdom: she swaps suffering with Christ so that the two become one—similar to Blandina, Felicitas, a slave mother-martyr, is united with Christ in death and in blood.[155] Again, but in a somewhat different way, pain and suffering in martyrdom are a medium of union with the suffering Christ. Furthermore, as in the story of Blandina, the martyrdom establishes a form of love that defies natural blood bonds or existing social categories to create a new community of cosufferers.[156] Blandina takes on a role of spiritual motherhood for her fellow martyrs, and Felicitas forgoes the joys and responsibilities of physical motherhood so that she and her fellow martyrs might, together, experience a higher union within both the

---

Martyrs," in *Credible, Incredible: The Miraculous in the Ancient Mediterranean*, ed. Tobias Nicklas and Janet E. Spittler (Tübingen: Mohr Siebeck, 2013), 299.

153. Hartmut Böhme, "The Conquest of the Real by the Imaginary: On the *Passio Perpetuae*," in Bremmer and Formisano, *Perpetua's Passions*, 226.

154. Böhme, "The Conquest of the Real," 226.

155. Referring to this union between Felicitas and Christ, Thomas Heffernan uses the phrase "a hypostatic union," in Thomas J. Heffernan, *Sacred Biography: Saints and Their Biographers in the Middle Ages* (New York: Oxford University Press, 1988), 228.

156. This social role and impact of pain is of critical importance to Asad in his thinking about the relationship between agency and pain as he follows Ludwig Wittgenstein's idea of pain as a public relationship as opposed to Elaine Scarry's notion of pain as an inscrutable, private experience. Pain is about inhabiting and enacting a social relationship as shown in the examples of Felicitas and Blandina and other martyrs and Christian sufferers. See Asad, *Formations of the Secular*, 81–83.

community (*communio*) of martyrs and those who are willing to follow their examples, all of whom in turn coalesce with Christ.[157]

How about Perpetua? When she is thrown into the arena, her tunic is torn. Despite her pain (*dolor*), she is more concerned about her feminine modesty and covers her thigh (20.3). When Perpetua is called back through the Gate of Life, she slips into an ecstatic state in the Spirit and does not feel an attack by the heifer despite some signs of struggle on her body and clothing (20.8–10). However, her painlessness is only temporary, as she has yet to "taste some pain" (*doloris gustaret*) (21.9). While her fellow martyrs die by the sword "without moving and in silence," Perpetua screams when the gladiator's sword strikes her bone (21.8, 9). Thomas Heffernan here compares Perpetua's scream of pain in the arena with Felicitas's claim about mutual suffering with Christ and her union with him—that in the arena another will be in her who will suffer for her as she will suffer for him (15.6).[158] Does Perpetua's scream in the arena mark "the precise moment that Perpetua has been joined (as Felicity predicted) by her Lord, who will suffer in her stead?"[159] If that is the case, Perpetua's pain can also be seen as her longing to be united with the suffering Christ, for whose sake she indeed cast off all earthly identities and relationships, including the relationship with her son and her father. Earlier in her "prison diary," Perpetua describes her tormenting anxiety (*sollicitudine*) for her child (3.6), and the hardship (*labore*) and anxiety (*sollicitudine*) that she endures over her separation from her child (3.9). She also repeatedly describes encounters with her "loving" father, who is in pains to dissuade her from the pain (*doluit*) of martyrdom (5.6; 6.5; 9.3); and Perpetua feels pain (*dolor*) on behalf of her suffering father. Perpetua's physical pain is in fact not extensive compared to that of Blandina, Felicitas, and many other martyrs; however, her psychological and relational pain due to her confession of Christian identity and embrace of martyrdom is just as poignant and meaningful; both kinds of pain together arouse her deep longing for union with God and Christ as "the beloved of God and a wife of Christ" (18.2). In fact, in her first and fourth visions before her contest, her acceptance by and intimacy with God (or Christ) through her martyrdom is intimated and foreshadowed in the figures of a gray-haired man who welcomes her ("child") to Paradise (4.8–9) and a gladiatorial trainer (*lanista*) who offers her ("daughter") a victory branch

---

157. Cf. Böhme, "The Conquest of the Real," 227.
158. Heffernan, *The Passion of Perpetua and Felicity*, 355.
159. Heffernan, *The Passion of Perpetua and Felicity*, 355.

with a kiss (10.12–13). Perpetua finally faces her martyrdom by redirecting the hand of the inexperienced gladiator to her throat "as if she herself willed it" (21.9). Perpetua is portrayed as having firm control over how she will die, not unlike Polycarp, who "played man" in his martyrdom, as we will see. This scene of Perpetua's death draws a perceptive comment by Craig Williams, regarding pain in Latin traditions: "whereas the Latin textual tradition is full of assertions that it is womanly to fear or avoid pain or to bear it badly, Perpetua takes action precisely in order to have some taste of pain (*ut aliquid doloris gustaret*), boldly bringing on death with a gesture that completes an ineffectual man's botched killing with a purposeful woman's suicidal thrust."[160] As Stephanie Cobb notes, if "Perpetua needs to experience pain in order to dismiss perceptions of womanly weakness, and if previous tortures did not result in pain, then here—at the narrative climax—we encounter Perpetua simultaneously feeling and overcoming pain in her quest for martyrdom."[161] This places the construction of Perpetua's pain in a larger cultural discourse of gender that takes a critical position in her very masculine martyrdom.

The Martyrdom of Polycarp (c. 150–250 CE) constructs a pedagogy of pain in relation to Polycarp's sacrificial and hypervirile martyrdom in imitation of Christ. At the outset, the text lays out Polycarp's martyrdom "in accordance with the Gospel" (1.1), that is, "in accordance with God's will" (2.1). Polycarp, the chosen vessel of God, reenacts in his own martyrdom the passion of Jesus as an efficacious and acceptable sacrifice to God. Thus, the details of his martyrdom are patterned on the gospel accounts of Jesus's passion: Polycarp's judge is called Herod (6.2; 8.2), and his betrayers are compared to Judas (6.2); the soldiers arrest him as though he were a brigand (7.1); he enters Smyrna riding a donkey (8.1); and when Polycarp enters his passion, he intercedes for "everyone and for all the churches" in the world (5.1; 8.1) and prays, "May God's will be done" (7.1). When the aged bishop enters the amphitheater, he hears a voice from heaven: "Be strong, Polycarp, and play the man (*andrizou*)" (9.1). After the governor orders him to be burned at the stake, he is bound like a sacrificial ram, "a holocaust prepared and made acceptable to God" (14.1). Polycarp stands steadfast without bounds while his body is inflamed (15.2); rather, his inflamed body has "such a delightful fragrance" of incense (15.2). In this way Polycarp's body becomes

---

160. Craig Williams, "Perpetua's Gender: A Latinist Reads the *Passio Perpetuae et Felicitatis*," in Bremmer and Formisano, *Perpetua's Passions*, 75; cf. Edwards, "The Suffering Body," 262.

161. Cobb, *Divine Deliverance*, 110.

a sensory reminder of Jesus's sacrifice as remembered in the Eucharist,[162] as Polycarp's sacrifice was prepared by God himself and performed through the "eternal and celestial high priest, Jesus Christ" (14.3). Just like Christ's sacrifice (i.e., martyrdom), Polycarp's sacrifice, in imitation of Christ, overcomes the Adversary (19.2) and saves not only himself but also all his brothers and sisters, and thus becomes an exemplar for others to imitate (1.2). Here the Martyrdom of Polycarp (as the *Letter of the Churches of Lyons and Vienne*) "cherishes the victim *qua* victim as a blameless sacrifice to God"—this is "the new style of 'active' suffering"[163] whose sanctity is approved beyond measure. In the end his body stays intact, unlike other martyrs, while he is pierced on his side (16.1); and his life is taken by the sword while he sheds so much blood that it extinguishes the flames (16.1). Along with the sacrificial imagery in imitation of Christ, by "playing the man" Polycarp endures pain as the man. This precisely puts him in the classical tradition as the endurance of pain is a quintessential male quality and trait in ancient medical and philosophical texts—so long as men are not soft and womanly.[164] Cicero gives an example of the "true man (*vir*)," one Gaius Marius, a countryman, who refused to be bound while undergoing an operation on his leg; "being a man he bore pain (*Ita et tulit dolorem ut vir*)" and demonstrated his self-mastery (*Tusc.* 2.22.53). As we have seen, Seneca mentions those men who have not uttered a moan while undergoing tortures (*Ep.* 78.19). Ultimately, however, it is Christ's strength given to him that enabled him to endure the flames without flinching (13.3). Polycarp's endurance of pain has multilayered meaning and significance in this text. Polycarp overcomes pain with Christ's strength and manly endurance, and his pain is reconfigured as a sacrifice to God in imitation of Christ. This mimetic identification then gives meaning to an otherwise horrible and frightening series of events by reinscribing his death within the larger context of Jesus's suffering according to the will of God.[165]

Suggested in the example of Polycarp, we see some martyr texts counteracting the link between torture and pain more explicitly; they express martyrs' immunity to pain in torture. In the Latin recension of the *Mar-*

---

162. Susan A. Harvey, *Scenting Salvation: Ancient Christianity and the Olfactory Imagination* (Berkeley and Los Angeles: University of California Press, 2006), 12, mentioned in Cobb, *Divine Deliverance*, 60.

163. Carole Straw, "'A Very Special Death': Christian Martyrdom in Its Classical Context," in *Sacrificing the Self: Perspectives on Martyrdom and Religion*, ed. Margaret Cormack (Oxford: Oxford University Press, 2002), 43.

164. E.g., Aristotle, *Eth. nic.* 7.7; Galen, *Temp.* 2.6 (625–626K).

165. Kelley, "Philosophy as Training for Death," 745.

tyrdom of Carpus, Papylus, and Agathonike, Pamfilius not only "uttered no
cry of pain" (*vocem non dedisset doloris*) while being scraped with claws but
also replied to the proconsul's plea to recant: "These torments are nothing.
I feel no pain" (*Haec vexationes nullae sunt. Ego autem nullum sentio dolorem*) (3.5–6). Unlike the silence that marked some martyrs' experiences
of pain,[166] here the text seems to draw a causal link between him not crying
out in pain and not feeling pain. This is reminiscent of the description of
Jesus's silence at the crucifixion, and the idea of Christ's not feeling pain as
conveyed in the Gospel of Peter. Pamfilius then provides a reason for his
insensitivity to pain: "I have someone to comfort me; one whom you do
not see suffers within me" (*patitur in me, quem tu videre non poteris*) (3.6).
Presumably Christ suffers within him, almost as a buffer to torture, and
comforts him. Several texts underscore the presence and suffering of Christ
with the martyrs giving them comfort, encouragement, and strength (as
we have seen in a couple of narratives), but this explicit link between
Christ's presence and suffering in (not just with) the martyred person,
and the martyred person's immunity to pain, is noteworthy.

This link is also present in the well-known story of Sanctus in *Letter of
the Churches of Lyons and Vienne* with a greater degree of complexity. After
Sanctus endures, extraordinarily and superhumanly, all the horrendous
tortures, he insistently confesses, "I am a Christian," while the governor
and the torturers press red-hot bronze plates to the tenderest part of his
body (i.e., his genitals) (1.20–21).

> And though these did burn him, he remained unbending and unyielding,
> firm in his confession of faith, refreshed and strengthened by the heavenly
> fountain of the water of life that flows from the side of Christ. But his body
> bore witness to his sufferings, being all one wound and bruise, drawn out
> of shape, and having lost any external human form (*kai apobeblēkos tēn an-
> thrōpeion exōthen morphēn*); but Christ suffering (*paschōn*) in him achieved
> great glory, overwhelming the Adversary, and showing as an example to all
> the others that nothing is fearful where the Father's love is, nothing painful
> (*mēde algeinon*) where there is Christ's glory. (1.21–23 [translation mine])

Sanctus, by all measures, endures the most sensational torture with un-
flinching self-mastery like a Stoic and a model of Roman masculinity. How-

---

166. E.g., *Mart. Lyons* 1.51; *Mart. Pion.* 18.10; 20.1; *Mart. Con.* 9.2. Silence in these
texts cannot be clearly linked to martyrs' insensitivity or immunity to pain, as Cobb
argues in *Divine Deliverance.*

ever, as Candida Moss points out, if Sanctus's masculinity is displayed in his unyielding tenacity and the firmness of his confession, it is immediately challenged by the newfound shapelessness of his body as a result of the torture.[167] Amorphousness or shapelessness could refer to a female body, to a trait proper, or according to the writings of Plato and Galen, to a women's sexual function as receptacle.[168] Or it could be christologically based on Isaiah 53:2 (LXX): "he has no form or comeliness." In fact, at the moment that Sanctus's amorphous wound blurs his identity, Christ suffers in him and overpowers the Adversary—and delivers Sanctus from pain.[169] Yet there is more. Some days later the torturers torment Sanctus again a second time while his body is still swollen and inflamed greatly. But "to the men's complete amazement, his body arose and became straight under the subsequent tortures; he resumed his former appearance and the use of his limbs so that by the grace of Christ the second suffering proved to be not a torture but a healing" (1.24 [translation mine]). Rather than breaking his resolve, the additional torture only results in the miraculous reconstruction of Sanctus's body. This narrative progression of Christ's suffering in Sanctus—Christ delivering him from pain, and miraculously restoring and healing his body—leads to the encouragement of those who were fearful of martyrdom, or who had already turned away from the prospect of it, not to fear the possibility of torture and pain. Indeed, Biblis, who had once denied Christ, was brought back to the rack; but while on the rack, she awoke "as it were from a deep sleep"—as though she did not feel pain; she renewed her confession in Christ during her second torture and joined the martyrdom (1.25–26). Both Sanctus and Biblis suffer for Christ but are spared from pain—this text distinguishes pain and suffering, as Hilary of Poitiers does for Christ's experience of the passion. Here the objectives of torture and pain "are constantly frustrated and circumvented. Torture becomes formative, it becomes a painless, exalting communion with Christ."[170]

Our last example of the martyrs' imperceptivity to pain comes from the *Martyrdom of Montanus and Lucius*, which purports to be an epistle of the martyrs in prison to the church in Carthage and also an account of their execution during the emperor Valerian's persecution (258 CE). When

---

167. Moss, *Ancient Christian Martyrdom*, 109; Moss, "Miraculous Events," 293.

168. Plato, *Tim.* 50c–51b; Galen, *Temp.* 2.576, both mentioned in Moss, "Miraculous Events," 293nn29–30.

169. Cf. Moss, "Miraculous Events," 109.

170. Moss, "Miraculous Events," 297.

Flavian received a sentence, he was overjoyed at the prospect of his martyrdom. In a vision he asks the Carthaginian bishop-martyr Cyprian if "the final deathblow was painful (*doleret*), for as a martyr-to-be," he explains, "I wished to ask his advice on bearing the pain (*passionis*)" (21.3). Notice his implied anxiety about torture and pain, an anxiety expected from the audience/readers as well. Cyprian replies: "The body does not feel this (*nequaquam corpus hoc sentit*) at all when the mind is entirely devoted to God" (21.4). Cyprian's perhaps surprising answer comes as a consolation and encouragement to Flavian; the reason given for the martyr's painlessness is a mind-body dualism; it resembles a classic Stoic principle in Seneca and Epictetus. The narrator then extols this exchange: "What an exchange, with one martyr encouraging the other! The one denied that enduring the deathblow was painful (*dolorem*) in order that the other who was about to die, might be filled with more courage since he would not fear the slightest sense of pain (*quod nec paruum sensum doloris in passionis*) in the final blow" (21.5). Similar to the case of Sanctus and Biblis, one martyr's insensitivity to pain encourages a would-be martyr not to fear and to strengthen his resolve for martyrdom as community formation. In this way, martyrs' immunity to pain in these texts, rather than showing "that pain is merely an illusion" or rejecting pain as a locus of meaning for martyrdom, forms an important part of pain narratives so that it might function as a useful and potent pedagogy of pain, not only for the martyrs themselves but also for their admirers and imitators to be (including the audience and readers).[171]

Besides the martyr texts, other texts were also engaged in the valuation of pain and suffering in martyrdom. Tertullian makes much of theological and practical value and meaning of pain in his polemical work *Scorpiace* (*Antidote for the Scorpion's Sting*), which denounces "heretics" ("gnostics and Valentinians") who oppose martyrdom on theological or ethical grounds as "poisons of the scorpion's sting" (e.g., 1.5–7; 4.2; 10.1; 15.3). God does will martyrdom for the faithful as necessary and good because it combats the evil of idolatry; God and God's will are good (4.2–5; 5.1–2). Just because martyrdom is painful and challenging, that does not make it any less good (5.5b–6.11). In defending martyrdom, Tertullian uses two culturally familiar analogies about how short-term pain leads to long-term benefit: medical (5.6–13) and athletic (6.1–11). First, God uses the martyrs' suffering as a doctor's scalpel, a hot iron, and a tonic; it is inevitably painful to the

---

171. Contra Cobb, *Divine Deliverance*, x, 18, 20, 64.

person who is cut, burned, and stretched but heals one's sickness with its "useful pains" (*dolores utiles*) (5.6).[172] Tertullian highlights the benefit of medically induced pain, which is absolutely necessary to healing (5.7).[173] Furthermore, the doctor employs medicines that are equal in character to the patient's symptom and disease, "restraining fever by applying more heat, [and] quenching a burning thirst by tormenting the thirst more," for example (5.8). Likewise, God the Great Physician heals and grants everlasting life "through fires and swords, and anything sharp" that produces pain (5.7). God also applies the principle of likes in healing—"to abolish death by death, to scatter slaying by slaying, to shatter the instruments of torture by the instruments of torture"—and the principle of opposites—"to confer life by removing [it], to help the flesh by damaging [it], [and] to serve the soul by tearing [it] out" (5.9). These are reminiscent of the Hippocratics' principles of likes and opposites for curing pain and illness, which had become popular by the time of Tertullian. Indeed, God himself accomplished this for our salvation by assuming our illness, an illness caused by Adam, through Jesus Christ (5.10–12); thus, how can one be "annoyed to suffer (*pati*) from a remedy that which he or she was not previously annoyed to suffer (*pati*) from an offence" (5.13)? Second, this age has long believed that the best way to test and evaluate the skills of bodies and voice for excellence with the glorious reward was to hold athletic contests as a spectacle (e.g., the Pythian Games) (6.2–3). The contests (*agonis*) are full of inflicted violence (*uiolentiae*) and wounds (*iniuriarum*) like bruises, gashes, and swellings; but no spectator accuses the judge of enacting those wounds, and neither does any contestant complain of the pain (*dolere*) to himself because he wins no crown of victory, glory, or gifts without the experience of painful wounding (*iniuria*) (6.3–5). Then, how much more would it be appropriate for God "to bring forth skills and teachings into public view, into this 'secular stage'" through martyrdom "as a spectacle for people and angels and the universal powers" not in the name of the struggle but also

---

172. See Pliny the Elder, *Nat.* 20.22. On the medical element/imagery in *Scorpiace*, see Thomas Heyne, "Tertullian and Medicine," *Studia Patristica* 50 (2011): 148–49.

173. Tertullian uses a similar analogy of medical care demanding severe pains in case of penance after postbaptismal sin as disease: "Or rather, when penance is to be performed, there is no longer any question of suffering, since it is become a means of salvation. It is painful to be cut (*secari*) and to be cauterized (*cauterio exuri*) and to be tortured by some medicinal caustic (*pulveris alicuius mordacitate*). Nevertheless, remedies (*medentur*) which are unpleasant justify the pain they give by the cure they effect, and they render present suffering agreeable because of the advantage which is to come in the future" (*Paen.* 10.9–10).

for its own benefit (6.6–7)? Painful martyrdom is God's gracious remedy providing second atonements for the faithful in accommodation for human weakness (6.9). In these analogies, Tertullian magnifies the necessary efficacy of pain and regards it as a fundamental identity marker for constructing a theology of martyrdom, and thus "orthodox" Christianity.

In the post-Constantinian context of "peace," Augustine, on the anniversaries of martyrs' deaths, often reflected on their suffering in his sermons. However, against the backdrop of the Donatist enthusiasm for martyrdom and Pelagian teaching on free will, which he strongly denounced, Augustine rather emphasized the divine help and grace of God, and de-emphasized the will and heroism of martyrs.[174] He underscores Christ's assistance and suffering in the martyr (like some of the martyr texts) in a series of sermons on the martyr Vincent, whose patience "must all be referred to the glory, not of man, but of God" (*Serm.* 274). In *Sermon* 276, he writes, "such hideous cruelty was being unleashed on the martyr's body, and such calm serenity being displayed in his voice; . . . while Vincent was suffering, it was someone else, not the speaker, that was being tortured" (2). Elsewhere he downplays the martyr's bravado in suffering and even masochism—enduring torture and punishment per se does not make one a martyr since many bandits endure them with "the greatest patience" (*Serm.* 299.8; 274); what distinguishes the martyr from the bandits is his cause—"because he [Vincent] contended for the truth, for justice, for God, for Christ, for the faith, for the unity of the Church, for undivided charity" (274; *Enarrat. Ps.* 34.13). Thus, unlike Tertullian, Augustine does not view pain or suffering as a constitutive or meaningful part of martyrdom per se; rather, "martyrs must be united with the Church, obedient members of the Body of Christ [i.e., the Catholic Church, as opposed to the Donatist church] where this charity is found. *Extra ecclesiam nulla salus* [there is no salvation outside the Church]!"[175]

Ascetic Text (Syncletica):
Shareability of Pain and Community of Cosufferers

We now move to a pain narrative and pedagogy written from an ascetic and hagiographical perspective, *The Life and Regimen of Blessed and Holy*

---

174. Cf. Carole Straw, "Martyrdom and Christian Identity: Gregory the Great, Augustine, and Tradition," in *The Limits of Ancient Christianity: Essays on Late Antique Culture and Thought in Honor of Robert Markus*, ed. William Kligsheim and Mark Vessey (Ann Arbor: University of Michigan Press, 1999), 251.

175. Straw, "Martyrdom and Christian Identity," 251.

Teacher Syncletica (*Vita*). As addressed in chapter 2, it showcases Synclet-ica's holiness not only through her rigorous ascetic regimen (*askēsis*) and teaching but also through her bearing of a decaying illness and bodily pain that the text describes in graphic detail. Whereas martyrdom as "second baptism" achieves Christian perfection through (relatively) instant victory in a public spectacle, asceticism as "bloodless martyrdom" pursues ever-continuing perfection through private, lifelong discipline (*askēsis*) with no guarantee of its attainment (*Vita* 15, 19, 26, 38, 47, 53). Yet, as both martyrs and ascetics are the imitators of Christ par excellence, ascetic texts also share the familiar motifs of the martyr texts such as the *agōn* (contest/struggle) motif with athletic and military metaphors (8, 18), the cosmic nature of that *agōn* with the devil (8, 18, 106–107), constructions of mas-culinity and virtue (30, 54–58, 72–73, 111), and the subjective heroism of a particular ascetic. All these characteristics are exhibited fully in the *Life*.

This *Life* portrays the saint's illness and pain "as a shared way of liv-ing rather than as solitary suffering" for her and her ascetic cosufferers; Syncletica's pain and illness become a means of constructing communal subjectivities.[176] As such, this text illustrates pain "as a social relationship," as pain is part of what creates the particular conditions of action and expe-rience, according to Talel Asad.[177] First, in her teaching, Amma Syncletica sees a pedagogical value in the pain of ascetic austerities (e.g., fasting) "be-cause every sprig of virtue grow[s] straight as a result of pain" (41); but the pain of *askēsis* can also be harmful and even evil, as it can lead the ascetic to be self-deceived and prideful (52–54). The painful *askēsis* is a remedy (*pharmaka*) and virtue, but it must be accompanied by humility, which is the critical virtue and remedy of the soul, particularly with regard to pride (54–58). Furthermore, "the evil originating from God," such as famines, droughts, plagues, poverty, and other misfortunes, is for the salvation of the soul and the training of the body (84). The determinists falsely regard these as evils of the soul; however, they are in fact saving remedies offered to those who believe in free will for conversion by the Almighty (84). Thus it is the self-mastery of will through lifelong *askēsis* in humility that benefits from the seemingly evil afflictions coming from God. However, the devil

---

176. Peter A. Mena, "Scenting Saintliness: The Ailing Body, Chicana Feminism, and Communal Identity in Ancient Christianity," *JFSR* 33, no. 2 (2017): 7, 8 (5–20). Mena in this study reads the *Life* using the Chicana feminist theorists' construction of communal subjectivity in their theories of health, pain, illness, and the body.

177. See Asad, *Formations of the Secular*, 85.

also uses suffering such as illness to weaken the ascetic's resolve (98). Still, in that case, illness functions as a medication and remedy, helpful to the ascetic because it is allowed for the destruction of pleasures. Just like martyr texts, the suffering, pained body becomes a site of spiritual battle and healing; and beyond body-soul dualism, the attainment of spiritual health "demands the complete engagement with the human body," particularly in pain and suffering.[178] As we discussed in chapter 2, the great *askēsis* and marker of holiness is then "to persevere in illness and to keep sending up hymns of thanksgiving to the Almighty" (99). Asceticism here goes beyond renunciation but is about "self-formation"[179] worthy of transformation, not only of the sufferer but also of the community.

Second, Syncletica indeed exemplifies this transformative "self-formation" in the bearing of her terrible disease, not alone, but in community. The devil first strikes her lung, making her cough up pieces of her lungs bit by bit and suffer unremitting fevers for seven years (105); he again strikes her inner organs with unceasing torment for another three and a half years to the extent that her suffering is said to surpass that of the martyrs (106). While nobly enduring her disease, she fights against the enemy through her teaching—that is, by her teaching she keeps healing those in her community wounded by the enemy. Here, her teaching, as it has always been, is a means of saving remedy for her ascetic community, so the enemy then afflicts her speaking organ so she cannot teach. Nevertheless, "as the women *contemplated* her suffering *with their own eyes*, they were strengthened in their will; the wounds in her body healed their afflicted souls" (110 [emphasis added]). Here her pained body becomes a "spectacle," and as such, an instrument in the healing of her community; and while it is still she herself who is suffering, her disciples are joined to her suffering through their contemplation and observation.

The enemy then hits a final blow to her mouth, inflicting pain in one tooth and causing infection in her gum. "The bone deteriorated and the sore spread through the whole jaw and became the source of infection for the adjacent body parts. Within forty days the bone decayed, and after a period of two months a hole appeared. . . . Putrefaction (*sēpsis*) and a very foul-smelling stench overpowered her body throughout so that those

---

178. Cf. Elizabeth A. Castelli, "Mortifying the Body, Curing the Soul: Beyond Ascetic Dualism in *The Life of Saint Syncletica*," *Differences: A Journal of Feminist Cultural Studies* 4, no. 2 (1992): 137.

179. Castelli, "Mortifying the Body," 140.

women who tended her *suffered more than she did*. They used to withdraw for quite long periods of time, unable to endure the inhuman stench (*dys-ōdias apanthrōpon*)" (111 [emphasis added]). Notice the progression in the depth of the community's cosuffering; as Syncletica's pain and suffering grow ever intolerable and unbearable, so does her community's. A particular cause of their cosuffering is the "inhuman stench"; this is significant. Earlier in the text, the same women smelled "the sweet fragrance (*euōdia*) of her most glorious sufferings (*euklestatōn autēs ponōn*)"—that is, her stringent *askēsis* (fasting, praying, etc.) (21). Now the community suffers with her and even more than her, not only in seeing her transformed (i.e., decayed and decomposing) body, but also in smelling her "inhuman stench" as she suffers through the ultimate *askēsis* (i.e., persevering in her illness). As in the case of Sanctus earlier, we see here Syncletica's female body with ascetic virility becoming (further) porous and permeable in her illness, and in the emission of her scent and stench; and her disciples also experience the permeability of their own bodies by becoming receptacles of Syncletica's emission. As Peter Mena notes, "the permeability of bodies" to bodily traumas and suffering—"Syncletica's and those who attend to her—is the central feature" of how they suffer together and thus "how holiness is transmitted in the *Life*."[180] Here suffering and sanctity are united in the cosuffering of Amma Syncletica and her disciples. Thus her apparent solitary illness and her experience of pain indeed mark "a construction of communal ascetic identity" as her caring disciples become a part of her and she a part of them through the experience of their shared suffering.[181]

Gregory of Nazianzus:
Ontological Pain and Community of Cosufferers

We now move to yet another kind of Christian construction of pain by returning to the writings of Gregory of Nazianzus, himself a bishop-ascetic and Trinitarian theologian and poet. In the following autobiographical poem, Gregory likens himself to a Christian Philoctetes, a Greek tragic hero we encountered earlier: "Folklore has it that when the bitter serpent has sunk his malicious fang into someone, that man will reveal the pestilent pain (*algos*) only to those who have been likewise wounded by the same hateful creature's feverish venom. They alone can appreciate its painful evil

180. Mena, "Scenting Saintliness," 13.
181. Mena, "Scenting Saintliness," 15.

(*kakon algos*). So with me. I shall recount my pain (*ponon*) to people united with me, by the same love (*erōs*), the same evil (*kakon*), and the similar pain (*algos*)" (*Carm. de se ipso* 235–241).

We have discussed how the unspeakable inexpressibility and utter isolation and abandonment of Philoctetes's experience of pain captured the inshareability of pain, in that those who had no experience of similar pain simply remained indifferent to his suffering. Gregory's audience, however, stands in contrast to those unsympathetic witnesses. They share with Gregory the same desire, evil, and pain that he experiences for himself, which unite them with him: "because they alone could hear the tale with sympathy. They are capable of insight into the mysteries of a downcast heart, who yearn to take upon their shoulders the burden of the cross, who have their portion in the fold of the great King, who love the path of rectitude and treat the fallen with compassion" (242–247). Together with Gregory, they yearn to take on the burden of the cross and have "known in their inmost being the sharp stab of desire for the King" (249). In his poetry, Gregory often writes about a "piercing longing" or "the common desires of all creation" addressed to the Creator and Christ. Thus, drawing on the classic notion of the inshareability of pain in Philoctetes's experience, Gregory in fact inverts it and highlights his experience of pain as a shared reality in communion with his community joined by the common burden of the cross and the same painful longing for the divine.

Indeed, the shared "piercing longing" for the divine arises from a life in the mortal body as a pain when one yearns to be united with the immortal God.[182] Gregory then understands this pain as an ontological one, a disposition that arises from the tension or disparity between the longing of our finite nature and the longing of our mutable nature for the immutable divine, God, who is incomprehensible and "who always was, is, and will be—rather who always is" (*Or.* 38.7). As Antigone Samellas notes, the ontological notion of pain has a precedent in a Neoplatonist, Plotinus:

> When two things aspire to unity, since the unity which they have is an extraneous one, the origin of pain (*algein*), it is reasonable to expect, lies in their not being permitted to be one. I do not mean "two" as if there were two bodies, for they would have one and the same nature; but when nature wants to share [itself] with another one and of a different kind . . .

---

182. Samellas, "Public Aspects of Pain," 275. For this insight and the connection to Plotinus, I am indebted to Samellas, 276–79.

and becomes thus two and one, having come to be in between what it was
and what it could not grasp . . . it swings up and down, and as it comes
down it proclaims its pain and as it goes up is longing for communion.
(*Enneads* 4.4.18 [2.73.25–29])

This is in fact Plotinus's Neoplatonist spin on the Hippocratic observation we have seen. The Hippocratic Corpus was concerned about the
physical composition of a human being. However, Plotinus speaks about
the pain generated by the mixed nature of humanity between "body and
a trace of the soul versus higher soul and intellect" in the sense that the
soul feels the pain because it comes into contact with a lower nature (cf.
Enneads 1.8.15 [1.125.12–14]).[183]

How does Gregory construct a pedagogy of pain with regard to this universal yet isolating metaphysical pain? "Those who have a common Spirit,
have also a common *pathos*; those who suffer equally, also believe equally.
For one would not believe another regarding something that one has not
experienced oneself; yet, the one who has the experience is more eager to
give his consent; he is an invisible witness of an invisible *pathos*, an intimate mirror [*oikeion esoptron*] of the other's form [*morphēs allotrias*]" (*Or.*
26.1). This passage, in a sense, brings us back to our initial passage because
his ontological pain, which left him isolated and separated from society,
becomes the foundation of a Christian community. Here *pathos* may have
a double meaning: it may indicate the suffering that Gregory and his audience undergo because of their common faith—the common "burden of
the cross"; and it may also be their common longing for what is incomprehensible and immortal, which has a transcendent aspect—it is a mirroring
of the otherness of those whose elusive and uncertain knowledge of the
invisible God brings them together by way of a common Spirit. As this
intimate mirroring through the Spirit brings together a community of co-

---

183. Samellas, "Public Aspects of Pain," 277. Svetla Slaveva-Griffin also points out
that Plotinus views the presence or lack of pain and pleasure as "a physical or biological 'sign' (*sêma*) of the ontological truth that the soul and body are heterogeneous in
nature" and that his "argument for the heterogeneous nature of the psycho-somatic
compound follows the logic of the leading Hippocratic argument, espoused in the
Nature of Man, that the composition of man is not homogeneous, consisting of one
substance, but heterogeneous, consisting of substances of different natures," in Svetla
Slaveva-Griffin and Ilaria L. E. Ramelli, eds., *Lovers of the Soul, Lovers of the Body:
Philosophical and Religious Perspectives in Late Antiquity* (Cambridge, MA: Harvard
University Press, 2020), chap. 6.

sufferers in Christ, Gregory's pedagogy of pain mirrors the Hippocratic cure of pain by the principle of likes—one cures pain by creating another pain. For Gregory and his cosufferers, this longing can be cultivated and refined through concrete practices or habits—*askēsis*. Ascetic imitation of Christ is painful but more pleasant than the ordinary joys of the crowd: "the nails are sweet, even though painful. For to suffer for and with Christ is more to be desired than a life of ease with others" (*Or.* 45). Moreover, any affliction and pain such as sickness can give believers the opportunity to draw near to God and gain an intimate knowledge of God (16.4), "since a troubled soul is next to God and their need turn them to the one who can provide, him who is even perchance despised for his unstinted generosity" (17.5). Thus there are times "when pain is preferable to health, patience to relief, visitation to neglect, punishment to forgiveness" (17.5), so that through their yearning God might purify them and in purifying them might make them like God (38.7)—this is the *telos* of Gregory's pedagogy of pain.

Augustine on Pain in This Age and Solidarity in (Affective) Pain

As is the case with just about every topic in Augustine's theology, Augustine sees pain within his grand vision of the creation, fall, redemption, and the eschaton. While absent in the original creation, pain (*dolor*) emerges from original sin as its manifestation and punishment (e.g., *Mor.* 1.35; *Civ.* 1.10; *Ep.* 130.12; *Enarrat. Ps.* 136.14).[184] God's punishment of Eve, by giving her pain in childbirth (Gen. 3:16), indicates how a woman, as a result of her sin, comes to the mortal condition of her body from the early state of immortality (*Gen. litt.* 19.29).[185] Even in the fallen state of this world, however, pain, "whether it is in the mind or the body, cannot exist except in good natures," as evil can exist only in something good (*Nat. bon.* 20). Pain is a sign of life in this age (*Civ.* 21.3). Only a healthy enough body can feel pain when it is hit or hurts (21.3), and the pain that body has can be poten-

---

184. Contra Julian of Aeclanum, who regarded (moderate) pain as "part of the human nature as created by God in the beginning," in Josef Lössl, "Julian of Aeclanum on Pain," *JECS* 10, no. 2 (2000): 221, 223, 233.

185. Augustine is silent on God "increasing pain" and its implied preexistent pain in childbirth, but he acknowledges that female animals bring forth their offspring in pain as a natural consequence of their mortal nature; so it could be a case for female human beings as well (*Gen. litt.* 19.29). For Julian of Aeclanum, woman's labor pain is a "normal" phenomenon and therefore "essentially (*naturaliter*) good," in Lössl, "Julian of Aeclanum on Pain," 236.

tially beneficial or harmful. When "it is forced to something better" (such as producing spiritual fruit or repentance), pain is beneficial (*Enarrat. Ps.* 55.6; *Serm.* 397.4); thus, commenting on a woman's desire mentioned in the same Genesis text, Augustine symbolically interprets that pain is a constitutive but necessary part of "restraining the will from any desire of the flesh" until one forms a better habit (*Gen. litt.* 19.29). However, when "it is forced to something worse" (such as suffering punishment for theft or injustice), it is harmful (cf. *Enarrat. Ps.* 55.6). Yet there is no benefit in an unconscious body that feels no pain, and the worse its condition, the less it feels, since there is no life in it (*Enarrat. Ps.* 55.6). Thus, in the body a wound with pain is better than decay without pain, which indicates the body's corruption (*Nat. bon.* 20). However, an immortal body in the age to come feels no pain either, because all decay has been swallowed up and the mortal and perishable body has been clothed with immortality and imperishability; in the heavenly homeland filled with peace and everlasting felicity, "no pain will there be then, for no cause for pain will exist" (*Enarrat. Ps.* 55.6). Therefore, there is no pain in an unconscious body and no pain in an immortal body; but the health of a person in pain is closer to immortality than is the insensitivity of one who cannot feel anything (55.6). Hence, the Holy Spirit works not necessarily to free Christians from pain but to use it for God's purpose. In the words of Augustine:

> Yes, he [the psalmist] says to God, you shape our pain as your precept [Ps. 93(94):20], you fashion pain into a precept laid on me. You give form to my pain; you do not leave it shapeless but mold it to your purpose, and this carefully formed pain inflicted on me will be for me a commandment from you, so that you may set me free. You form pain, scripture says, you shape our pain, you mold our pain, you do not send any unreal pain to us. As an earthenware pot is so called because it is a potter's work, so do you like a potter mold our pain into shape. (*Enarrat. Ps.* 38.17)

Pain for Augustine is not a bodily phenomenon per se, but it links the body and soul in that even if pain's origin is bodily, it is the immortal soul (*anima*) that experiences pain (bodily and affective), just as it experiences all sensation (*Civ.* 21.3). What we call bodily pain always leaves the soul affected because bodily pain is invariably preceded by sadness, which is the pain of the soul (*Enarrat. Ps.* 87.3). The soul's role and capacity to feel both physical and affective pains as a sign of life are important for Augustine as he constructs pain's shareability through Christ's pain.

Christ the Lord and God himself felt *in his soul* a deep and anguishing sorrow to the point of death (Matt. 26:38; Mark 15:34)—in carrying our own weakness, Christ bore our sorrows (*Enarrat. Ps.* 55.6; 63.18; 87.3; cf. Isa. 53:4). When he prayed, "Father, if it is possible, let this cup pass from me" (Matt. 26:39), he identified us with himself (*Enarrat. Ps.* 63.18). Although the Father and the Son are one, Christ, in his capacity as the servant, bore our hearts within his own to instruct it by his own example (63.18). Born from Mary, "scourged was He who expelled from the bodies of man the scourges of all distresses; crucified was He who put an end to our crucial pains (*omnium dolorum*); dead did He become who raised the dead" (*Catech.* 40). Buried, risen, ascended into heaven, and now sitting at the Father's right hand, he is the head and we are his body. As such, Augustine speaks of the *totus Christus*, "the whole Christ," with Christ as head and the church as body, sharing in one life and one identity on account of the incarnation, the Word of God uniting himself to humanity even to the point of embracing pain and suffering in order to transform humanity into himself (e.g., *Enarrat. Ps.* 30.3).[186] Therefore, Augustine says the following in a mystical manner:

> Whatever he suffered, we too suffered in him, and whatever we suffer, he too suffers in us (*quia quidquid passus est, in illo et nos passi sumus; quia et nos quod patimur, in nobis et ipse patitur*). Think of an analogy: if your head suffers some injury, can your hand be unaffected? Or if your hand is hurt, can your head be free from pain? . . . When any one of our members suffers, all the other members hasten to help the one that is in pain. This solidarity meant that when Christ suffered, we suffered in him; and it follows that now that he has ascended into heaven, and is seated at the Father's right hand, he still undergoes in the person of his Church what-

---

186. On Augustine's doctrine of *totus Christus*, see Kimberly Baker, "Augustine's Doctrine of the *Totus Christus*: Reflecting on the Church as Sacrament of Unity," *Hor* 37, no. 1 (2010): 7–24; Tarsicius van Babel, "The 'Christus Totus' Idea: A Forgotten Aspect of Augustine's Spirituality," in *Studies in Patristic Christology: Proceedings of the Third Maynooth Patristic Conference*, ed. Thomas Finan and Vincent Twomey (Portland, OR: Four Courts, 1998), 84–94; Bernard Bruning, "Die Einheit des Totus Christus bei Augustinus," in *Scientia Augustiniana: Studien über Augustinus, den Augustinismus und den Augustinorden*, ed. Cornelius Petrus Mayer and Willigis Eckermann (Würzburg: Augustinus-Verlag, 1975), 43–75; David Vincent Meconi, *The One Christ: St. Augustine's Theology of Deification* (Washington, DC: Catholic University of America Press, 2013), 194–216.

ever it may suffer amid the troubles of this world, whether temptations or hardship, or oppression (for all these are the necessary means of our instruction, and through them the Church is purified, as gold is by fire). (*Enarrat. Ps.* 62.2)

As Gregory of Nazianzus has also noted earlier, Christ's pain, that is, his assumption of our weakness by identifying us with himself,[187] is the basis of our pain in mimetic identification with Christ. The suffering Christ is the model of our suffering and, at the same time, the healer of our pain. We are united with Christ in our mutual suffering with Christ, which requires our ability to feel pain in this life; for Christ, our head, continues to experience the pains and sufferings of the world through his body, the church, until the coming of the new age. Thus Augustine describes Christ's complete solidarity with us as follows: "This is like the tongue speaking in the foot's name. It may happen that someone's foot is trodden on in a crowd, and it hurts: the tongue cries out, 'You are treading on me!' It does not say, 'You are treading on my foot'; it says it is being trodden on. Nobody has touched it, but the crushed foot is not severed from the tongue" (30.3). In the same way, Christ cries out in pain with the foot that directly experiences the pain as the whole body experiences the suffering together. Then our solidarity with Christ in suffering necessarily binds the members of Christ's body in cosuffering with one another, just like the tongue and foot in the quote. Again, pain's shareability lies in a particular relationship that enables and mediates sociality.[188] Since our union with Christ through our mutual suffering with Christ is contingent upon our capacity to feel pain, this capacity to feel one another's pain is the vital sign of the collective health and sanctification of the body—and even the health of the head. This is what Paul was speaking about, according to Augustine. Quoting Paul, Augustine writes: "Is anyone weak, and I am not weak too? Is anyone tripped up, without my being afire with indignation?" (2 Cor. 11:29). If Paul "had been unmoved and immune to pain" by some scandal, or the ruin of some weak person, that would have been callousness, not tranquility (i.e., *apatheia*) (*Enarrat. Ps.* 55.6).

In the *City of God*, Augustine pursues this horizontal shareability and pedagogy of pain through compassion (*misericordia*) against the Stoic denigration of it (9.5). The Stoics consistently regarded compassion (*eleos; misericordia*) as a morally bad species of emotion (e.g., Cicero, *Tusc.* 3.9.20).

---

187. Cf. John Chrysostom, *Hom. Matt.* 26.39; *Hom. Gen.* 58.13.
188. Cf. Asad, *Formations of the Secular*, 89.

Within our [Christian] discipline, then, we do not so much ask whether a pious soul is angry, as why he is angry; not whether he is sad, but whence comes his sadness; not whether he is afraid, but what he fears. For I do not think that any right-minded person would condemn anger (*irasci*) directed at a sinner in order to correct him; or sadness (*contristari*) on behalf of one who is afflicted in order to liberate him, or fear (*timere*) for one in peril, lest he perish. The Stoics, indeed, are wont to reproach even compassion (*misericordia*); but how much more honorable it would have been for the Stoic in Aulus Gellius' story [*Noct. att.* 19.1] to be moved by compassion (*misericordia*) for a man in order to liberate him, than by [pre-liminary] fear of shipwreck. . . . What is compassion but a kind of fellow-feeling in our hearts for the misery of another which compels us to help him if we can (*Quid est autem misericordia nisi alienae miseriae quaedam in nostro corde compassio qua utique si possumus subvenire compellimur*)? This impulse is the servant of right reason (*rationi*) when compassion (*misericordia*) is displayed in such a way as to preserve justice (*iustitia*), as when alms are distributed to the needy or forgiveness extended to the penitent (*sive cum indigenti tribuitur, sive cum ignoscitur paenitenti*). (*Civ.* 9.5)

Augustine here endorses compassion as the distinctive feature of his Christian ethics, in the context of endorsing other emotions such as anger and fear, all of which the Stoics regarded as reactionary and arising from false judgments. Defined as "a kind of fellow-feeling in our hearts for the misery of another," compassion's goal is to "liberate" (from vice) and fear lest someone "perish" (cf. 14.9). Thus it is good to commiserate (*contristari, conpati*) with one who is afflicted, judged by reason in preserving justice. This combination of reason and justice is significant for Augustine. Stoics (Seneca in particular) reproach compassion as something that acts from feelings of pity, and clouds judgment, without regard for justice; for example, sentimental people, wanting to throw open the prison doors, would let out the most harmful criminals, if once they see them cry, without any proper judgment or regard for justice (Seneca, *Clem.* 2.4.1).[189] Against that Seneca praises mercy or clemency (*clementia*), a tendency or act to pardon from punishment in such a way as to preserve the principles of justice without any

189. Comparing Augustine and Seneca in this paragraph, I am indebted to Sarah Byers, "The Psychology of Compassion: Stoicism in *City of God* 9.5," in *Augustine's City of God: A Critical Guide*, ed. James Wetzel (Cambridge: Cambridge University Press, 2012), 135–37.

affective involvement (1.18.1–2; 2.6.1). Augustine, while agreeing with Seneca's justice-oriented *clementia* to an extent, redefines it as compassion—a good emotion (*eupatheia*) spurring or accompanying good works; and this compassion is a proper, just response to bodily and moral afflictions—this is a Christian pedagogy of pain. Then, Augustine's reference to the Stoic in Gellius's *Attic Nights* points out that the Stoic's response of momentary fear of shipwreck was an "indifferent," involuntary ascent to external stimuli according to the Stoics but was in fact not a true (moral) good in reference to virtue (i.e., to liberate someone from vice). Augustine is saying that "a concern for virtue as such will entail that the sage has emotional reactions to others' virtue or lack thereof"; that is, "the sage *ought to* have concern for virtue as such—but that the Stoic sage does *not* actually have this."[190] As we have seen in Augustine, pain is indeed an affect[191] and is predicated in social relationship. Whereas the Stoics are trying to avoid affective pain caused by involvement with the other in the name of self-sufficiency and self-mastery, Christians not only acknowledge but also embrace that affective pain as a regular, necessary part of life (even for the sage)—Christians "feel fear and desire, pain and gladness (*metuunt, cupiuntque, dolent gaudentque*), but in a manner consistent with the Holy Scripture and wholesome doctrine; and because their love is righteous, all these emotions are righteous in them (*istas omnes affectiones rectas habent*)" (*Civ.* 14.9).

That is why Christians feel sad "on behalf of one who is afflicted" and fear for one in peril, lest he perish" (9.5; 14.9); they feel pain for their own sins (*dolent in peccatis*) and gladness in their own good works and also on account of those whose liberation they desire (14.9); Christians are troubled and anxious by their friends' famine, war, disease, or captivity, fearing that in slavery they may suffer evils beyond their powers of imagination (19.8). A Christian love demands from us solidarity in pain: not only to grieve *for* the ills and impairments in our neighbor's well-being, but also to grieve *with* our neighbor in his or her grief (cf. 19.8; 19.4)[192]—"if we felt no [negative] emotions at all while subject to the infirmity of this life, we should then certainly not be living rightly" (14.9). A Christian's pain and suffering are a condition of her relationship with Christ because, by virtue of Christ's pain and suffering, she always lives in tangible, inseparable vertical and horizon-

---

190. Byers, "The Psychology of Compassion," 138; cf. *Civ.* 14.9.

191. Also for Julian of Aeclanum, in Lössl, "Julian of Aeclanum on Pain," 217.

192. Cf. Nicholas Wolterstorff, "Augustine's Rejection of Eudaimonism," in Wetzel, *Augustine's City of God*, 161.

tal relationships; it is something that includes her ability to respond sympathetically to the pain of other sufferers, both vertically and horizontally.[193] For Augustine, sadness, anger, fear, grief, pain, and the like are not goods in one's life as such, for there will be no *pathē* in the life to come. However, in this present life with its "misery" and "great mass of evils" (14.9; 19.8), sharing one another's pain and suffering through compassion delivers us from an "inhuman mind and insensitive body," and unites us as cosufferers with Christ as Christ himself feels pain, anger, fear, and grief when we suffer from the effects of the Fall (14.9; cf. *Enarrat. Ps.* 62.2).

Conclusion

This chapter has examined how early Christians constructed their notions of pain by way of their pain narratives and pedagogies, and in close engagement with larger cultural discourses of pain and suffering. They actively appropriated existing concepts, languages, and forms of communication from the medical, literary, and philosophical traditions to understand and explain their precarious existence and their experiences of suffering based on the founding narrative of Jesus's passion. On the one hand, this appropriation placed the narratives of Christian pain and suffering squarely within the Roman cultural and sociopolitical tradition, particularly depicting Christian martyrs as Stoic sages, athletes, and gladiators who exhibit the masculine virtue of endurance in the Roman penal system. On the other hand, the unmistakable foundation of their founder's pain and suffering not only gave them the christological framework and means of resisting Roman value and power but also made them embrace their pain and suffering in reference to, as an extension of, and in union with their founder. Their particular openness to pain constituted itself as "a form of agency,"[194] and thus transformed the uniquely subjective unshareable pain narrative of Jesus into something repeatedly shareable by his followers. In this process, they expanded the application of christological pain and suffering beyond the pain experienced under torture and violence to the pain experienced by illnesses and other afflictions, all of which required moral agency, as already seen in the Greco-Roman literature and philosophy. On the basis of these narratives and pedagogies of pain, early Christians forged their unique communal identity as a community of cosufferers in mimetic

---

193. Cf. Asad, *Formations of the Secular*, 82.
194. Asad, *Formations of the Secular*, 87.

identification with Christ. Pain and suffering became a constitutive part of sustaining that relationship and communal identity. Thus their pedagogy of pain would always accompany or compel tangible decisions or actions by sufferers, whether they may be martyrdom, ascetic discipline, longing for the divine, enduring illness, caring for the sick, or showing compassion to the grieving and the suffering. Finally, while they understood that different degrees of meaningfulness in pain existed, they paradoxically established the significance of pain and suffering as an expected and meaningful part of the human experience.

4

HEALTH CARE
IN THE GRECO-ROMAN WORLD

In this chapter and the next we will focus on health care. This chapter addresses health care in the Greco-Roman world, while the next chapter explores early Christianity as it adopts, reformulates, and interacts with Greco-Roman health care. Since illness is by and large a sociocultural phenomenon, so is health care;[1] health care includes people's beliefs about sickness, "norms governing choice and evaluation of treatment," "socially-legitimated statuses, roles, power relationships, interaction settings, and institutions."[2] In this sense, health care is "the nexus of adaptive responses to the human problems created by sickness"; as such, "the issue of 'efficacy' is central to it" in response to the various etiologies we encountered in chapter 2.[3] As already seen in chapters 1 and 2, healing includes a biomedical response to illness but also involves the attempts of the health-care system to provide culturally (and religiously) relevant meaning(s) for the disruptions caused by illness.[4] Therefore, health care in this chapter and the next will encompass a number of approaches that were thought efficacious and meaningful by the patients and healers: from religious healings (temple medicine), rational medicine, hospitals, and popular medicine, to remedies for the soul in medico-philosophical treatises. The selective

1. Arthur Kleinman, *Patients and Healers in the Context of Culture: An Exploration of the Borderland between Anthropology, Medicine, and Psychiatry* (Berkeley and Los Angeles: University of California Press, 1980), 27, 360.

2. Kleinman, *Patients and Healers*, 24.

3. Kleinman, *Patients and Healers*, 26–27.

4. Kleinman, *Patients and Healers*, 72–80.

treatment of these various forms of health care will represent the comprehensive and symbiotic nature of the diverse means of health care available
in the Greco-Roman world and will also show us the socially constructed
nature and shape of health care.

Greco-Roman Health Care

In chapter 1 I examined the Greco-Roman construction of illness narratives in light of their culture and religion, including Aristides's own illness
narrative through his complex engagement with both temple medicine
and rational medicine. There I treated the complementary relationship
between religious healing and rational medicine from the perspective of
a patient in the Greco-Roman world. Here I extend the discussion further
and focus on the therapeutics of religious healing based on the common
cultural understanding that gods can both cause and heal sickness.

Religious Healing (Temple Medicine)

A close correlation between religious healing and rational medicine is seen
in the fact that the rise of healing shrines, including the cult of Asclepius in
the fifth century BCE, went hand in hand with the rise of Hippocratic medicine[5] and, in Athens, with the appearance of public physicians. As seen in
chapter 1, in Greek religion, potentially every Greek deity could cause and
heal human sickness.[6] Apollo and Artemis, for example, came to the acropolis of Aegialea and rescued the community from the deadly plague (Pausanias, *Descr.* 2.7.7–8). When the people of Troezen suffered from a plague,
Pan Lyterius ("the one who releases from disease") appeared to magistrates
while they slept and provided healing from the epidemic (2.32.6). Hermes
also averted a pestilence from the city of Tanagra "by carrying a ram round
the walls; to commemorate this Calamis made an image of Hermes carrying a ram upon his shoulders" (9.22.1). In Oropos, the cult of Amphiareus,
a mythical hero who, like Asclepius, was later worshiped as a god, became

5. On the close relationship between Hippocratic medicine and healing cults,
especially that of Asclepius, see Maria Elena Gorrini, "The Hippocratic Impact on
Healing Cults: The Archaeological Evidence in Attica," in *Hippocrates in Context:
Papers Read at the XIth International Hippocrates Colloquium*, ed. Philip J. van der
Eijk (Leiden: Brill, 2005), 135–56.

6. See Vivian Nutton, *Ancient Medicine*, 2nd ed. (London: Routledge, 2013),
108–9.

the leading healing cult.[7] During the Hellenistic and Roman periods, the popular imported Egyptian deities such as Isis[8] and Serapis were renowned as healing deities. Diodorus of Sicily writes: "For practically the entire inhabited world is their witness, in that it eagerly contributes to the honors of Isis because she manifests herself in healings. For standing above the sick in their sleep she gives them aid for their diseases and works remarkable cures upon them as they submit themselves to her" (*Library of History* 1.25.2–5). Cicero also places Serapis on par with Asclepius in healing a supplicant through incubation: "It can be either Asclepius or Serapis who is able to prescribe a treatment for health in sleep" (*Div.* 2.123).[9] Healed of his blindness by Serapis, Demetrius of Phalerum "wrote five books about dreams and cures involving Serapis" (Artemidorus, *Onir.* 2.44).[10] Other Greco-Roman deities like Zeus/Jupiter, Artemis/Diana, Juno, Hercules, Mars, and Mercury healed at their own sacred temples or sanctuaries.

However, by 300 BCE, it is Asclepius who came to be seen as "the healing god par excellence" with his cultic center at Epidaurus.[11] The Asclepian temple medicine in particular was "an integral part of ancient medicine"[12] and "the most popular of the treatment options available during the High Roman Empire."[13] As his temples[14] attracted sick worshipers, pilgrims, and visitors from all over the Greco-Roman world, the asclepieia in Epidaurus,

---

7. Georgia Petridou, "Asclepius the Divine Healer, Asclepius the Divine Physician: Epiphanies as Diagnostic and Therapeutic Tools," in *Medicine and Healing in the Ancient Mediterranean*, ed. Demetrios Michaelides (Oxford: Oxbow Books, 2014), 292.

8. On the curative activities of Isis, see Hector Avalos, *Health Care and the Rise of Christianity* (Peabody, MA: Hendrikson, 1999), 51–53.

9. Cf. John E. Stambaugh. *Serapis under the Early Ptolemies* (Leiden: Brill, 1972), 76: "Inscriptions in the Delos Serapea frequently allude to dream appearances, which parallel those in which Asclepius appeared at his shrines."

10. Stambaugh. *Serapis under the Early Ptolemies*, 76.

11. Nutton, *Ancient Medicine*, 104. For a concise yet thorough study of the Asclepian medicine and cult, see Florian Steger, *Asclepius: Medicine and Cult*, trans. Margot M. Saar (Stuttgart: Franz Steiner Verlag, 2018).

12. H. F. J. Horstmanshoff, "Did the God Learn Medicine? Asclepius and Temple Medicine in Aelius Aristides' *Sacred Tales*," in *Magic and Rationality in Ancient Near Eastern and Graeco-Roman Medicine*, ed. H. F. J. Horstmanshoff and M. Stol (Leiden: Brill, 2004), 337.

13. Ido Israelowich, *Patients and Healers in the High Roman Empire* (Baltimore: Johns Hopkins University Press, 2015), 111.

14. For a detailed study of the temples of Asclepius and their role in healing, see Hector Avalos, *Illness and Health Care in the Ancient Near East: The Role of the Temple in Greece, Mesopotamia, and Israel*, HSM 54 (Atlanta: Scholars Press, 1995), 37–98.

Cos, Athens, Tricca, and Pergamum became among the most famous because the god revealed himself to the worshipers in those places and healed them (e.g., Strabo, *Geogr.* 8.6.15). Among the preserved testimonies and cure inscriptions, sickness of a chronic nature features prominently: infertility, paralysis, blindness, swellings, headaches, insomnia, and facial blemishes, although we also read about those suffering from ulcers, abscesses, or intestinal worms. As in other temples, representations of the body parts that were healed such as eyes, ears, hands, legs, feet, breasts, and genitalia were carved in terra-cotta, stone, ivory, bronze, silver, or gold, and were hung on the walls of the shrines or displayed on shelves.[15]

At the Asclepian shrines, desperate supplicants who were prepared to meet the god would undergo a purification rite at a sacred spring, offer a suitable sacrifice, and then, in white robes, purify themselves again before entering the *abaton* ("impassable place"), a place of incubation ("healing sleep"), in order to ensure the efficacy of the incubation.[16] In the *abaton* the supplicants, while asleep, would receive a dream-vision from Asclepius himself as a physician or surgeon, and would receive either direct healing on the spot or a remedy for subsequent healing;[17] sometimes one of the god's accompanying snakes or dogs would appear to lick or enter the patient for healing;[18] sometimes, as we saw in chapter 1, the dream-vision would require further explication and would be interpreted by the temple wardens, as in the case of Aristides. While the patients themselves would typically sleep in the temple in hopes of getting a healing dream-vision, stand-ins could also experience the incubation on their behalf.[19] For example, for the sake of Arata from Lacedaemon, who suffered from dropsy, her mother slept in the asclepieion in Epidaurus and saw a dream: "It seemed

15. According to Steven M. Oberhelman, "Anatomical Votive Reliefs as Evidence for Specialization at Healing Sanctuaries in the Ancient Mediterranean World," *Athens Journal of Health* 1, no. 1 (2014): 47–54, the Asclepian sanctuaries in different cities specialized in the treatment of different diseases based on the evidence of votive reliefs. For example, the asclepieion at Athens specialized in treating eye disease, and Corinth specialized in limbs, appendages, and venereal disease.

16. Cf. Nutton, *Ancient Medicine*, 110; Avalos, *Illness and Health Care*, 75–76.

17. For Aristides's self-report of incubation, see *Or.* 23.16 and 48.31–33.

18. Nutton, *Ancient Medicine*, 110. Cf. Emma J. Edelstein and Ludwig Edelstein, *Asclepius: Collection and Interpretation of the Testimonies*, 2 vols. (Baltimore: Johns Hopkins University Press, 1945), T. 423, II 39, 42.

19. Georgia Petridou, "The Healing Shrines," in *A Companion to Greek Science, Technology, and Medicine in Ancient Greece and Rome*, ed. Georgia L. Irby (Malden, MA, and Oxford: Wiley Blackwell, 2016), 1:437.

to her [Arata's mother] the god cut off the head of her daughter and hung the body neck downwards. After much fluid had run out, he united the body and put the head back on the neck. Having seen this dream she returned to Lacedaemon and found on her arrival that her daughter was well and that she had seen the same dream."[20] Here we see the god even capable of healing at a distance. As in other healing cults, the dreams in the Asclepian cult were "the therapeutic means of treating an illness," and not merely a diagnostic tool, as had been common in the rational medicinal practices of the Hippocratic Corpus and Galen.[21]

As is well known, those dream remedies from the god would often find parallels with contemporary rational medicine such as bloodletting, drugs, poultices, emetics, diet, exercise, and baths from Hippocratic medicine. To be more specific, Galen, led by two clear dreams from Asclepius, treated a patient with bloodletting (*How to Cure People through Bloodletting* 23.11), which played a significant role in Galen's own medical system.[22] According to Galen, and the dedication to Asclepius in Pergamum by Aristides's near contemporary Publius Aelius Theon, Asclepius's prescription to consume half an onion and fifteen pepper seeds every morning closely corresponded to the common regimen of warming agents prescribed by Greco-Roman doctors.[23] Galen also relates a story about a wealthy man from Thrace who was led to Pergamum by a dream of Asclepius's, "prescribing that he should drink every day of the drug produced from the vipers and should anoint the body from the outside" (*Outline of Empiricism* 10).[24] His disease, elephantiasis, "after a few days turned into leprosy; and this disease, in turn, was cured by the drugs which the god commanded" (10). First, an incurable disease has turned into a curable disease by following the divine instruction, and then Galen and his team apparently cured the curable disease with the drugs commanded by the god. Teun Tieleman further points out that Asclepius was also familiar with the Platonic notion of the tripartite and tri-located soul, to which Galen also subscribed, in treating mental and

---

20. Edelstein and Edelstein, *Asclepius*, T. 423, II 21.

21. Petridou, "The Healing Shrines," 439. See also Israelowich, *Patients and Healers,* 53–55.

22. Teun Tieleman, "Religion and Therapy in Galen," in *Religion and Illness,* ed. Annette Weissenrieder and Gregor Etzelmüller (Eugene, OR: Wipf & Stock, 2016), 25. On Galen's practice of bloodletting, see p. 11.

23. Israelowich, *Patients and Healers,* 56–57.

24. Edelstein and Edelstein, *Asclepius*, T. 436.

psychosomatic disorders:[25] "A very important witness is also our ancestral god, Asclepius, who often prescribed the composition of poems, mimes and songs to people in whom the movements of the spirited part [of the soul] had become too vehement and had made their bodily temperament hotter than it should be" (*On the Preservation of Health* 1.8 [41K]).

The dreams experienced by patients in asclepieia were thus comparable with the practices of rational medicine. Nevertheless, other dream therapies by the god were clearly divergent from contemporary medicine, as in the case of Aristides, or were apparently random, thereby accentuating the superiority of Asclepius's treatment to the physician's. For a man suffering from gout, Asclepius intervened through a goose, which bit the patient's feet "and by making them bleed, made him well."[26] In fact, the healing encounter that took place in the asclepieion was about much more than the god's medical techniques in dreams; the temple with the altar, its physical setting, the sacred spring, the sacred grove, the gymnasium or stadium, a theater for various performances such as songs, mimetic dances, and processions—all these strongly testify to the fact that the healing event in the asclepieion was a holistic experience for mind and body, one that went beyond physical cure only.[27]

In this sense, the asclepieion was the place of holistic healing, especially for patients who did not have any alternative but to seek Asclepius when doctors using rational medicine could do no more for them, or when doctors refused to treat them due to their incurable sickness. The Greek orator Aeschines left the following votive inscription: "Having despaired of the skill of mortals, but with every hope in the divine, forsaking Athens, blessed with children, coming to your sacred grove, Asclepius, I was healed in three months of a festering wound which I had had on my head for a whole year."[28] Additional cases considered hopeless (*aphēlpismenos hypo pantos anthrōpou*) included Lucius, who suffered from lung disease (pleurisy), and Julian, who was bringing up blood. During their separate visits to the temple, Asclepius prescribed that Lucius "should go and from the threefold altar lift ashes and mix them thoroughly with wine and lay them on his side"; and he prescribed that Julian "should go and from the threefold altar take the seeds of a pine cone and eat them with honey for three

---

25. Tieleman, "Religion and Therapy in Galen," 25.
26. Edelstein and Edelstein, *Asclepius*, T. 423, II 43.
27. Nutton, *Ancient Medicine*, 111.
28. Edelstein and Edelstein, *Asclepius*, T. 404.

days."²⁹ They were both cured by those prescriptions, and both publicly offered thanks to the god before the people. Finally, the following story presents Asclepius's supernatural power beyond his superhuman power by healing an inconceivable disability beyond the dictates of natural law:

> A man came as a suppliant to the god. He was so blind that of one of his eyes he had only the eyelids left—within them was nothing, but they were entirely empty. Some of those in the Temple laughed at his silliness to think that he could recover his sight when one of his eyes had not even a trace of the ball, but only the socket. As he slept a vision appeared to him. It seemed to him that the god prepared some drug, then, opening his eyelids, poured it into them. When day came he departed with the sight of both eyes restored.³⁰

Was there any criterion as to what kind of patient to whom Asclepius revealed himself? According to Philostratus's *Life of Apollonius of Tyana*, an Assyrian youth, who had dropsy, came to visit Asclepius but indulged himself with drink even on his sickbed, neglecting a dry diet, which he was presumably prescribed (1.9). So Asclepius neglected him and refused to visit the youth in his sleep. In response to the young man's complaint about the god not showing up, Asclepius then appeared to him and directed him to consult with Apollonius for his relief. Apollonius rebuked the youth for blaspheming the god and explained that the god reveals himself to those who want healing (*tois gar boulomenois didōsi*), not to those who act contrary to their illness. Philostratus also records that Apollonius told the Asclepian priest to refuse entry into the sanctuary of the god to a man with only one eye who had, before he obtained anything from the god, offered extravagant sacrifices to the god at the altar. (1.10). Asclepius appeared to the priest by night and confirmed Apollonius's warning about this man, telling him that the man should keep his goods and leave, for he did not deserve even to have one eye. Upon investigation, the priest found out that the man had slept with his wife's daughter from her previous marriage; but the mother caught them in bed and put out both of the girl's eyes and one of the man's with her brooch pins (1.10). It seems that for Asclepius the right attitude (wanting healing), the right action (making an active effort to improve one's lifestyle and following the prescribed regimen from the

---

29. Edelstein and Edelstein, *Asclepius*, T. 438.
30. Edelstein and Edelstein, *Asclepius*, T. 423, I 9.

god),[31] and moral uprightness were important criteria for his manifestation to and healing of a patient.[32] These stories and criteria "were intended to assign responsibility for the failure to be healed to the patients, not to Asclepius' inefficiency," as many incubants probably would have left "the *abaton* without having received any divine dream or vision and without any health improvement."[33]

While Asclepius was the divine healer par excellence, Apollonius of Tyana, the itinerant Pythagorean of the first century, was the human healer par excellence. His well-known biography, Philostratus's *Life* in the early third century (c. 220 CE), colorfully portrays him as an ascetic holy man. In contact with the gods through his travels, sayings, and miracles, he was often accused of being a sorcerer. And this biography spurred debates between Christians and non-Christians ("Greeks") in late antiquity.[34] Although it is difficult to recover the "historical" Apollonius, our focus here is the recorded healing miracles in the *Life*. For example, while in India, Apollonius healed several Indians in a row (3.39). First, a lame man whose hip was dislocated by a lion immediately recovered when Apollonius ("the Wise Man") massaged the man's hip with his hands. Second, Apollonius restored the eyesight of a blind man, and third, he stretched out a man's withered hand. Finally, he helped a woman who had seven miscarriages to deliver a baby successfully by advising her husband to walk around her with a hare while she was giving birth and to release the hare. In Ephesus, as a visionary, Apollonius warned the Ephesians of an approaching epidemic, but they ignored his prophetic warning (4.4). So he moved on

31. Cf. Edelstein and Edelstein, *Asclepius*, T. 423, II 37: "Cleimenes of Argus, paralyzed in body. He came to the Abaton and slept there and saw a vision. It seemed to him that the god wound a red woolen fillet around his body and led him for a bath a short distance away from the Temple to a lake of which the water was exceedingly cold. When he behaved in a cowardly way Asclepius said he would not heal those people who were too cowardly for that, but those who came up to him into his Temple, full of hope that he would do no harm to such a man, but would send him away well."

32. Cf. Petridou, "The Healing Shrines," 438.

33. Olympia Panagiotidou, "Asclepius: A Divine Doctor, A Popular Healer," in *Popular Medicine in Graeco-Roman Antiquity: Explorations*, ed. William V. Harris, Columbia Studies in the Classical Tradition (Leiden: Brill, 2016), 103.

34. On this see Christopher P. Jones, "Apollonius of Tyana in Late Antiquity," in *Greek Literature in Late Antiquity: Dynamism, Didactism, Classicism*, ed. Scott Fitzgerald Johnson (London: Routledge, 2006), 49–64, where the author discusses the divergent views on Apollonius between the Christians in the East and the West in response to the non-Christian polemics against Christians using the story of Apollonius.

to Smyrna; but, when the plague actually swept Ephesus, the Ephesians sought out Apollonius to "make him the physician of their misfortunes" (4.10.1). Declaring an end to the plague, Apollonius assembled the Ephesians in the theater and had them stone an old wretched beggar; to the puzzlement of the Ephesians, Apollonius urged them even more (4.10.2). It turned out that the beggar was a demon and was transformed into a dog, the size of the largest lion—dead and spewing foam "as maniacs do" (4.10.2). As illustrated in the previous paragraph, Apollonius is portrayed as "Asclepius's servant and companion" (1.12.1), and in Aegeae, Asclepius "told the priest that he was glad to cure the sick with Apollonius as his witness" (1.8.2). Thus, when Apollonius arrived at the asclepieion in Pergamum, "after cleansing the Ephesians of the plague" (4.11.1), "he made suggestions to the god's supplicants about all they should do in order to get auspicious dreams, and he cured many of them (*pollous de kai iasamenos*)" (4.11.1). Moreover, in Athens, Apollonius exorcised a demon-possessed boy while lecturing on libations to the gods (4.20.1–2); this not only amazed the people present in the lecture but led the boy to "convert" to philosophy (4.20.3). Finally, in Tarsus, Apollonius healed a youth's rabies from a mad dog and also restored the dog itself (6.43.1–2).

As we can see, religious healings by the gods and itinerant holy men like Apollonius were an integral part of the Greco-Roman medical market. These stories of healing not only legitimized the patients' experiences of healing, but also normalized their expectations of healing from the gods or their human "servants"—an ascetic holy man, as in the case of Apollonius of Tyana, as well as the physicians themselves. What follows is the therapeutics of the latter.

(Rational) Medicine

As in the case of Aristides, people who visited the asclepieia did not ignore physicians or exclude rational medicine. Already in the Hippocratic *Oath*, the author establishes the intimate connection between Asclepius (among other gods and goddesses) and doctors, even as Hippocrates himself was recognized as a descendant of Asclepius (1). Still, Hippocratic physicians operated in other ways than through temple healings. Their self-stated aim as practitioners of medicine is captured in this famous saying: "To help, or least to do no harm" (*Epidemics I* 11). To be more specific, they defined the goal of medicine as a patient's total or relative recovery; the author of *The Arts* says that the purpose of medicine is "to do away with the sufferings of the sick, to lessen the violence of their diseases" (3). How would they achieve the goal of recovery for their patient? They expressed the two fundamental Hippocratic principles of treatment toward that goal. First, "one must know that diseases due to repletion are cured by evacuation, and those due to evacuation are cured by repletion; those due to exercise are cured by rest, and those due to idleness are cured by exercise. To know the whole matter, the physician must set himself against the established character of diseases, of constitutions, of seasons and of ages; he must relax what is tense and make tense what is relaxed. For in this way the diseased part would rest most, and this, in my opinion, constitutes treatment" (*Nature of Man* 9). Here we find allopathy, treatment by opposites, as the fundamental principle of treatment. We see the same principle in *Aphorisms* ("Diseases caused by repletion are cured by depletion; those caused by depletion are cured by repletion, and in general contraries are cured by contraries" [*II* 22]) and also in *Breaths*:

> For knowledge of the cause of a disease will enable one to administer to the body what things are advantageous [by using opposites to combat (*ek tōn enantiōn epistanienos*) the disease]. Indeed this sort of medicine is quite natural. For example, hunger is a disease, as everything is called a disease which makes a man suffer. What then is the remedy for hunger? That which makes hunger to cease. This is eating; so that by eating must hunger be cured. Again drink stays thirst; and again repletion is cured by depletion, depletion by repletion, fatigue by rest. To sum up in a single sentence, opposites are cures for opposites. (1)

This does not mean that the Hippocratic doctors excluded homeopathy, therapy by like agents (*Places in Man* 42), which was relatively well known in the fifth century BCE; however, in general, they envisioned medical treatment as "combating" the disease and its cause with opposite force that would produce the opposite result.[35] In this framework, the role of the doctor was to "'struggle' (*antagonisasthai*) against the disease thanks to his art" (cf. *Prognostic I*).[36]

If allopathy was the first principle of Hippocratic treatment, the second principle was change: "Everything that changes from the existing state ben-

---

35. Jacques Jouanna, *Hippocrates*, trans. M. B. DeBevoise (Baltimore: Johns Hopkins University Press, 1999), 342–43.

36. Jouanna, *Hippocrates*, 343.

efits what is ill, for if you do not change what is ill, it increases" (*Places in Man* 45). Since change was the basis of pathology in the Hippocratic Corpus,[37] treatments should also change the state of the sick body. However, this change should be carried out in the right way: "Whoever knows how to cause in men by regimen moist or dry, hot or cold, he can cure this disease also, if he distinguishes the seasons for useful (*kairous*) treatment, without having recourse to purification and magic" (*The Sacred Disease* 21). For the change to be carried out appropriately and correctly, it must be gradual, since every abrupt, major change is harmful to patients. Therefore, "in seeking to combat the disease, the physician had to know how to measure the proper degree of change to be introduced into the body, and to seize upon the appropriate moment for introducing it, if he wished to restore health without causing damage."[38] In Greek, *kairos* denotes the ideas of suitable quantity or measure and right timing or moment; the skilled physician should adjust to the condition of the sick body and "administer to the body neither too much nor too little, neither too soon nor too late."[39] Hence, the most difficult aspect of medicine for the doctors is to discern the right interventions at the right times. As we saw in the treatment of pain in chapter 3, and as we will see later in this chapter, this notion of adjusting to the patients' conditions and treating them according to their specific needs in the right way and at the right time will also be important for medico-philosophical treatises in their therapies for the soul.

Galen, who drew up a close connection between the authority of Asclepius and that of physicians,[40] also defines the primary objective of therapeutics as "the elimination of disease" (*Meth. med.* 9.10 [636K]) and regards allopathy and change as the two important principles of his treatment. In *The Art of Medicine*, Galen writes: "Thus, if everything unbalanced is contrary to nature and everything balanced is in accord with nature, it is necessary to restore everything that is unbalanced to a balance by the due measure of the opposite" (28 [381K]).[41] In the case of the imbalance of humors (*dyskrasis*), which is the cause of all kinds of diseases, Galen names

37. See chap. 1, pp. 39–41.

38. Jouanna, *Hippocrates*, 344.

39. Jouanna, *Hippocrates*, 344.

40. On Galen's belief in Asclepius and Asclepius's dream-cures, see Fridolf Kudlien, "Galen's Religious Belief," in *Galen: Problems and Prospects*, ed. Vivian Nutton (London: Wellcome Institute for the History of Medicine, 1981), 117–30.

41. See also *Meth. med. Glaucon* 1.10 (32K): "For opposites are cures of opposites, curbing what is excessive and reintroducing what is lacking"; cf. *Meth. med.*

two types of cures: change and evacuation (36 [404K]). Change[42] can be brought about by the body itself cooking (concocting) these humors or by certain medication; evacuation[43] takes place by means of medication. Again, in the case of a fever kindled by putrefying humors, "the indication (*endeixis*) is for change and evacuation," says Galen. Here "indication" is the reflection of the body of the patient under examination that indicates or shows what is wrong with it and how it should be treated.[44] As Galen observes the helpful and relevant signs of the patient's fever, he again mentions a therapeutic principle of change (i.e., concoction through opposites) and evacuation. Galen notes, "change puts an end to the putrefaction; if the cause remains in existence, evacuation removes the whole cause from the body" (28 [382K]). As Galen explains how to effect change and evacuation in the patient, he further stresses the importance of knowing when to apply evacuation with regard to time, quality, quantity, and how to use it (28 [383K]).

Having established how the Hippocratic Corpus and Galen define the purpose of medicine and its treatment principles, we can delve into the specific methods of its therapeutics. According to Jacques Jouanna, the therapeutic triad in classical Greek literature, including the Hippocratic Corpus, consists of medicine (*pharmakon*), incisions, and cauterization.[45] In *Aphorisms*, the physician author ranks these three kinds of treatment on the basis of their degree of effectiveness: "Those diseases that medicines do not cure are cured by the knife. Those that the knife does not cure are cured by fire. Those that fire does not cure must be considered incurable" (*VII* 87).

First, concerning medicines, Jouanna notes that most aimed at evacuating the "cavities" of the body.[46] Because the Hippocratic doctors regarded humans to have two main cavities—the upper cavity, that is, the chest, and the lower cavity, that is, the stomach—they recommended either upward evacuation via vomiting or downward evacuation via bowel movement as

---

3.8 (214K); *The Art of Medicine* 34 (400K): "we shall effect a cure by introducing the opposite quality."

42. On therapeutics as changing the state of the body, see also Galen, *Hygiene* 1.1 (1K).

43. See below under the Hippocratic medicine.

44. Philip J. van der Eijk, "Therapeutics," in *The Cambridge Companion to Galen*, ed. R. J. Hankinson (Cambridge: Cambridge University Press, 2008), 292.

45. This and following paragraphs on the Hippocratic therapeutics are based on Jouanna, *Hippocrates*, 156–62.

46. Jouanna, *Hippocrates*, 156.

preventative or curative measures.[47] *Regimen in Good Health* advises inducing vomiting in winter and bowel evacuations in summer, especially in the case of diseases: "Emetics and clysters for the bowels should be used thus. Use emetics during the six winter months, for this period engenders more phlegm than does the summer, and in it occur the diseases that attack the head and the region above the diaphragm. But when the weather is hot use clysters, for the season is burning, the body bilious, heaviness is felt in the loins and the knees, feverishness comes on and colic in the belly. So the body must be cooled, and the humours that rise must be drawn downwards from these regions" (5).

With the development of a humoral theory, the Hippocratic Corpus used medicines as the main way to evacuate humoral excess (for example, of phlegm or bile), which was thought to be the cause of disease.[48] *Nature of Man* describes the direct linkage between specific medication and the evacuation of any of the four humors that constitute human nature: "If you were to give a man a medicine which withdraws phlegm, he will vomit you phlegm; if you give him one which withdraws bile, he will vomit you bile. Similarly too black bile is purged away if you give a medicine which withdraws black bile. And if you wound a man's body so as to cause a wound, blood will flow from him" (20).

Downward evacuations were aided by clysters, as mentioned, or purgatives, ranging from boiled donkey's milk (mild) to black hellebore (strong) (*Regimen in Acute Diseases* 23). Discerning the appropriate dosage of these evacuants was challenging to the physicians, and they were fully aware of the danger of evacuants: "medications that clean bile or phlegm are a source of danger [for those who are treated] and of blame for the person treating" (*Affections* 33).

For Galen, medications acted on the mixture (*krasis*) of the body also to produce an effect, especially evacuation, according to the allopathic principle[49] (e.g., a cold *pharmakon* cools an organism and thus helps in curing diseases provoked by heat or vice versa).[50] As already mentioned, Galen recommended evacuation induced by means of medication with strong heating effect, cathartics, clysters, sweating and vomiting, in order

---

47. Jouanna, *Hippocrates*, 156.

48. Jouanna, *Hippocrates*, 157.

49. See Galen, *On the Powers of Simple Drugs* 1.1: XI 380: "Everything that has some power (*dynamis*) to alter our nature we call a *pharmakon*.... A *dynamis* is some active cause, whether in actuality or in prospect." Quoted in Sabine Vogt, "Drug and Pharmacology," in Hankinson, *The Cambridge Companion to Galen*, 307.

50. Cf. Vogt, "Drug and Pharmacology," 307.

to treat any imbalance of the humors (*The Art of Medicine* 36 [404K]). Also, in treating a fever through putrefying humors, evacuation is to be done "by phlebotomy, clysters, the urine and transpiration through the skin, as well as revulsion and diversion to other places" (28 [382K]). This is according to his principle that the cure of any illness involving moisture is "complete evacuation"—whether in the case of excessive blood in the vein, pus or blood in the stomach, intestines, rough artery or lungs, or an excess of food or drink in the lungs and chest.[51] Thus, in treating tertian fevers, Galen also advises a double evacuation of yellow bile through vomiting and bowel movement with gentle clysters (*Meth. med. Glaucon* 1.10 [32K]). And one recipe for a purgative includes a mixture of aloes, scammony, colocynth, agaric, bdellium, and gum arabic (*Meth. med.* 5.14 [375K]). Like the Hippocratics, Galen also exhorts other physicians to discern an appropriate time and amount of dispersed medications for evacuation, considering the temperament (*krasis*) of the body, the severity of the imbalance (*dyskrasis*), and the position of the organs.[52]

Second, an incision made by knife shared the same goal as evacuation—curing sickness via the elimination of impure liquids.[53] The Hippocratic *Nature of Man*, which presents a complete account of the blood vessels, lays down the principles of bleeding or venesection as follows: "The habit should be cultivated of cutting as far as possible from the places where the pains are wont to occur and the blood to collect. In this way the change will be least sudden and violent, and you will change the habit so that the blood no longer collects in the same place" (11). Physicians thought bleeding was helpful for many diseases and identified various parts of the body as suitable places of bleeding: "arms at the bend of the elbow and the legs behind the knee or at the ankle," as well as the head and beneath the tongue (11).[54] Nonetheless, they also acknowledged the necessity of caution in advising venesection as a treatment for acute diseases: "The acute affections you treat with phlebotomy, if the disease seems to be severe, and patients are at the height of their youth and strength" (*Regimen in Acute Diseases [Appendix]* 3). Still, *Epidemics V* describes the successful cure of a man's upset stomach by bleeding: "He drank various drugs to

51. *The Art of Medicine* 33 (392K); cf. *On the Constitution of the Art of Medicine* 299–300K.

52. *On the Constitution of the Art of Medicine* 287K; *Meth. med. Glaucon* 1.1 (5K); 2.4 (94K–95K).

53. Jouanna, *Hippocrates*, 159.

54. Jouanna, *Hippocrates*, 160.

purge upward and downward, and was not benefited. But when he was bled in each arm in turn until he was bloodless, then he was benefited and freed from the trouble" (6).

By the time of Galen, bleeding was a well-known therapy, and Galen also considered it a proper treatment for "virtually any serious illness."[55] As Scribonius Largus reported in the first century CE, most people were understandably terrified of the knife (for incision) or hot iron (for cauterization) (*Intr.* 2).[56] As a skilled physician, Galen was well aware of the severe reaction of his patients when he took a large amount of blood; but in the case of fever with an excess of blood (*plēthos*), he believed the drastic measure of phlebotomy delivered the patient from his sickness:

> I deliberately took from him an amount [of blood] sufficient to bring about fainting, since I had been taught that both on theoretical grounds and from experience this is a most important remedy for continuous fevers in someone of strong capacity. For first, a body being cooled in the fainting very quickly comes to the opposite state. Nobody would be able to discover anything more pleasant or useful than this, either for patients or for the actual nature governing animals. Since, of necessity in such bodies gastric excretion follows, while there is sometimes also vomiting of bile, the result is that moisture is immediately removed from the whole body, or there are sweats. Therefore, when these things all happened in sequence in that man [patient], I quenched the fever right away so that some of those present said, "You, sir, have slain the fever!," upon which we all laughed. (*Meth. med.* 9.4 [612K])

Despite this success, however, Galen clearly knew the potential danger of venesection and cautioned his peers about conducting it in the right way, and at the right time: "Indeed, in the case of phlebotomy, which I said a little earlier needs to be carried out to the point of fainting for the sake of quenching the fire of the continuous fever due to stoppage of the pores, no little harm is likely to follow, if it is not done at the appropriate time and with due moderation" (*Meth. med.* 9.10 [637K]).

55. Susan P. Mattern, *The Prince of Medicine: Galen in the Roman Empire* (Oxford: Oxford University Press, 2013), 234. See also *Galen on Bloodletting: A Study of the Origins, Development, and Validity of His Opinions, with a Translation of the Three Works*, trans. Peter Brain (Cambridge: Cambridge University Press, 1986), 122.

56. Mentioned in Mark Grant, *Galen on Food and Diet* (London: Routledge, 2000), 6.

Finally, the strongest form of ancient medicine was cauterization (burns). According to the Hippocratic treatise *The Arts*, "of the caustics employed in medicine fire is the most powerful, though there are many others less powerful than it. Now affections that are too strong for the less powerful caustics plainly are not for this reason incurable; but those which are too strong for the most powerful plainly are incurable" (2). Given the seriousness of this treatment, fire was considered the treatment of last resort.[57] When confronted with a recurring disease of the head, *Diseases II* recommends a total of nine cauterizations in addition to three types of evacuants: "two beside the ears, two on the temples, two behind the head . . . , and two on the nose. . . . And burn the vessels beside the ears until they no longer throb" (12). The doctor then anticipates a positive result: if the patient "does these things, he recovers" (12). In the case of a disease caused by the flow of phlegm and bile through the blood vessels, *Internal Affections* suggests a total of thirteen cauterizations to be performed: four beside the right shoulder blade, three into the socket of the right hip joint, two under the buttock, two in the middle of the thigh, and one each above the knee and ankle" (18)! Similarly, the author believes that these cauterizations will be effective and block the advance of disease: if the patient "is cauterized in this way, it will not allow the disease to migrate either upwards or downwards" (18). However, *Epidemics V* also records the unfortunate case of a man whose severe pains in his right hip and groin and adjacent to them were successively treated with evacuation, incision, and cauterization as the last resort (7). After his cauterization, "the scars were numerous, large, and close together"; although much thick pus ran out, "he died a few days after that, from the size and number of the wounds and from weakness of his body" (7). The physician author then offers an alternative treatment of one or two incisions to drain the pus, which would have been successful in his mind (7).

In *A Method of Medicine to Glaucon*, Galen applies all three methods of therapeutics in treating "the cancerous tumors (*tōn karkinōdōn*) occurring in all the parts [of the body], but especially in the breasts of women" whose menstruation is either abnormal or nonexistent (2.12 [139K]). He thinks this disease is caused by the accumulation of residues of black bile in the liver during the formation of blood, but it is left behind when the bile is purged through the spleen (2.12 [139K]). These residues are created when the liver is weak, when thick and muddy blood is generated by the diet, and

57. Jouanna, *Hippocrates*, 160.

when the spleen is weak in attracting the humor; then "the blood becomes turbid and thickened in the veins" (2.12 [140K]). In terms of treatment, the disease is curable only at the beginning; otherwise the physician should resort to surgery (2.12 [141K]). Before the operation, the physician should attempt to "thin" the blood first with the help of purgative medications. Then the surgery should excise the tumor, cutting around the whole mass (2.12 [141K]). Galen cautions the physician to be extremely careful since there is a great danger of hemorrhage, and of spreading the neighboring area with cancer. In this context, Galen mentions cauterization for burning the roots of the tumor and warns of the inherent danger of that as well (2.12 [141K]).

In addition to the therapeutic triad, the regimen (*diaita*) of the patient, a notion developed within the Hippocratic Corpus, played an important therapeutic role for the diseased, as well as a preventative role for the healthy. Among Hippocratic texts, while *On Ancient Medicine* understands *diaita* as the alimentary diet of food and drink, *Airs, Waters, Places* defines it in terms of food, drink, and exercise.[58] In chapter 1 we observed that for the Hippocratic doctor, good health comes from the balance between the two, that is, food and exercise (*Regimen III* 69). Accordingly, *Regimen II* lays down a basic rule for the physician to know the "power of various foods and drinks, both what they are by nature and what by art," in order to prescribe an alimentary regimen for the patient (39); it goes on to state the properties of various cereals, meats and poultry, fish, eggs, cheese, vegetables, fruits, and drinks specified as water, wine, and vinegar (40–46). According to *On Ancient Medicine*, Greeks have a superior diet to the barbarians because it (especially its quantity) is adapted to the different states of the person, whether one is healthy or sick (3–5). Indeed, diet should be adapted to human nature (*trophēn harmozousan tē physei*) to be beneficial and useful (5). In fact, when a physician recognizes a disease rising from a regimen, the physician should change the regimen of the patient "to suit the several conditions of age, season, physique and disease," as well as the habits of the patient (*Nature of Man* 9; *Aphorisms I* 17), according to three degrees of the patient's state of strength/weakness: a diet based on solid food (*sitia*); an in-between diet between solids and liquids (*rhyphēmata*); and a liquids-only diet (*pōmata*) (*On Ancient Medicine* 5). However, "even when all this is known, the care of a man is not yet complete," says the author of *Regimen I*, "because eating alone will not keep a man well; he must also take exercise"

58. *Airs, Waters, Places* 1; cf. *Regimen I* 2; *Regimen II* 61; *Regimen III* 69.

(2). Since health consists of a due proportion of food and exercise, the physician should "discern the power of the various exercises, both natural exercises and artificial, . . . know which of them tends to increase flesh and which to lessen it; and not only this, but also . . . proportion (*symmetrias*) exercises to bulk of food, to the constitution of the patient, to the age of the individual, to the season of the year, to the changes of the winds, to the situation of the region in which the patient resides, and to the constitution of the year" (2). Thus *Regimen II* discusses and recommends various exercises such as those of sight, hearing, voice, and thought, and walking, running, and swinging the arms in varying degrees (61–66), along with baths (52).

In *Hygiene*, where Galen discusses the various measures of health, he lays down the three objectives of a healthy regimen, which consists of diet (food and drink), exercise, baths, massage, and sexual activity: first, replacement of things emptied out; second, separation of superfluities;[59] and third, avoidance of premature aging (1.3 [9K]). As we saw in chapter 1, since there are different states of health among people, there need to be different regimens for different people in different stages of life (5.1 [306–7K]). Diet has to do with all three objectives. It is important since it can affect the mixture (*krasis*) of the body, which is the key to health, and facilitate concoction. Galen's principle is that "wherever a result can be achieved purely by regimen [here diet in particular], it is preferable to refrain from pharmacological prescriptions" (*The Thinning Diet* 1). For example, the thinning diet[60] counteracts the thickening effect of the humors of our bodies, the enlargement of the spleen, and the hardening of the liver; it helps those suffering from chronic breathing difficulties and "completely" cures minor or incipient cases of epilepsy (1). Also, hot and moistening foods (such as wine) counter the coldness and dryness of old age (*Hygiene* 5.3 [319K]). Another example involves treating different types of fatigue as classified by Galen:

> Thus, the wound-like fatigue, if it is to be treated adequately with apotherapy,[61] requires the customary nourishment or slightly less, and less of those

---

59. Superfluidity is "what is over and above especially after the digestion of food": Ian Johnston and G. H. R. Horsley, introduction to *Galen: Method of Medicine*, ed. and trans. Ian Johnston and G. H. R. Horsley, LCL (Cambridge, MA: Harvard University Press, 2011), 1:xciv.

60. Some examples of the foods with thinning qualities include garlic, onions, cress, leeks, and mustard among vegetables, poppy seeds, barley, rockfish, vinegar, honey, sweet wines, and the whey of milk (*The Thinning Diet* 2–11).

61. A form of physical therapy that combines massage, suppression of breath (*pneuma*), and binding by the masseur. See Ian Johnston, "General Introduction,"

nutrients that are more moist. . . . However, the tensive fatigue needs even more reduction of nutriment, while the inflammation-like fatigue above all needs nutriment that is very moist and very small in amount, and which also contains something cold. All those who are fatigued have a similar need for *euchymous* ("healthy") nutriment of the kind. . . . It is appropriate to guard against what is viscous in the nutriment, both in the wound-like and inflammation-like fatigues, as it would prevent the superfluities being dispersed. However, in the tensive fatigue, one must give such foods while keeping away from an excess of them. (3.8 [203K])

These diets are designed to help manage an excess of blood in the veins (*plethora*) and the unhealthy state of the four humors (*kakochymia*), which are the major causes of fatigues; and we see again the significance of a specific diet adapted to the different types of fatigues. In *Hygiene* Galen also mentions several common foods and drinks that are important in preserving health: milk in infancy usually from a mother or a nurse, and also milk in general depending on its sources (1.7 [36K]); breads, depending on how they are prepared and what they are taken with—for example, "those that are oven baked, leavened and moderately pure";[62] and various wines with their respective benefits and harms—for example, "the thinnest in consistency, and in color what Hippocrates was wont to call 'orange tawny,'" yellow, or pale yellow for old age.[63]

Moreover, in the same work Galen presents his principles and descriptions of various exercises involving speed, vigor, and violence, from the simple to the complex (2.8 [136K]; 2.9 [139–147K]). For optimal hygiene, as exercise increases innate heat, it should be followed by apotherapy, which counteracts the drying effect of exercise, and then by food and rest. The general factors for considering certain kinds of exercises include: first, the condition and constitution of the whole body; second, age; third, customs; and then affections (*pathē*) (5.10 [321–325K]). Old people should rest the weak parts and engage exercises that are customary and moderate, but young people should attempt to change their habitual exercises moderately (5.10 [360K]). As mentioned in chapter 1, in *On the Exercise with a Small Ball*, Galen promotes what he thinks is the most beneficial exercise: exercise with a small ball! One can easily procure a small ball, and this exercise

---

in *Galen: Hygiene; Books 1–4*, ed. and trans. Ian Johnston, LCL (Cambridge, MA: Harvard University Press, 2018), 1:xxxiv.

62. 4.4 (261K); 5.7 (342–343K).

63. 4.6 (275–276K); 5.5 (335–336K); 6.3 (392K).

moves all the parts of the body equally (although Galen says nothing about what the exercise is like); and it benefits both body and soul with pleasure when it is done in moderation (2–3 [901–906K]). These are some major ways that Galen promoted a healthy regimen based on moderation.

Thus, rational medicine as represented by the Hippocratic Corpus and Galen, while not incompatible with temple medicine, provided a natural means of therapeutic treatment and prophylactic measures. As we will see in the next chapter, early church leaders, in their development of Christian health care, generally accepted and were familiar with the therapeutics of Hippocratic and Galenic medicine.

Roman Hospitals (Valetudinaria)

For a relatively short period, the Romans built hospitals (*valetudinaria*)[64] for slaves and soldiers—the two types of workers considered most necessary to the political economy of the empire.[65] Wealthy owners favored slave hospitals (special shelters or dormitories usually in the countryside) to treat exhausted or injured slaves, mainly from the first century BCE to the end of the first century CE. In times of diminishing supply of slave labor and price hikes for slaves, maintaining healthy slaves by providing medical and rehabilitative care was, as expressed by the prominent agricultural writer Columella, advantageous to the owners: "Attention of this kind is a source of kindly feeling and also obedience. Moreover, those who have recovered their health, after careful attention has been given them when they were ill, are eager to give more faithful service than before."[66] However, when this main labor force became free tenant farmers (*coloni*), living in their peasant households, slave hospitals became no longer advantageous or necessary; we no longer hear about them after 80 CE.[67]

---

64. See, for examples, Guenter B. Risse, *Mending Bodies, Saving Souls: A History of Hospitals* (Oxford: Oxford University Press, 1999), 38–68; Ido Israelowich, "Medical Care in the Roman Army during the High Empire," in *Perspectives on Popular Medicine in Classical Antiquity*, ed. William V. Harris (Leiden: Brill, 2016), 215–30; Peregrine Hordern, "The Earliest Hospitals in Byzantium, Western Europe, and Islam," *Journal of Interdisciplinary History* 35, no. 3 (2005): 372–73. This section draws on these sources.

65. Hordern, "The Earliest Hospitals," 372.

66. Columella, *Rust.* 1.8.

67. Hordern, "The Earliest Hospitals," 372.

Then came military hospitals, from the time of Emperor Augustus to the mid-third century CE. During this period, military *valetudinaria* were mainly located in the northern imperial frontier at the rivers Rhine and Danube (now Germany and Hungary), where the Roman legions tried to mitigate adverse climatic conditions and make up for the lack of other local infrastructure by erecting more permanent fortresses, which included buildings for sick and injured soldiers to recover.[68] These hospitals had "lavatories, baths, a kitchen, rooms for medical supplies, and treatment rooms" for the patients and the military physicians (*medici*).[69] According to the former army officer and historian Velleius Paterculus, medical care in the hospitals included "treatment on the battlefield itself, safe evacuation of the wounded, a healthy diet, and sanitary facilities."[70] Nonetheless, "in the third century, the army was reorganized, and a local militia, supported by a mobile field army, defended the frontier, thus rendering the construction of fortress hospitals unnecessary."[71] Thereafter, sick soldiers were mainly treated by their family members, just as sick tenant farmers were typically cared for by their family or relatives; thus there exists no evidence of a military hospital after 250 CE.[72]

As we can see, at certain periods of time, these Roman hospitals had a specific purpose of offering (rational) medicine only for targeted groups who served a utilitarian function in the empire; there were no hospitals available for ordinary people, let alone the poor. While these hospitals employed military physicians, physicians in general did not operate out of hospitals as a rule. Everyday folks, however, had still other means to seek health care for themselves.

Popular Medicine and Magic (Medical Pluralism)

As significant as they were, religious healing (temple medicine) and formal medicine practiced by physicians (*iatroi*) like the Hippocratics and Galen were only a part of the Greco-Roman health-care system. As medical historians and medical anthropologists note, Greco-Roman patients partic-

---

68. Risse, *Mending Bodies, Saving Souls*, 48.
69. Israelowich, *Patients and Healers*, 102.
70. As mentioned in Israelowich, "Medical Care," 218.
71. Hordern, "The Earliest Hospitals," 372.
72. Hordern, "The Earliest Hospitals," 372.

ipated in "an open medical marketplace" consisting of various religious, empirical, medical, and magical practitioners.[73] They could start with self-treatment, utilizing the many herbs and plants available in gardens, fields, and mountains; they could also consult root cutters and drug sellers, especially if they lived in the countryside.[74] If they were female, they would most likely receive maternity care from midwives.[75] In addition to visiting temples or seeing doctors, depending on the patients' interpretations of the etiologies of their illness, they could resort to "other healers" such as exorcists, diviners, or astrologers; they could also, in their hopes for healing, obtain magical amulets, charms, and incantations readily available from "other healers" or itinerant sellers who traveled from one village to another.[76] In the world of Aristides, besides physicians, gymnastic trainers (*paidotribai*) and the temple wardens of the Pergamene asclepieion also served as appropriate health-care workers whom Aristides himself consulted—the former for regimen, and the latter for dream interpretations, particularly interpretations of the god's prescription (e.g., *Or.* 48.69; 48.30, 31, 35, 47). To a certain Thrasybulus, Galen[77] addressed the question whether hygiene (health) belonged to the physician or the gymnastic trainer (Thrasybulus). Despite his expected conclusion that hygiene not only falls within the purview of the doctor but also that the methods of gymnastics trainers are often counterproductive to health, what this demonstrates is that the gymnastic trainers, among many health-care au-

73. Israelowich, *Patients and Healers*, 32. On popular medicine, see, for example, William V. Harris, ed., *Popular Medicine in Graeco-Roman Antiquity: Explorations*, Columbia Studies in the Classical Tradition (Leiden: Brill, 2016); Steven M. Oberhelman, ed., *Dreams, Healing, and Medicine in Greece: From Antiquity to the Present* (Surrey, UK, and Burlington, VT: Ashgate, 2013), 8–30; Nutton, *Ancient Medicine*, 274–77; Vivian Nutton, "From Medical Certainty to Medical Amulets: Three Aspects of Ancient Therapeutics," in *Essays in the History of Therapeutics*, ed. W. F. Bynum and V. Nutton (Amsterdam: Rodopi, 1991), 13–22; Vivian Nutton, "Healers in the Medical Marketplace: Towards a Social History of Graeco-Roman Medicine," in *Medicine in Society: Historical Essays*, ed. Andrew Wear (Cambridge: Cambridge University Press, 1992), 15–58; Richard Gordon, "The Healing Event in Graeco-Roman Folk-Medicine," in *Ancient Medicine in Its Socio-Cultural Context*, ed. Philip J. van der Eijk, H. F. J. Horstmanshoff, and P. H. Schrijvers (Amsterdam: Rodopi, 1995), 2:363–76.

74. Oberhelman, "Anatomical Votive Reliefs," 48.

75. Oberhelman, "Anatomical Votive Reliefs," 48.

76. Cf. Oberhelman, "Anatomical Votive Reliefs," 48.

77. For popular medicine in Galen, see Danielle Gourevitch, "Popular Medicine and Practices in Galen," in Harris, *Popular Medicine in Graeco-Roman Antiquity*, 251–71.

thorities, were dispensing health advice regarding regimen at that time. Hence, in this "medical pluralism," many divergent and overlapping but sometimes competing types of healing were mixed together, "with a range as broad as the social status of their practitioners and as the variety of doctrines that they held."[78]

Among these popular remedies, magical amulets, spells, and chants against disease were widespread. A papyrus of the third century CE (P. Oxy. 42.3068) testifies to the use of an amulet against a disease (tonsillitis) with specific instructions to the recipient in Roman Egypt: "The amulet against tonsillitis, for the gold plate, send it to Sarmates, having copied it on a slip of papyrus word by word."[79] The Roman historian Marcellus of Bordeaux, in the fourth century, described some sixty-five incantations, charms, and magical formulas as intrinsic parts of medical therapy, ranging from prayers, simple commands, and nonsensical rhymes to ceremonies accompanied by recitation, spitting, or physical action (e.g., *De medicamentis* 12.24; 8.171, 191).[80] Moreover, Marcellus included guidelines for various amulets, phylacteries, and rings in averting or curing a disease (1.41; 8.27, 45; 1.85; 2.7; 28.21). Marcellus's contemporary, the orator Libanius, consulted various soothsayers (*manteis*) for his own sickness in addition to the treatment by doctors in consultation with Asclepius; and he also attributed one of his severe attacks of disease to malicious magic used against him through a dead chameleon (*Or.* 1.139–143, 243–250). Thus, even in the period of increasing Christianization, these two committed pagan authors showed their responses to diseases in traditional and "popular" ways.

If you recall, in chapter 1, we covered the Hippocratic diagnosis of "uterine suffocation" by a "wandering womb" in the section on women's diseases.[81] Although Galen and Soranus rejected the "wandering womb" of the Hippocratic Corpus (since the womb was securely anchored in the abdomen by ligaments), they still accepted the diagnosis of uterine suffocation, whose symptoms were very much like epilepsy. According to Christopher Faraone, this belief of uterine suffocation and a wandering womb was also circulating among the popular healers, who attributed its cause to gods, demons, and other supernatural forces infiltrating the human body;

---

78. Nutton, *Ancient Medicine*, 277.

79. I. Andorlini, "Crossing the Borders between Egyptian and Greek Medical Practice," in Harris, *Popular Medicine in Graeco-Roman Antiquity*, 171.

80. Cf. Jerry Stannard, "Marcellus of Bordeaux and the Beginnings of Medieval Materia Medica," *Pharmacy in History* 15, no. 2 (1973): 50.

81. See p. 47.

this etiology then generated a new therapeutic method of exorcism through magical amulets and spells.[82] For example, the earliest Greek magical inscription against the wandering womb from Beirut (*Greek Magical Amulets*, no. 51) comes from the first century BCE or the first century CE: "I adjure (*exorkizo*) you, womb of *Ipsa* [herself], whom *Ipsa* [herself] bore, in order that you never abandon your place, in the name of the lord god (*kyrios theos*), the living, the unconquerable: remain in your spot."[83]

Another recipe against uterine suffocation from a Greek magical handbook from Upper Egypt comes from the third or fourth century CE (*PGM* 7.260–71):

For the ascent (*anadromê*) of the womb:

I adjure you, womb, [by the] one established over the abyss, before heaven, earth, sea, light or darkness came to be, who created the angels, foremost of whom is AMCHAMCHOU and CHOUCHAÔ CHERÔEI OUEIACHÔ ODOU PROSEIOGGÊES, and who sits over the Cherubim, who bears his own throne: return again to your seat and do not lean into the right part of the ribs nor the left part of the ribs, nor bite into the heart, like a dog, but stop and remain in your proper place without chewing as long as I adjure you by the one who in the beginning made heaven and earth and all that is therein. Hallelujah! Amen![84]

As one can see, these amulets share similar features with contemporary medical texts that discussed uterine suffocation. The first amulet from Beirut seems to reflect the Hippocratic understanding of the "mobile" or "wandering womb" in its command; the second amulet from Upper Egypt, which appears to invoke Jewish Yahweh as the creator of the universe, seems to reflect the more recent perspectives of Galen and Soranus, which limited the movement of the womb within the abdomen area.[85] Therefore,

82. Christopher A. Faraone, "Magic and Medicine in the Roman Imperial Period: Two Case Studies," in *Continuity and Innovation in the Magical Tradition*, ed. Gideon Bohak, Yuval Harari, and Shaul Shaked (Leiden: Brill, 2011), 138–39.

83. Faraone, "Magic and Medicine," 139. The names in the quotation, by themselves, do not indicate that the amulet is Jewish or Christian, as they often appear in pagan magical texts.

84. Faraone, "Magic and Medicine," 142.

85. Cf. Faraone, "Magic and Medicine," 143–44.

in contrast to the assumption that these magical recipes were evidence of "superstition" antithetical to the Greek medical tradition, the exorcists who devised these amulets rather "seem to be literate persons, who share with their medical rivals a number of key ideas and strategies."[86] This indicates a more blurred boundary between formal (rational) medicine and popular medicine in the Greco-Roman medical marketplace.

Therapy for the Soul (Medico-Philosophical Treatises)

As we now discuss the medico-philosophical therapy for the soul, we recall, from chapter 1, Galen's approach to the psychological affections and the cure (*therapeia*) of emotions in the manner of both Stoic-Epicurean and Platonic-Aristotelian ways. These schools had contrasting ideals. The Stoic-Epicurean school, which originally developed the genre of therapy of emotions, argued that one should seek total freedom from emotions (*apatheia*). The Platonic-Aristotelian school, adopting the genre, argued not for the elimination of emotions but for their moderation (*metriopatheia*). However, Galen converged these two approaches, rather uneasily, by blending Stoic-Epicurean rational or cognitive methods (i.e., replacing false beliefs with true beliefs or knowledge) with a Platonic-Aristotelian psychological framework, a framework in which emotions are located in a nonrational part of the psyche.[87] Our task now is to examine further the medico-philosophical therapy for the soul/emotions based on the prevalent analogy between medicine and philosophy.

The notion that philosophers provide treatment for the soul, which is analogous to the therapy provided by physicians for the body, features prominently in Plato (e.g., *Gorg.* 475d, 505c; *Rep.* 444c–e; *Tim.* 86d–90d).[88] Epicurus also describes his philosophy as a therapy for the soul: "Vain is the word of the philosopher by which no human suffering is healed. Just as medicine confers no benefit if it does not cast out bodily disease, neither is there from philosophy if it does not drive away suffering from the soul" (*Fr.* 54; cf. *Fr.* 221). Based on this analogy, the Stoics and Epicureans developed philosophical therapy for the soul thoroughly during the Hellenistic era.

---

86. Faraone, "Magic and Medicine," 143, 144.

87. Christopher Gill, *Naturalistic Psychology in Galen and Stoicism* (Oxford: Oxford University Press, 2010), 264.

88. Cf. Gill, "Philosophical Therapy," 343.

According to the Stoic Chrysippus in his (lost) fourth book of *On Passions* (*Peri pathōn*),[89] philosophical health and therapy parallel medical health and therapy. In contrast to the dominant Platonic-Aristotelian view that emotions (passions)—with extreme emotions being an exception—are a normal aspect of the human experience,[90] he identifies emotion as a "'rejection' of reason, as a form of psychological weakness or 'lack of tension' (*atonia*), and madness or psychological blindness" or "feverishness."[91] Given this notion of emotion as a form of psychological disease due to false beliefs or judgments, he encourages his readers to see their emotions as being in need of cognitive therapy.[92] For Chrysippus, this understanding of emotion is in fact a part of the therapeutic process and the beginning of "acquiring a better and more 'healthy' belief-set";[93] for "changes in belief or understanding about key ethical ideas necessarily carry with them changes in motivational pattern or attitude."[94] Therefore, while the philosopher-therapist has to take account of the emotional state of the person treated,[95] one's recognition of the need for psychic cure and of the roots of one's sickness in itself "provides the basis for self-improvement and the development towards a more 'healthy' state."[96] The Epicurean Philodemus (first century BCE), in his guidelines for philosophical therapy, likens using frank speech (*logos*) to "applying a scalpel to those who are ill" (*Lib.* col. 12a) and compares its "suitable use" to the physician's discernment of disease from reasonable signs (*sēmeiōsamenon*) and the adaptation of remedies to specific cases (*Lib. Fr.* 57).[97] As in medical therapy, the same treatment would not

89. See Teun Tieleman, *Chrysippus'* On Affections*: Reconstruction and Interpretation* (Leiden: Brill, 2003).

90. Gill, *Naturalistic Psychology*, 285. On different understandings of *pathos* between the Platonic-Aristotelian tradition and the Stoic tradition, see Kallistos Ware, "The Meaning of 'Pathos' in Abba Isaias and Theodoret of Cyrus," *Studia Patristica* 20 (1989): 315–22.

91. Gill, *Naturalistic Psychology*, 281. See also William V. Harris, *Restraining Rage: The Ideology of Anger Control in Classical Antiquity* (Cambridge, MA: Harvard University Press, 2001), 370.

92. Gill, *Naturalistic Psychology*, 281; Harris, *Restraining Rage*, 370.

93. Gill, "Philosophical Therapy," 344.

94. Gill, *Naturalistic Psychology*, 288.

95. Gill, *Naturalistic Psychology*, 281.

96. Gill, *Naturalistic Psychology*, 288.

97. Cf. Paul R. Kolbet, *Augustine and the Cure of Souls: Revising a Classical Ideal* (Notre Dame: University of Notre Dame Press, 2010), 43.

be appropriate for every person; rather, treatment should be accommodated, at the appropriate moment, to the diverse needs of different types of souls: the strong, the weak, those of upper status, those of lower status, women, and the elderly (7, 10).

Philodemus's emphasis on frank speech as a philosophical therapy highlights speech or writing (*logos*) as a central tool in remedying the sick soul. Already in Plato's *Phaedrus*, Socrates defines rhetoric as an art of guiding souls (psychogogy) by means of words (*psychagōgia tis dia logōn*) in the service of philosophy (261a), and likens rhetoric to "the method of the art of healing" (270b). In physical healing, a doctor analyzes the body in order to prescribe proper medicine and diet for health and strength; in rhetoric, an orator/philosopher analyzes the soul in order to give it "the desired belief and virtue" through "proper discourses (*logous*) and training" (270b). Therefore, the orator/philosopher should be able to "describe the souls with perfect accuracy . . . and classify the speeches (*logōn*) and the souls (*psychēs*) and . . . adapt each to the other, showing the causes of the effects produced and why one kind of soul is necessarily persuaded by certain classes of speeches (*logōn*), and another is not" (271b). According to Martha Nussbaum, this analogy between *logos* and medical treatment, with *logos* being understood as a powerful and sufficient treatment of the diseases of the soul, goes all the way back to Homer; and by the Hellenistic era, therapeutic *logos* came to mean philosophical orations and writings in particular, and, with degrees of adaptability, was used as a drug (*pharmakon*) or surgical procedure for the healing of the soul.[98]

This significance of *logos* in changing false beliefs in the therapeutic process of the diseased soul corresponds to Christopher Gill's threefold distinction in the writings of ancient philosophical therapy: protreptic, therapy, and advice.[99] They are typically "seen as having interrelated functions" and form a well-designed therapeutic process toward health. "Protreptic offers encouragement to undergo therapy"[100] by "demonstrating [the] value and [the] benefit of a particular philosophy and defending objections of the adversaries in refutation."[101] Therapy "removes false beliefs that

---

98. Martha C. Nussbaum, *The Therapy of Desire: Theory and Practice in Hellenistic Ethics* (Princeton: Princeton University Press, 1994), 49–53.

99. Gill, "Philosophical Therapy," 342.

100. Gill, "Philosophical Therapy," 342.

101. Helen Rhee, *Early Christian Literature: Christ and Culture in the Second and Third Centuries* (London: Routledge, 2005), 24.

produce psychological sicknesses" and is strongly tied to medicine. Finally, "advice replaces the false beliefs with true, or at least better-grounded, ones."[102] These three activities function like regimen (*diaita*) of medical practices, that is, "life-style management" that is amenable to personal control, particularly diet, exercise, choice of environment, and state of mind or mood;[103] and we can see Chrysippus's writing above already having these three elements as a form of cognitive therapy. By the first and second century CE then, the notion of philosophy as therapy was well established in the Greco-Roman world.

In light of this threefold distinction, we can better understand Seneca's *On Anger* (*De ira*) in the early imperial period. First of all, using protreptic, Seneca presents anger as a diseased state in need of a cure at the outset and throughout the book: a "wholly violent," inherently "ugly and horrible picture of distorted and swollen frenzy," and a "cruel" phenomenon with barbaric behavior, from which no one is safe.[104] In that process, he explicitly challenges and undermines the competing Aristotelian understanding of emotions (anger), which sees their (its) usefulness with moderation (*metriopatheia*).[105] Having established the need to see anger as the sickness of the soul, something contrary to nature, Seneca provides the prospect of therapy (*remedia*): "The best course is to reject at once the first incitement to anger, to resist even its small beginnings, and to take pains to avoid falling into anger" (1.8.1). Throughout the book, Seneca counters false beliefs that see anger as necessary or useful with the examples of the evil of anger from Roman history.[106] Furthermore, Seneca provides an extensive therapy (*remedia*), which advises the reader "not to fall into anger, and in anger to do no wrong," advice which, for Seneca, should be applied throughout one's life (2.18.1). Doing so requires the formation of habit especially during the period of education, and is dependent on one's temperament. For example, the fiery temperament should avoid wine and overeating; exercise vigorously, to a point just short of exhaustion; and play games. The more moist, drier, and colder natures must beware of faults such as fear, moroseness, discouragement, and suspicion; they need encouragement, indulgences, and calls to cheerfulness (2.20). Here Seneca advises

---

102. Gill, "Philosophical Therapy," 342–43.
103. Gill, "Philosophical Therapy," 340, 346–47.
104. E.g., Seneca, *Ira* 1.1–4; 2.35–36; 3.17–21.
105. E.g., Seneca, *Ira* 1.7, 9–10.4; 1.11.1; 1.12.1–3; 1.17.1; 3.3.
106. E.g., Seneca, *Ira* 1.11, 12, 14, 16, 18, 20; 2.2, 10, 12, 14, 15.

different, even contrary, remedies dependent on the type of soul (2.20). Moreover, he identifies credulity as "a source of very great mischief" and offers extended advice on how to shape one's own mind: "believe only what is thrust under [your] eyes and becomes unmistakable" and "develop the habit of being slow to believe"; don't be annoyed by trifles, things, or people that have no ability to harm you (2.24, 22–27; cf. 2.29). Since "the best corrective of anger lies in delay," delay angry reactions in order to completely conquer anger (2.29). In book 3, Seneca advises readers to discern what irritates them since "people are not all wounded at the same spot"; therefore, he says, "you ought to know what your weak spot is in order that you may especially protect it" (3.10). However, they should understand that all the causes of anger, including foods, drinks, insulting words, disrespectful gestures, etc., really amount to little; "not one of them, though we take them so tragically, is a serious matter, and not one is important," after all (3.34). Hence, we should "give to the soul that peace which is afforded by constant meditation on wholesome instruction (*praeceptorum salutarium*), by noble deeds, and a mind (*mens*) intent upon the desire for only what is honorable" (3.41); for before you know it, death is at hand (3.43).

If Seneca's *On Anger* is an example of the Stoic approach to philosophical therapy of the soul, Galen's *Avoiding Distress* (*De indolentia*)[107] presents a more Platonic-Aristotelian approach while still incorporating the Stoic-Epicurean elements in the first half of the essay. It is in the form of a letter to an unknown man who wants to know "what sort of training (*askēsis*) or arguments (*logoi*) or doctrines (*dogmata*)" prepared Galen never to be distressed (*lypeisthai*), especially in light of the magnificent loss of his personal possessions (books, drugs, instruments, and other resources needed for his medical practice) in the great fire at Rome in 192 CE (1). Galen responds to this query first by recounting those losses in detail and then by explaining the rational or cognitive basis of his magnanimity in Stoic-Epicurean

---

107. For studies on Galen's *Avoiding Distress*, see, for instance, Christopher Gill, "Galen's Περὶ Ἀλυπίας as Philosophical Therapy: How Coherent Is It?" in *A Tale of Resilience: Galen's* De Indolentia *in Context*, ed. C. Petit (Leiden: Brill, 2018), 135–54; Teun Tieleman, "Wisdom and Emotion: Galen's Philosophical Position in *Avoiding Distress*," in Petit, *A Tale of Resilience*, 199–216; L. Michael White, "The Pathology and Cure of Grief (λύπη): Galen's *De indolentia* in Context," in *Galen's* De indolentia*: Essays on a Newly Discovered Letter*, ed. Clare K. Rothschild and Trevor W. Thompson (Tübingen: Mohr Siebeck, 2014), 221–50; David H. Kaufman, "Galen on the Therapy of Distress and the Limits of Emotional Therapy," *Oxford Studies on Ancient Philosophy* 47 (2014): 275–96.

terms. Galen goes on to describe a courageous indifference to misfortune (*apatheia*) or absence of disturbance (*ataraxia*) on the one hand, and in Platonic-Aristotelian terms, a sense of moderation (*metriopatheia*) of passion, on the other. In the first part of therapy, Galen stresses "enduring everything that happened" by focusing not on what has been lost but on what is sufficient to meet one's expenses (38, 44, 46) as a way to critique and avoid insatiability (*aplēstia*) (39–48, 79–84). Quoting Euripides, Galen says he has trained his imagination to prepare for the total loss of everything he owns, and he advises, "train your soul's imagination to cope with almost any turn of events" (54–56). However, it is at this point that Galen moves to endorse the Platonic-Aristotelian framework of the role of other less rational or nonrational factors such as inborn nature (*physis*), family upbringing, and education as he says that "this training is only an option for those predisposed to courage by nature and beneficiaries of the finest education" like himself in his own family background (57, 58–65). In fact, Galen distances himself from the Stoic-Epicurean ideal: "Some suppose that freedom from disturbance is the good, although I know that neither I, nor any other human being, nor any living creature exhibits" (68; cf. 73–74); no complete invulnerability to external losses is possible, even for a philosopher (70, 71). Here he comes back to the issue of insatiability (*aplēstia*) and argues for developing a realistic picture of having a sufficient level of physical and psychological health with adequate material possessions, not nonpossessions, to avoid distress (79–84). On the basis of his experience, he prefers the Platonic-Aristotelian moderation of one's emotions through natural talent, habituation, and education (81, 83). In this way Galen provides a way of thinking and resources that enable his readers "to strengthen [their] emotional resilience and the capacity to withstand what he sees as disasters."[108]

Conclusion

When it comes to health care in the Greco-Roman world, the sick used multiple approaches, beginning with self-treatment and moving on to alternatives of religious medicine, rational medicine, and popular medicine based on the nature and etiology of their illness, "prior personal or family experience, beliefs, local opportunities, as well as financial consid-

---

108. Gill, "Philosophical Therapy," 352.

erations."[109] Religious healings through the cult of Asclepius in particular provided the fundamental and steady context for the hope and expectation of extraordinary healing in a basically religious society. Rational medicine's therapeutic triad of medicine, incision, and cauterization offered specialized treatments of diseases in conjunction with temple medicine, and its regimen presented prophylactic measures for the preservation of health. Then, at a certain period of time, Roman hospitals proffered collective medical care to sick slaves and soldiers, who were indispensable to the operation of the empire. Popular medicine such as magical amulets filled any gap in "the open medical marketplace" of the Greco-Roman world. Finally, medico-philosophical treatises, based on medical analogies, afforded therapies of the soul by applying the appropriate *logos* in the forms of protreptic, therapy, and advice. This plurality of therapeutics corresponded to the plurality of etiologies available in Greco-Roman culture and reflected both socially constructed and efficacious means within Greco-Roman society. In this very context early Christians would establish their adopted and transformed health care for their own identity formation, and also for the shaping of the larger late antique society. To this, we now turn.

---

109. Guenter B. Risse, "Asclepius at Epidaurus: The Divine Power of Healing" (lecture delivered on May 13, 2008, updated March 10, 2015), 17. For all types of healing options in Greco-Roman society, see Avalos, *Health Care*, 75, 77–79, 81.

HEALTH CARE IN EARLY CHRISTIANITY

In this final chapter we will examine Christian health care in the context of Greco-Roman health care. I will divide the development of Christian health care into two periods: the second and third centuries, when Christians developed their engagement with Greek medicine, identified the care of the sick as one of their primary ministries, and sought religious healing; and the fourth and fifth centuries, when Christianity, having become known as the "religion of healing" par excellence, established hospitals and monastic infirmaries with limited medical care and practiced private health care through both religious healing and rational medicine. In my treatment of these developments, I will show both the embedded nature of Christian health care within Greco-Roman health care and the distinct contributions of Christian health care to Christian identity formation in the late antique period. Early Christian health care was thoroughly imbued with Greco-Roman health care, but at the same time, Greco-Roman health care was transformed by and within early Christianity.

Christian Health Care in the Second and Third Centuries

From the rich context of the variety of Greco-Roman health care, we turn to Christian health care before the time of Constantine, starting with Christian attitudes toward (Greek) medicine.

(Rational) Medicine

Some Christian apologists of the second and third centuries regarded rational medicine positively as they engaged Greco-Roman culture and philosophy alongside their own developing Christian theology. In North Africa, Tertullian saw medicine as God's gift: "Let Asclepius have been the first who sought and discovered the cures. Isaiah mentions that he ordered medicine for Hezekiah when he was sick. Paul also knew that a little wine was good for the stomach" (*Cor.* 8.2).[1] According to Clement of Alexandria, "health by medicine" had its origin and existence as a consequence of divine Providence as well as human cooperation (*Strom.* 6.17). He also considered medicine, "the healing art" of curing bodily diseases, as the highest of the arts acquired by human skill (*Paed.* 1.2.6), and, as mentioned in chapter 2, presented many examples of regimen throughout his writings. ⟶ ?SPECULATIVE THEOLOGIAN

Origen believed that the knowledge of medicine was a direct gift from God to humanity, one that existed alongside the gift of spiritual healing (*Hom. Num.* 18.3). Just as God provided medicines for humanity's bodily illnesses, so did he provide medicines for the soul through the healing truths of Scripture (*Hom. Ps.* 37.1.1). Christ was then "the physician (*iatros*) of soul and of body," healing the leper in Matthew (cf. Matt. 8:3), for example, in two ways, not only from physical leprosy but also from spiritual leprosy "by his truly divine touch" (*Hom. Lev.* 7.1; *Cels.* 1.48). However, when it came to using medicine for healing, Origen established a two-tiered system: "A man ought to use medical means to live in the simple and ordinary way. If he wishes to live in a way superior to that of the multitude, he should do this by devotion to the supreme God and by praying to Him" (*Cels.* 8.60). Origen often distinguished the ordinary and the perfect Christian in his writings through physical and spiritual parallels. He thought that the more spiritual Christians should only resort to God for healing and refrain from using human medicines and physicians.[2] As Gary Ferngren notes, here Origen pioneered a principle later used by other Christian writers and ascetics: "namely, that medicine, however lawful (and indeed efficacious) for all Christians, might freely be renounced by those who seek a closer de-

---

1. For a more in-depth study of Tertullian and medicine, see Thomas Heyne, "Tertullian and Medicine," *Studia Patristica* 50 (2011): 131–74.

2. Gary B. Ferngren, *Medicine and Health Care in Early Christianity* (Baltimore: Johns Hopkins University Press, 2009), 27.

pendence on God and who look for bodily healing through prayer alone."[3] We will see multiple examples of this attitude later in this chapter.

Care of the Sick[4]

During this period, Christians provided sacrificial care for the sick as they knew how, even as they prayed for healing and exercised the gift of healing, which we will discuss shortly. According to the *Apostolic Tradition*, visiting the sick was one of the tests for every catechumen, measuring one's readiness for baptism (*Trad. ap.* 15; 20.1); providing meals for the sick, along with widows and the poor, was a common duty of all the faithful (24). Among the clergy, deacons (and deaconesses, especially in the East), in particular, were charged with the care of the sick and the infirm, informing the bishops about their needs so that they could be visited and prayed over by the "high priests," that is, bishops (34).[5] We also have Bishop Polycarp exhorting presbyters to be compassionate and to "visit all the sick" in his letter to the Philippians (*Phil.* 6.1).

These overlapping practices and their diverse participants created concentric circles of care for the sick. These practices prepared Christians to meet the overwhelming challenge of an epidemic in the mid-third century in the most impressive way. Several contemporary accounts from Carthage and Alexandria enlighten us with what the infected endured and how Christians cared for them even in the midst of their own suffering. The deadly plague, which had originated in Ethiopia in 250 CE, spread quickly throughout Egypt and North Africa, and from there to Italy and to the West (as far as Scotland), as the pandemic lasted fifteen to twenty years in intervals.[6] The symptoms of the plague included diarrhea, sores on the jaws, continual vomiting, redness of the eyes, and in some cases, the loss of limbs and the impairment of hearing or eyesight (Cyprian, *Mort.* 14).

---

3. Ferngren, *Medicine and Health Care*, 27.

4. This section is drawn from Helen Rhee, *Loving the Poor, Saving the Rich: Wealth, Poverty, and Early Christian Formation* (Grand Rapids: Baker Academic, 2012), 128–31.

5. Cf. *Didasc.* 3.4; 16.3; cf. *Pseudo-Clem. Ep.* 12; *Ap. Const.* 3.19.

6. Michael M. Sage, *Cyprian*, Patristic Monograph Series (Cambridge, MA: Philadelphia Patristic Foundation, 1975), 269; Gary B. Ferngren, "The Organisation of the Care of the Sick in Early Christianity," in *Actes/Proceedings of the XXX International Congress of the History of Medicine*, ed. H. Schadewaldt and K.-H. Leven (Düsseldorf: Vicom KG, 1988), 193.

Its effects were extensive and destructive, with an estimated mortality rate higher than that of any epidemic previously known (Zosimus, *Hist. Nova* 1.26, 37; cf. 1.36, 46). At its height, it is recorded that more than five thousand succumbed to death in one day in Rome alone (*Hist. Aug. Vita Gall.* 5.6). While Cyprian reported from Carthage in 252 CE that "many of our people are dying from this mortality" (*Mort.* 15), his biographer Pontius further illustrated the ravaging and horrifying scene: "Countless people were seized daily in their own homes by a sudden attack; one after another the homes of the trembling crowd were invaded. Everyone shuddered, fled to avoid contagion, wickedly exposed their dear ones, as if along with the person who was about to die from the plague one could also shut out death itself" (*Vita Cypriani* 9).

Streets were filled not only with the corpses but also with the cries of the diseased and dying who were begging for help (*Vita Cypriani* 9). Dionysius also recounted the terror from Alexandria around 260 CE: "Now indeed all is lamentation, and all men mourn, and wailings resound throughout the city because of the number of dead and of those that are dying day by day" (Eusebius, *Hist. eccl.* 7.22.2). It affected Christians and non-Christians alike, though its fuller impact fell on non-Christians, as Dionysius reports (Eusebius, *Hist. eccl.* 7.22.6). Modern calculations based on Dionysius's account suggest that two-thirds of ancient Alexandria's population may have perished.[7]

The civil authorities called on the traditional gods by making sacrifices and customary supplications to placate their anger, but hardly did anything tangible to improve the situation.[8] Classical society lacked any organized program for the treatment of the sick on either a regular or an emergency basis[9] except for the case of slaves and soldiers (as mentioned in chap. 4). The Christian churches, on the other hand, established a rather systematic means of caring for the sick; these efforts were under the leadership of bishops who directed relief efforts. According to Pontius, Cyprian gathered his people for the whole city and gave them a theological grounding for their works of charity and mercy; based on Christ's examples and Scripture, their care of the diseased and the dying would gain merit with God

---

7. A. E. R. Boak, *A History of Rome to 565 A.D.*, 3rd ed. (New York: Macmillan, 1947), cited in Rodney Stark, *The Rise of Christianity: A Sociologist Reconsiders History* (Princeton: Princeton University Press, 1996), 77; cf. Eusebius, *Hist. eccl.* 7.21.9.

8. Gary B. Ferngren, "Medicine and Compassion in Early Christianity," *Theology Digest* 46 (1999): 318.

9. Ferngren, "Medicine and Compassion," 319.

and should be extended to Christians *and* non-Christians alike (*Vita Cypriani* 9). Cyprian instructed them to go beyond the practice of the "publican or heathen" and to seek perfection "by overcoming evil with good and by the exercise of a divine-like clemency, loving even [their] enemies, and by further praying for the salvation of [their] persecutors, as the Lord advises and encourages" (9). As Christians of all social ranks assembled to help, their care was offered "according to the nature of the men and their rank" (10). Many who, on account of their poverty, could not contribute wealth as a means of helping the poor provided their own precious labor, while the wealthy donated money (10). Under Cyprian's effective leadership, there was the generosity of overflowing works accomplished for all people, both Christians and non-Christians (10), and these activities of looking after the victims of the plague continued until Cyprian's exile during Valerian's persecution five years later (258 CE) (11).

"What Cyprian encouraged Christians to do, Dionysius said they did,"[10] especially stressing their all-inclusive and sacrificial care: "very many, indeed, of our brothers and sisters through their exceeding love and merciful kindness did not spare themselves but kept close to one another and cared for the sick without taking thought for themselves" (Eusebius, *Hist. eccl.* 7.22.7). These Christians ministered to the victims—again, both Christians and non-Christians—so earnestly and sacrificially that "many who tended the sick and restored them to health died themselves, having transferred to themselves the death that lay upon others" (7.22.7). As a result, Dionysius concludes, "the very best of the brothers and sisters among us departed life in this manner—presbyters, deacons, and some of the laity. They are exceedingly worthy of praise, since this kind of death, occurring because of great piety and a strong faith, seems in no way to be inferior to martyrdom" (7.22.8). Conversely, "the pagans acted the very opposite" by fleeing from and thrusting away those in the first stages of disease, even their own families, and by casting the half dead to the roads and treating them as refuse (7.22.10). Pontus describes similar actions in Carthage, as people attempted to avoid contagion.

When another plague ravaged the eastern part of the empire in 312, in addition to a famine, an outbreak of another disease ("anthrax"), and a war against Armenia by Maximin—all in the midst of the Great Persecution for Christians (at this point only in the East intermittently)—Christians served

10. Everett Ferguson, *Early Christians Speak: Faith and Life in the First Three Centuries*, vol. 2 (Abilene, TX: Abilene Christian University Press, 2002), 131.

the general population in a similarly selfless manner (9.8). According to Eusebius, "countless was the number of those who were dying in the cities, and still larger of those in the country parts and villages, . . . the entire population perished all at once through lack of food and through plague" (9.8.4–5). In that tragic state of affairs, Christians alone, Eusebius records, exhibited concrete evidence of their sympathy and humanity (*philanthrōpia*): "all day long" some would "diligently persevere in performing the last offices for the dying and burying them (for there were countless numbers, and no one to look after them); while others would assemble the multitude" who were famished all throughout the city and distribute bread to all of them "so that their action was on all men's lips, and they glorified the God of the Christians, and convinced by the deeds themselves, acknowledged that they alone were truly pious and God-fearing" (9.8.14).

The compassionate actions of the laity and clergy all accord with instructions about their respective roles and responsibilities found in other sources; and this instruction and conduct taken together highlight even more a portrait of contrasting conduct between Christians and non-Christians toward the victims of the epidemics (though not without idealizations and apologetic purposes). As shown by Rodney Stark, the Christian-organized care of the sick (however rudimentary it might have been), motivated by a higher love, was a factor in the greater survival rate of Christians versus non-Christians during the time of epidemics and other crises, which ultimately contributed to the growth of Christianity.[11] And as argued by Gary Ferngren, this Christian response to the plagues might have spurred the church to advance in its organized medical charity for those in need, eventually leading to the founding of permanent institutions such as hospitals in the late fourth century.[12]

Healing

We covered, in chapter 2, the theological significance of Jesus's healings in the Gospels and the healings by his disciples in Acts, and the phenomenon of religious healing in the Greco-Roman world in the previous chapter. Here we will explore the extent to which Christianity was seen and functioned as a "religion of healing" in its common Greco-Roman context, and how such healing functioned as a part of Christian health care in the second

---

11. Stark, *The Rise of Christianity*, 73–94.
12. See Ferngren, *Medicine and Health Care*, 113–39.

and third centuries. According to Ferngren, Christianity was not a "religion of healing" in the second and third centuries, as reflected in the paucity of specific healing accounts in apologetic literature.[13] While that remains largely true, the following "general" comments by Irenaeus, Tertullian, and the apocryphal Acts cannot easily be discounted.

Irenaeus makes explicit references to the contemporary practice of healing by the church in contrast to the "magical deceptions" of the Simonians and Carpocratians (*Haer.* 1.23.1). They "can neither confer sight on the blind, nor hearing on the deaf, nor chase away all sorts of demons. . . . Nor can they cure (*curare*) the weak, or the lame, or the paralytic, or those who are distressed by any other part of the body, as has often been done in regard to bodily infirmity" (2.31.2). In contrast, in the church "sympathy, and compassion, and steadfastness, and truth, for the aid and encouragement of [hu]mankind, are not only displayed without fee or reward, but we ourselves lay out for the benefit of others our own means; and inasmuch that those who are cured (*curantur*) very frequently do not possess the things they require, they receive them from us" (2.31.3). Indeed, the true Jesus followers (i.e., the church), over against heretics and non-Christians, perform miracles, drive out devils, see visions, and prophesy (2.32.4). "Others still, heal the sick (*iaomai*)[14] by laying their hands upon them, and they are made whole" and even raise the dead (2.32.4). "For as [church] has received [the gifts: *charismata*] freely from God, freely also does she minister [to others]" (2.32.4).[15] In the larger context of this treatise, Irenaus here points to these contemporary miraculous healings as evidence of the authenticity of Jesus's own miraculous healings.[16] Tertullian also confirms the contemporary efficacy of healing in North Africa with his rhetorical statement: "How many men of rank (to say nothing of common people) have been delivered from devils, and healed of diseases!" (*Scap.* 4). It indeed implies many were exorcised and healed. In fact, even the emperor

13. Ferngren, *Medicine and Health Care*, 64–75. See also Thomas Heyne, "Were Second-Century Christians 'Preoccupied' with Physical Healing and the Asclepian Cult?" *Studia Patristica* 44 (2010): 63–69.

14. Cf. Justin, *Dial.* 39.2; Origen, *Cels.* 3.24; Hippolytus, *Trad. ap.* 14.

15. See also *Haer.* 2.32.5: "If therefore the name of our Lord Jesus Christ even now confers benefits [upon people], and cures (*curat*) thoroughly and effectively all who anywhere believe on Him."

16. See *Haer.* 5.12.5–6; 2.17.9; 4.8.2. Cf. Andrew Daunton-Fear, *Healing in the Early Church: The Church's Ministry of Healing and Exorcism from the First to the Fifth Century*, Studies in Christian History and Thought (Eugene, OR: Wipf & Stock, 2009), 59.

Septimius Severus himself "was graciously mindful of the Christians; for he sought out the Christian Proculus, surnamed Torpacion, the steward of Euhodias, and in gratitude for his having once cured him by anointing, he kept him in his palace till the day of his death" (4).

These intentional witnesses to contemporary Christian healings are significant, as these apologists were consciously critical of Asclepius and his healings. Tertullian, using Pindar, criticized Asclepius's avarice and love of gain in his "unlawful use of his medical art" (*Apol.* 14.5–6); his older contemporary Justin Martyr, while acknowledging similarities between the healings by Jesus and Asclepius, attributed the latter to the devil (*1 Apol.* 1.22.6; *Dial.* 69.3); his younger contemporary Origen also attributed Asclepius's healings to evil spirits, aided by "Egyptian magic and spells" (*Cels.* 3.24–25, 36).

We encountered the significance of healing (and exorcism) in the apocryphal Acts when we discussed demonic etiologies in chapter 2. Contemporaneous with Irenaeus and Tertullian, the Acts of Paul highlights the following complex healing episode: at Myra Hermocrates, a man with dropsy, shares his confidence in the God of healing with Paul, and is made whole when Paul promises healing through the name of Jesus Christ. While he and his wife "receive the grace of the seal" (i.e., baptism), their son Hermippus and his companions rise up against Paul for having cured his father; for he had hoped to inherit his father's property after his death. In contrast to his sibling, Hermippus's younger brother Dion hears Paul gladly, but dies; upon the plea of his mother, Nympha, Paul raises Dion from the dead. Hermippus, as punishment for his hostility against Paul, becomes blind. As he repents, Paul and the others pray for him, and his sight is restored by a vision in which Paul comes and lays his hands on him (4). Notable in this account is healing in the name of Jesus and healing also by the laying on of hands, both of which lead to the conversion of those healed. In a summary fashion, the Acts of Peter states that a multitude of Romans "brought the sick to him [Peter] on the Sabbath and asked him to treat them. And many paralytics and podagrous were healed, and those who had two- and four-day fevers and other diseases, and believed in the name of Jesus Christ" (31). Again, the Acts of Thomas speaks of Thomas's fame that "spread over all the cities and villages, and all who had sick persons or such as were troubled by unclean spirits brought them to him . . . and he healed all by the power of the Lord" (59). Then those healed and freed from demons praised Jesus and decided to follow him: "Glory to you, Jesus who in like manner has given healing to all through your servant and apostle Thomas . . . we pray that we may become members of your flock and be counted among your sheep. Receive us, therefore, O Lord" (59). In the polemical context of these acts against paganism, Jesus Christ, whose power is mediated through the apostles, overcomes demons and diseases, and people respond to this power by worshiping Jesus as the only true God.[17]

What is more, a healing account at the end of the Acts of Thomas is connected with Thomas's grave (170). King Misdaeus directed the killing of Thomas. A long time after Thomas's death, Misdaeus's son becomes a demoniac, and no one can cure him. Misdaeus considers taking Thomas's bone from his grave for the healing of his son. However, when he opens the grave, he finds the bones gone (since a Christian took them away to Mesopotamia)! Then Misdaeus takes dust from the same grave and puts it around his son's neck, saying: "I believe in you, Jesus Christ, now that he has left me who troubles men and opposes them lest they should see you" (170). The boy then becomes healed. Here we see that, as early as the third century, a belief in the miraculous power of martyrs' relics emerged, and that with the confession of faith in Christ, this belief brings about healing and anticipates important development in subsequent centuries.

Christian Health Care in the Fourth and Fifth Centuries

Beginning in the fourth century, Christianity, the "religion of healing," practiced private health care through both religious healing and rational medicine and established hospitals and monastic infirmaries. We take up these phenomena in turn.

Healing (and Magic)

However removed the worldview of the apocryphal Acts was from mainstream Christianity in the second and third centuries, their perspective in fact became commonplace in the fourth and fifth centuries. As Ferngren acknowledges, "more instances of miraculous healing are reported from the fourth century than from the three preceding centuries combined";[18] and he attributes this to the rise of asceticism, the cult of saints,[19] and the credu-

17. At the same time, in these healing stories, including the conversions of those who were healed, the parallel between the main apostles in the respective apocryphal acts and Apollonius of Tyana, who were also both accused of magic/sorcery, is unmistaken.

18. Ferngren, *Medicine and Health Care*, 76.

19. On this topic, see Peter R. L. Brown, *The Cult of the Saints: Its Rise and Function in Latin Christianity* (Chicago: University of Chicago Press, 1982). On holy men as

lity of the age in the process of Christianization of the Roman Empire that began with Constantine.[20] As Owsei Temkin notes, the healing power of the ascetic holy men was "perhaps the commonest manifestation of Holiness."[21] We see this aspect of healing by holy man in Athanasius's *Life of Antony*. Through Antony "the Lord healed many of those present who suffered from bodily ailments" (14), writes Athanasius, setting a certain expectation of the relationship between Antony's spiritual prowess and healings early on. And yet he also lays down a fundamental principle in Antony's healing miracles: "He [Antony] encouraged those suffering (*tous paschontas*) to have patience and to know that healing belonged neither to him nor to men at all, but only to God who acts whenever he wishes and for whomever he wills. . . . And the ones who were cured (*hoi therapeuomenoi*) were taught not to give thanks to Antony, but to God alone" (56). "Antony's healing" is, as a matter of fact, God's healing, not Antony's; and this profession is a further proof of his holiness. Again, "Antony did, in fact, heal without issuing commands, but by praying and calling on the name of Christ, so it was clear to all that it was not he who did this, but the Lord bringing his benevolence to effect through Antony and curing those who were afflicted (*therapeuōn tous paschontas*)" (84). Enclosed in this principle front and back are specific case stories of healing: Fronto, who "bit his own tongue and was about to lose his eyes" (57); a girl with a terrible ailment and paralysis (58); Polycratia, a female ascetic who had grown weak and developed severe stomach pains as a result of her stringent *askēsis* (61)—all were freed from their ailments from a distance (by Antony's prophetic pronouncements on the first two cases). Therefore, Athanasius concludes about Antony that "it was as if he were a physician given to Egypt by God" (87).

As Athanasius portrayed Antony as the God-given physician, the emerging saints' lives followed suit in their underscoring of healing miracles. Among Pachomius's healing accounts, in a story reminiscent of the healing of a woman with a flow of blood in Mark 5, the author of the *Life of St.*

the agents of syncretism and the regional prophets in late antique Egypt, see David Frankfurter, "Syncretism and the Holy Man in Late Antique Egypt," *JECS* 11, no. 3 (2003): 339–85.

20. Ferngren, *Medicine and Health Care*, 76–79.

21. Quoted in Owsei Temkin, *Hippocrates in a World of Pagans and Christians* (Baltimore: Johns Hopkins University Press, 1991), 164n92. Ramsay MacMullen simply states, "[all holy men] healed . . . *going about their usual practices, praying, healing diseases, driving out demons*" (quoting Socrates, *Hist. eccl.* 4.24): "The Place of the Holy Man in the Later Roman Empire," *HTR* 112, no. 1 (2019): 15.

Pachomius writes about the healing of a woman, the wife of the councilor of Nitentori, who had been suffering from a flow of blood for a long time. With confidence in her healing by Apa Pachomius, she asks Apa Dionysios to lead her to him. As the two saints sat down together, "the woman came up behind our father and, in consequence of her great faith, as soon as she had merely touched him and his clothing, she was cured" (41). While the woman's gesture of faith and healing closely mirror those of the woman in Mark 5, in the fourth-century monastic context, this episode also indicates the healing power not only of the holy man Pachomius but also of his clothing. In the late antique belief, "holiness could be communicated through physical touch, such that oil, clothes, and other objects can be made holy by the holy men's physical contact."[22] In the West, one of the first biographies of a monastic hero is *The Life of St. Martin of Tours* by Sulpitius Severus in the early fifth century. Among the accounts of numerous healings by Saint Martin, Severus records the instantaneous healing of a leper by Saint Martin's kiss, followed by the thanksgiving of the healed man in the church (18). Then Severus emphasizes that "this fact, too, ought not to be passed over in silence, that threads from Martin's garment, or such as had been plucked from the sackcloth which he wore, wrought frequent miracles upon those who were sick. For, by either being tied round the fingers or placed about the neck, they very often drove away diseases from the afflicted" (18). This is one of the clearest statements about the healing power of physical objects through "sacred contact" with the holy men,[23] which functioned as de facto amulets,[24] that is, objects endowed with supernatural force; as such, the use of these sacred relics boomed, especially in the West.

22. Christopher Sweeney, "The Function of the Holy Men in Context: Inscriptions, *Eulogiai*, and *Vitae*" (paper presented at the North American Patristic Society Annual Meeting, Chicago, May 2014), 6, drawing on Gary Vikan's "Icons and Icon Piety in Early Byzantium," in *Sacred Images and Sacred Power in Byzantium* (New York: Aldershot, 2003), 7.

23. Vikan, "Icons and Icon Piety," 7. Cf. Sweeney: "In the 5th century, the materiality that counts, is the tactile materiality of the holy man—miracles are performed either through the holy man's direct contact, or through an object to which his sacred contact has been transferred," in "The Function of the Holy Men," 13. See also Theodoret, *Phil. hist.* 26.11–12.

24. On Christian amulets, see, for example, David Frankfurter, "Amuletic Invocations of Christ for Health and Fortune," in *Religions of Late Antiquity in Practice*, ed. R. Valantasis (Princeton: Princeton University Press, 2000), 340–42; Theodore De Bruyn, *Making Amulets Christian: Artefacts, Scribes, and Contexts* (Oxford: Oxford University Press, 2017); Theodore De Bruyn, "Appeals to Jesus as the One 'Who Heals

The healing power of relics is not confined to hagiographical sources. Ambrose draws attention to the healing of a blind man during the translation of the bones of the martyrs Gervasius and Protasius to the new basilica in Milan in 386 (*Ep.* 22.2) and further attests to the exorcism and healing of "many, having touched with their hands the robe of those saints, [who are] freed from those ailments which oppress them" (22.9). Augustine is the cowitness of the healing of the blind man with Ambrose and records another healing of a demon-possessed boy with a damaged eye at a shrine dedicated to the same martyrs (*Civ.* 22.8). Augustine further confirms a series of miraculous healings of various people—a blind woman, a bishop with fistula, a Spanish presbyter, three victims of the gout, etc.—and even the resurrection of three recently dead people through the relics of Saint Stephen (22.8). Likewise, Theodoret, in his *Curatio*, appeals to the curative power of the martyrs' shrines by reporting the prayers of those with ailments and "the *ex-voto* offerings [of eyes, feet, and hands] of those who are cured as the proof" of the success of martyrs' intercession (8.63–64).

If the ascetic holy men were physicians given by God, their healing power far exceeded that of ordinary physicians.[25] According to Jerome, a woman who had been blind for ten years was brought to Saint Hilarion and confessed that "she had spent all her substance on physicians" (*Vit. Hil.* 15). Hilarion then replied: "If you had given to the poor (*pauperes*) what you have wasted on physicians, the true physician Jesus would have cured you" (15). When she pleaded with the saint with a cry, however, "he spat into her eyes, in imitation of the Savior, and with similar instant effect" (15). Here Saint Hilarion is a stand-in for the True Physician, Jesus, with the same effect of healing, in contrast to the earthly physicians' impotence. Theodoret of Cyrrhus also records a woman whose eye was afflicted with a disease that had no medical remedy, coming to Peter the Galatian. Peter then demanded that she bid "farewell to doctors and medicines" and accept the medicine given by God as a sign of her faith (*Phil. hist.* 9.7). Then "he placed his hand on her eye and forming the sign of the saving cross drove out the disease" (9.7). Moreover, a woman brought her sick son to a certain Syrian ascetic Maësymas after "the physicians had written him off and said explicitly that

Every Illness and Every Infirmity' (MATT 4:23, 9:35) in Amulets in Late Antiquity," in *The Reception and Interpretation of the Bible in Late Antiquity: Proceedings of the Montréal Colloquium in Honour of Charles Kannengiesser, 11–13 October 2006*, ed. Lorenzo DiTommaso and Lucian Turcescu (Leiden: Brill, 2008), 65–81.

25. Here we see a clear parallel with Asclepius's superior healing power to the physician's.

the child would die" (14.3). The holy man then took the child in his hands and placed it at the foot of the altar; he "lay face downwards as he entreated the Physician of the souls and bodies" and restored the son to good health, and to his mother (14.3). These are just a few examples of the saints' healings used as last resorts in light of the inadequacy of the doctors.[26]

The papyrological evidence further attests to seeking after the healing powers of the holy men on the part of the sick people. Among a series of papyrological texts studied by Barrett-Lennard,[27] in P.Lond VI 1926, a Christian woman named Valeria writes of her "serious disease in the form of a terrible difficulty in breathing" and, showing a high degree of confidence in the efficacy of his prayers even at a distance, asks a certain Apa Paphnutius to pray for her healing. In another papyrus, P.Lond. VI 1928, Heraclides mentions his oppressive sickness (*nosos*), also clearly expresses his belief in the efficacy of Apa Paphnutius's prayers, and sees "our Lord Christ" as the eventual source of his healing. In P.Lond. VI 1929, a certain Athanasius is primarily concerned for Didyma and his mother's "bad health," despite his own "very weak health" (*atonōtata*); based on his confidence in the Savior, he requests a prayer from Apa Paphnutius and even suggests that the prospect of their continued good health may depend upon the continued prayers and piety of Paphnutius. In P.Neph. 1, a certain Paul and Tapiam state that their previously sick children recovered through the prayers of the recipient of the letter, Nepheros the ascetic. Finally, in P.Neph. 10, Horon also states that Nepheros's prayers have made him healthy (*hygiaivō*) and recognizes that God is the source of such healing. Together as a whole, as Barrett-Lennard stresses, these letters appear "to contain convincing evidence that there was a definite tradition of Christian healing in existence in Egypt and that even well-to-do Christians sometimes, at least, responded to their illness by seeking a cure from Christ through the medium of the prayers of a holy person."[28]

● Indeed, Christians resorted to all means of healing, just like their non-Christian neighbors, during this period, including amulets and spells. We

---

26. This theme would become accentuated in the hagiographies of the later medical saints, such as Saints Cyrus and John. See Anastasia D. Vakaloudi, "Illnesses, Curative Methods and Supernatural Forces in the Early Byzantine Empire (4th–7th C. A. D.)," *Byzantion* 73, no. 1 (2003): 189–94.

27. See R. J. S. Barrett-Lennard, *Christian Healing after the New Testament: Some Approaches to Illness in the Second, Third, and Fourth Centuries* (Lanham, MD: University Press of America, 1994), 43–86.

28. Barrett-Lennard, *Christian Healing*, 62.

observed the use of Christian magical amulets when we discussed demonic etiology in chapter 2. Christian amulets included not only holy men's relics (as mentioned earlier) but also various items such as "figurines, engraved stones, and texts written on papyrus, parchment, potsherds, wood, and metal," some of which survived in Egypt.[29] A papyrus amulet from the fifth or sixth century seeks to heal a woman's sickness by appealing to Christ's healing acts in the Gospels and through the intercession of the Virgin Mary (*PGM* 2.227): "And the one who healed again, who raised Lazarus from the dead already on the fourth day, who healed Peter's mother-in-law, who also performed many untold healings—beyond what are recounted in the holy Gospels, heal the bearer of this divine amulet from the illness crushing her, by prayers and interception of the ever-virgin Mother, the Theotokos, and all." According to David Frankfurter, the wording of this amulet, with references to the "holy Gospels" and the intercession of the Theotokos, suggests the work of ecclesiastical scribes; the preparation of amulets became one of their dominant functions along with holy men and Christian shrines.[30] Another papyrus amulet[31] from the fifth century, folded to fit into a pouch or tube, starts with creedal declarations: "Christ was born, amen. Christ was crucified, amen. Christ was buried, amen. Christ arose, amen. He has woken to judge the living and the dead. Holy stele and mighty *charaktēres*, chase away the fever with shivering from Kale, who wears this amulet, now now now, quickly quickly quickly." By these creedal declarations the fever and chills are to flee from a woman named Kale as she carries the amulet with her. Given the orthodox Christian creed reflected in this amulet, it might also have come from ecclesiastical scribes. "Such features of the spells reflect an earnest attempt to remain orthodox while calling the Christian pantheon into service for human affliction,"[32] even as the ecclesiastical authorities like Athanasius harshly rebuked the faithful for using amulets.[33]

29. Theodore De Bruyn, "Archaeological Views: Christian Amulets—a Bit of Old, a Bit of New," *BAR* 44, no. 5 (2018).

30. Frankfurter, "Amuletic Invocations of Christ," 341. On scribal features of scriptural amulets, see De Bruyn, *Making Amulets Christian*, chap. 5.

31. *SM* I 23 (*Papyri Haunienses Collection* III 51), in De Bruyn, *Making Amulets Christian*, 208–9.

32. Frankfurter, "Amuletic Invocations of Christ," 341.

33. Athanasius, *De amuletis* (PG 26:1320): "Let anyone who should get seriously ill recite the psalm 'I said, "O Lord, be gracious to me; heal me, for I have sinned against you" [Ps. 41:4],' because by recurring to the prayer and imploring the divine

The tradition of Christian healing in Egypt is further confirmed in the *Canons of Hippolytus*, a church order from the fourth century, and the *Sacramentary of Sarapion*, a collection of thirty liturgical prayers ascribed to the bishop of Thmuis, also from the fourth century.[34] The *Canons of Hippolytus* is an adaptation of the earlier-mentioned third-century *Apostolic Tradition* and therefore mandates a similar concern for and care of the sick by the catechumen and widows (19, 16). However, whereas someone with a gift of healing is not to be ordained in the *Apostolic Tradition* (14), if a person with a gift of healing asks for his ordination, he is to be ordained with the confirmation of healing in the *Canons* (8). More significantly, unlike the *Apostolic Tradition*, the ordination prayers for the bishop and the presbyter in the *Canons* (3–4) highlight curing the sick (and exorcism) as one of the key roles of the bishop and presbyter: "Give him power to loosen every bond of the oppression of demons, to cure the sick and crush Satan under his feet swiftly." Thus we see that the expectation of healing power was not just confined to ascetic holy men, but also extended to the official church leadership. Similarly, as a deacon is to inform the bishop of the sick person, and to facilitate the bishop's visit, canon 24 states the following as a matter of fact: "He [the sick person] is relieved of his sickness when the bishop goes to him, especially when he prays over him, because the shadow of Peter healed the sick [Acts 5:15]." Here the bishop's prayer for healing as the successor to Peter is expected to be efficacious. The *Canons* also testifies to the liturgical use of therapeutic elements in association

---

grace, he will follow the heavenly wisdom that states, 'My child, when you are ill, do not delay, but pray to the Lord, and he will heal you' [Sir. 38:9]. In fact, amulets and sorceries are useless in securing help. And if someone consulted these, let him know this clearly, that he has made himself instead of a believer, an unbeliever; instead of a Christian, a pagan; instead of an intelligent person, an unintelligent one; instead of a rational person, an irrational one . . . nullifying the seal of the cross that brought you salvation. Not only are the illnesses afraid of that seal, but also the whole crowd of demons fears and wonders at it." On ecclesiastical boundary drawing between licit and illicit ritual and healing practices, see Joseph E. Sanzo, "Early Christianity," in *Guide to the Study of Ancient Magic*, ed. David Frankfurter (Leiden: Brill, 2019), 226–37. Augustine, in several sermons, compares a sick Christian in bed, wracked with pain, who refuses to use an amulet against social pressure, to a martyr on one's sickbed, crowned by Christ—see *Serm.* 318.3; 328.8; 335D.3, 5; 360F.7.

34. For detailed studies, see R. J. S. Barrett-Lennard, "The Canons of Hippolytus and Christian Concern with Illness, Health, and Healing," *JECS* 13, no. 2 (2005): 137–64. This paragraph is largely drawn from this study; Barrett-Lennard, *Christian Healing*, 277–323.

with the bishop's prayer: "The sick also, it is a healing for them to go to the church to receive the water of prayer and oil[35] of prayer, unless the sick person is seriously ill and close to death: the clergy shall visit him each day, those who know him" (21). Barrett-Lennard notes that the *Canons'* contemporary, the *Sacramentary of Sarapion*, also contains a prayer "for oils and water that are offered" and a prayer "for oil of the sick or for bread or for water" for "a throwing off of every sickness and every infirmity . . . [and] a medicine of life (*pharmakon zōēs*) and salvation" (17, 5). In addition, another fourth-century document from Syria, the *Apostolic Constitutions*, also contains a prayer for the blessing of water and oil by the bishop: "Grant them the power to restore health, to drive away diseases, to put demons to flight" (8.29).[36] Therefore, such blessings and the use of those elements for therapeutic purposes had become pervasive in many (Eastern) Christian communities by the later fourth century.[37]

Hospitals, Monastic Health Care, and Private Health Care

Besides various forms of healing, late antique Christians practiced their health care through founding hospitals and monastery infirmaries with limited medical care, and pursuing both rational medicine (doctors) and religious healing as complementary measures in private contexts.

Organization and Funding of Hospitals

In the previous chapter, we studied the specific purposes and targeted clienteles of Roman hospitals, which operated for limited and specific periods of time. Christian hospitals also had a distinct purpose and target clientele, but for a contrary purpose and with a lasting impact. The Constantinian era that dawned with the legalization of Christianity in 313 CE brought about a watershed in every aspect of the church. Constantine's unprecedented imperial patronage of the church, including financial subsidies, tax exemptions, and clerical exemptions from compulsory public services,[38] along with "a system of gifts of food to churches, grain allowances to nuns,

---

35. Bishops exhorted their congregations to apply water or oil blessed by a bishop, priest, or a monk, or to make the sign of the cross instead of amulets.

36. Barrett-Lennard, "The Canons of Hippolytus," 153.

37. Barrett-Lennard, "The Canons of Hippolytus," 155.

38. Constantine's imperial patronage of the church did not exceed what was expected of the imperial patronage of the Roman state religion.

widows, and others in church services,"[39] was revolutionary in general but also in the specific area pertinent to our topic. It exponentially increased the scale of the church's charity and wealth and furthered their impact on Roman society as a whole. Now accountable to the imperial throne, the church no longer merely served the Christian poor with offerings from the faithful (especially the wealthy); the early church was now bound to serve the poor of the empire as a public service in return for public privileges.[40] This change linked Christian identity even more closely to the church's care of the poor in Roman society, and the bishop consolidated his position as the "lover of the poor" and the "governor of the poor" par excellence.[41] It did not necessarily prompt the development of any new theological base for the work that the church had been doing for centuries, but it highlighted the identification of the poor with Christ that was found in Matthew 25:31–45 in particular: in *every* poor person, Christ is fed, given to drink, and welcomed as a guest. As Ambrose had written: "Minister to a poor person and you have served Christ" (*Vid.* 9.54).[42]

By the end of the fourth century, various hostels (*xenodocheia*; *xenōnes*), poorhouses (*ptōchtropheia*; *ptōcheia*), and lodges/hospices (*katagōgia*) devoted to collecting, housing, feeding, and caring for the poor and the sick had sprung up in cities throughout the Eastern empire, cities such as Caesarea, Antioch, Constantinople, Alexandria, Jerusalem, and Edessa. These institutions, developing as the hospitals, were typically adjacent to churches and monasteries. For instance, Basil's famous *katagōgion*, just outside Caesarea, built in the early 370s (c. 372 CE), was modeled after that of Eustathius, a homoiousian bishop of Sebasteia in the mid-fourth century (357), who had designed his *ptōchtropheion* to serve persons afflicted with diseases and disabilities.[43] Basil's "New City" (Basileios, as it was called by Gregory of Nazianzus) housed strangers, the poor, the sick, and the

39. Ramsay MacMullen, *Christianizing the Roman Empire (AD 100–400)* (New Haven: Yale University Press, 1984), 49.

40. Cf. Peter R. L. Brown, *Poverty and Leadership in the Later Roman Empire* (Hanover, NH: University Press of New England, 2002), 31.

41. Rhee, *Loving the Poor, Saving the Rich*, 181; also Brown, *Poverty and Leadership*, 32, 45.

42. See also Jerome, *Epist.* 66.5: "in the persons of such [i.e., the poor and the disabled] [Pammachius] ministers to Christ Himself"; Basil, *Ep.* 150.3: "he who gives to the afflicted has given to the Lord"; Leo the Great, *Serm.* 9.3: "Rightly in the needy and poor do we recognize the person of Jesus Christ our Lord himself."

43. Timothy S. Miller, *The Birth of the Hospital in the Byzantine Empire* (Baltimore: Johns Hopkins University Press, 1985; reprinted with new introduction, 1997), 79; Epiphanius, *Pan.* 75.1 (*tous lelōbēmenous kai adynatous*)

elderly as well as lepers and the mutilated, many of whom needed medical treatments and care (*tois therapeias*).[44] It employed both physicians ("those who give medical care"—*tous iatreuontas*) and nursing staff (*tous nosokomountas*), in addition to his monastic staff.[45] It also provided patients with the means to reintegrate into society by training them in crafts, skills, and other occupations for their livelihood, presumably upon their recovery (Basil, *Ep.* 94).[46] Moreover, according to Basil himself, Cappadocia had a number of *ptōchtropheia* under the direction of country bishops by 373.[47] By 386 Antioch had a hostel (*katagōgion*) for beggars who were approaching the city, where the urgently sick could be seen;[48] in fact, already in the mid-fourth century Leontius (344–358), its bishop, had appointed three devout men to oversee the operation of hospices (*xenodocheia*) for the care of strangers.[49] In Constantinople, Macedonius, a semi-Arian bishop, founded several poorhouses (*ptōcheia*) that took in the sick and destitute in the mid-fourth century (Sozomen, *HE* 4.27). Shortly thereafter in the early 380s, the empress Placilla, the first wife of Theodosius I, conducted her rounds of *xenōnes* attached to the churches in the city, offering the patients meals (Theodoret, *Hist. eccl.* 5.19.2–3). John Chrysostom, upon his episcopal appointment in 398, not only financially supported the existing sick houses (*nosokomeia*) but also established several more, placing them under the control of his trusted clergy and hiring doctors (*iatrous*), cooks, and other staff to enact medical care for the sick (Palladius, *Dial.* 5). In Alexandria, in 388, the priest Macarius supervised the *ptōcheion* for lepers, with women on the upper floor and men on the ground floor (Palladius, *HL* 6).[50] In the mid-fifth century, the ascetic Theodosius established a *nosokomeion* near Jerusalem with three separate buildings staffed with physicians: one for monks, one for the "regular sick," and one for the poor, "who would be more permanent patients than the others" (*Vita Theodosii* 40).[51]

44. Basil, *Ep.* 94; Gregory of Nazianzus, *Or.* 43.63; Sozomen, *HE* 6.34.

45. Basil, *Ep.* 94; Gregory of Nazianzus, *Or.* 43.63.

46. Cf. Peregrine Horden, "Poverty, Charity, and the Invention of the Hospital," in *The Oxford Handbook of Late Antiquity*, ed. Scott F. Johnson (Oxford: Oxford University Press, 2012), 717.

47. Basil, *Ep.* 142.

48. John Chrysostom, *Ad Stagirum a daemone vexatum*, PG 47:490.

49. Horden, "Poverty, Charity," 721; see also Brown, *Poverty and Leadership*, 123n121; contra Timothy S. Miller, *Birth of the Hospital*, 21.

50. Cf. John Cassian, *Conf.* 14.4.

51. Peter van Minnen, "Medical Care in Late Antiquity," *Clio Medica: Acta Academia Internationalis Historiae Medicinae* 27 (1995): 160; see also Timothy S. Miller, *Birth of the Hospital*, 147.

In Edessa, Ephraem the deacon set up a temporary hostel with three hundred beds during the severe famine in circa 372–373 (Palladius, *HL* 40). In the early fifth century, Bishop Rabbula reestablished a *xenodocheion* for both the sick and the healthy, with clean linen and soft beds, operated by a deacon assisted by monks; he also built a separate *nosokomeion* for women, which was run by a deaconess who was assisted by nuns;[52] then he founded a third hospital for the lepers who lived in isolation outside the city.[53] In the years 500–501, when Edessa suffered a deadly famine and pestilence, with the care of Mar Noonus, the *xenodochos*, the people of the city assembled at the gate of the *xenodocheion* and went forth and buried the dead, from morning to morning.[54] Then two priests established an infirmary among the buildings attached to the Great Church of Edessa.[55] "Those who were very ill used to go in and lie down there; and many dead bodies were found in the infirmary, which they buried along with those at the *xenodocheian*."[56]

In the West, Fabiola, a Christian Roman noble lady and friend of Jerome, founded the first public hospital (*nosokomein*) with her own resources in Rome around 397 (Jerome, *Epist.* 77.6);[57] and Pammachius, Fabiola's acquaintance, established a similar institution in Ostia some years later.[58] Augustine, in the early fifth century, recorded the first establishment of *xenodocheion* in Hippo by his priest, Leporius (*Serm.* 356.10). As these examples show, Christian hospitals emerged as philanthropic institutions that provided public care of "the poor"—for example, the indigent, the sick, the infirm, migrants, strangers, and lepers—in inpatient facilities with basic physical, psychological, and nursing care, and medical treatments

---

52. *The Heroic Deeds of Mar Rabbula* 100–101. Cf. Timothy S. Miller, *Birth of the Hospital*, 100, 120.

53. *The Heroic Deeds of Mar Rabbula* 101.

54. William Wright, *The Chronicle of Joshua the Stylite*, translated with notes (Amsterdam: Philo, 1968), 42.

55. William Wright, *Chronicle of Joshua the Stylite*, 42.

56. William Wright, *Chronicle of Joshua the Stylite*, 42.

57. "She was the first person to found a hospital, into which she might gather sufferers out of the streets, and where she might nurse the unfortunate victims of sickness and want."

58. Jerome, *Epist.* 66.13: Pammachius personally served the poor in the hospital, "ranking himself among the poor, condescendingly entering the tenements of the needy, offering his eyes to the blind, his hands to the feeble, and his feet to the crippled"; Isabella Bonati, "The (Un)Healthy Poor: Wealth, Poverty, Medicine and Health-Care in the Greco-Roman World," *Akroterion* 64 (2019): 32n61.

in varying degrees.[59] Already in 362, the emperor Julian "the Apostate" inadvertently paid a memorable paradoxical tribute to Christian philanthropy through hospitals by writing to Arsacius, the high priest of Galatia, the following: "In every city establish frequent hospitals (*xenodocheia*) in order that strangers [the poor] may benefit from our benevolence (*philanthropia*) . . . whoever is in need. . . . For it is disgraceful that, when no Jew ever has to beg, and the impious Galilaeans [i.e., Christians] support not only their own poor but ours as well, all men see that our people lack aid from us" (*Ep.* 22.430c–d). Julian goes on to exhort Arsacius to teach the "Greeks" (non-Christians) to "contribute to the public service of this sort" (431a). In Julian's mind there was no non-Christian tradition on which to base hospitals as a public service.

As one can see, it was mainly the urban bishops who founded and oversaw these hospitals in the East. The Fourth Ecumenical Council at Chalcedon (451) stipulated that all *ptōcheia* and *xenodocheia* should come under the authority of the bishop in every city.[60] Syriac canons of the fifth century also required the local bishop to establish and maintain *xenodocheia* in each town and to appoint an administrator from among "the rank of monks."[61] A century later (544 and 545), Emperor Justinian issued legislation confirming the canons of Chalcedon by acknowledging the episcopal jurisdiction over hospitals and all other philanthropic institutions.[62] While under episcopal leadership, these hospitals were funded by both imperial and private resources in addition to episcopal resources. According to Theodoret, the Basileios itself was partly funded by a substantial grant of land from the emperor Valens, although Valens was at odds with Basil theologically (*Hist. eccl.* 4.19.13). Other *ptōcheia, xenones, gerokomeia* (houses for the elderly) in Constantinople and Jerusalem were financed by gifts of imperial estates by the empresses Pulcheria and Eudocia in the fifth century (Sozomen, *HE* 9.1.10). Fundamentally, the imperial legislations that started with Constantine and culminated with Justinian allowed all these charitable institutions to accept legacies and receive inheritances

---

59. "A hospital in its basic and, in world-historical terms, most widespread form is a place, an area, designated for the overnight care of the needy. Its ethos is nearly always charitable, and it is to be distinguished from 'outdoor' relief centers from which food or goods are distributed." Horden, "Poverty, Charity," 719.

60. Canons of Chalcedon 8 and 10.

61. Arthur Vöövus, *History of Asceticism in the Syrian Orient*, vol. 2, *A Contribution to the History of Culture in the Near East* (Leuven: Secrétariat du CorpusSCO, 1960), 371.

62. Timothy S. Miller, *Birth of the Hospital*, 101.

(in addition to offerings and rents).[63] As for other private funding, Basil petitioned a fiscal officer who was one of the patrons for *ptōchtropheia* at Amaseia (*Ep.* 143); and the city's wealthy people provided donations to finance Ephraem's hospital in Edessa (Palladius, *HL* 40). Rabbula's hospital was funded not only by the church's wealth but also from the legacies of "many."[64] According to one papyrological source, a gift of 371 *artabas* of wheat by a certain Apion family funded the *nosokomeion* of Abba Elias in Egypt; the same family gave seven *artabas* (of wheat) to the *nosokomeion* of Leukadios, according to another papyrological source.[65] In sum, in the words of Norman Underwood, "the hospital mediated a new reciprocal healthcare economy, in which almsgivers treated the disease of their souls by provisioning the care of the physically sick."[66]

Patients and Personnel of Hospitals

With the development of hospitals as public charities in the Eastern empire in the late fourth century, (limited) Christian sources reveal certain patterns of religious expectations for patients and their roles in these emerging institutions.[67] As institutions attached to churches (or monasteries), these hospitals provided religious spaces with ceremonies, routines, and expectations intended to affirm or reaffirm Christian faith among patients— almost exclusively the destitute and strangers who needed nursing and "doctoring."[68] If we can infer from the descriptions written by the church

63. Cf. Timothy S. Miller, *Birth of the Hospital*, 105. Cf. Horden, "Poverty, Charity," 729.

64. *The Heroic Deeds of Mar Rabbula* 100.

65. P. Oxy. 16.1898 (one *artaba* is equivalent to about thirty-nine liters) for the former, and P. Oxy. 61.4131 for the latter, in Adam Serfass, "Wine for Widows: Papyrological Evidence for Christian Charity in Late Antique Egypt," in *Wealth and Poverty in Early Church and Society*, ed. Susan R. Holman (Grand Rapids: Baker Academic, 2008), 98.

66. Norman Underwood, "Medicine, Money, and Christian Rhetoric: The Socio-Economic Dimensions of Healthcare in Late Antiquity," *Studies in Late Antiquity* 2, no. 3 (2018): 378.

67. This section on the patients of the hospitals is largely drawn from Helen Rhee, "Portrayal of Patients in Early Christian Writings," *Studia Patristica* 81 (2017): 129–31.

68. Guenter B. Risse, *Mending Bodies, Saving Souls: A History of Hospitals* (Oxford: Oxford University Press, 1999), 82. For a (memorial) chapel (*mnēmē*) attached to Basil's *ptōchotropheion*, see Basil, *Ep.* 176. The word "doctoring" is from Horden, "Poverty, Charity," 717.

fathers, the "admission rituals" for these institutions stressed the priority of spiritual well-being over bodily health.[69] For example, cleansing "the soul through confession of sins was to precede any efforts to care for the body," and rituals of prayer and vigils were part of the routine, especially during crises such as famines, epidemics, and wars.[70] Patients, who were segregated by their gender whenever possible, were also expected to uphold basic Christian character, moral instructions, and the discipline of the institutions as seen in the Basileios (cf. Gregory of Nazianzus, *Or.* 43.63).[71] Despite their general dependency and lack of control in that setting, some patients acted in recalcitrant ways, throwing their food, refusing to follow the physician's orders, and tearing the clothes of their caregivers.[72] Morally suspect and unruly patients were exhorted and admonished by the superior, but if they persisted in their conduct, condemnation and expulsion would follow, as would be expected for the monks in Basil's monastery (Basil, *SR* 155).

This fundamental religious expectation and the prioritizing of the patient's spiritual health was a natural corollary to the ideal of Christ as the Great Physician (*archiatros*), primarily of the soul, but also of the body.[73] A rare Greek inscription from Frîkyā, Syria, invokes Christ as the Physican (*iatros*).[74] Following Origen, the Cappadocian fathers and John Chrysostom are among the early Christian leaders who show

69. Risse, *Mending Bodies, Saving Souls,* 85.

70. See Fridolf Kudlien, "Beichte und Heilung," *Medizin Historisches Journal* 13 (1978): 1–14; Risse, *Mending Bodies, Saving Souls,* 85.

71. Andrew T. Crislip, *From Monastery to Hospital: Christian Monasticism and the Transformation of Health Care in Late Antiquity* (Ann Arbor: University of Michigan Press, 2005), 117.

72. Risse, *Mending Bodies, Saving Souls,* 85.

73. E.g., Tertullian, *Pat.* 15; Clement of Alexandria, *Paed.* 1.2.6; Origen, *Hom. Jer.* 3.4; *Hom. Luke* 13; Cyprian, *Eleem.* 1, 3; Gregory of Nyssa, *De oratione dominica* 4; Basil, *Hom.* 7.1; *LR* 55; Basil, *Quod Deus non est auctor malorum* 3; Gregory of Nazianzus, *Or.* 18.29; 32.25. For the idea of Christ as the physician in the writings of various church fathers, see Jan Nicolae, "'Christus Praedicator/Medicator': Homiletical, Patristic and Modern Elements of *Theologia Medicinalis,*" *European Journal of Science and Theology* 8, no. 2 (2012): 15–27.

74. Bonati, "The (Un)Healthy Poor," 30n51. Part of the fourth-century inscription from Timgad, the Roman African city of Thamugadi (Algeria), reads: "I beg you, Lord, come, Christ, you who only are the physician, to the aid of the saints and penitents" (*C<h>riste, tu | solus me|dicus, sa|nctis et | peniten|tibus*). See P. Monceaux, "Nouveau fragment de l'inscription chrétienne de Timgad relative au Christus medicus," CRAI 68.1:80, in Bonati, 30.

the greatest familiarity, knowledge, and positive attitudes toward Greek (Hippocratic/Galenic) medicine.[75] For them the medical art for somatic diseases is in harmony with Christian piety as "a model (*typos*) for the therapy of the soul," so long as both the patients and the doctors keep in mind the need to glorify God, eschew unwarranted human reliance, and uphold spiritual health as the highest priority (e.g., Basil, *LR* 55.2). Basil therefore praised a Christian physician, Eustathius, for his *philanthrōpia* in tending to the patients not just in their physical ills but also in their spiritual ills,[76] and this was a sentiment echoed by the other Cappadocian fathers and Chrysostom.[77] Consequently, it was only natural from their perspectives that the patients understood or were instructed about the priority of the soul over the body for their holistic treatment even as they were given palliative care and (limited) medical attention in those institutions. In fact, as Peregrine Horden reminds us, this outlook reflected the antique and late antique medical belief that attention to the patient's spiritual, emotional, and bodily needs and environment was an essential part of medicine.[78]

Along with religious expectations and routines, diet, rest, and nursing constituted the regular regimen offered in *nosokomeia, ptōcheia,* and *xenodocheia.*[79] Patients were fed the standard Byzantine diet of "bread, wine, and dried or fresh cooked vegetables dressed with olive oil," as well as barley soup and honey boiled in water (Theodoret, *Hist. eccl.* 5.19.2–3).[80] They received most of their nursing and care from nonmedical personnel, including monastics and lay volunteers, some of whom might have had some basic medical training, under the supervision of clergy, deacons, and deaconesses. Where physicians were available, those with internal ailments and external wounds would have received drugs; some counterirritation, surgical incisions; and cauterizations.[81] The chief therapeutic strategy, as in Galen's time, was evacu-

75. Cf. Jerome's praise of medicine as "the skill most useful to us mortals," in *Epist.* 53.6.

76. Basil, *Ep.* 189.1; Gregory of Nyssa, *On the Holy Trinity.*

77. Cf. Michael Dörnermann, "Einer ist Arzt, Christus: Medizinales Verständnis von Erlösung in der Theologie der griechischen Kirchenväter des zweiten Jahrhunderts," *ZAC* 17, no. 1 (2013): 121. See also Mary E. Keenan, "St. Gregory of Nyssa and the Medical Profession," *BHM* 15, no. 2 (1944): 150–61.

78. Horden, "Poverty, Charity," 728.

79. Risse, *Mending Bodies, Saving Souls,* 85.

80. Risse, *Mending Bodies, Saving Souls,* 85.

81. Risse, *Mending Bodies, Saving Souls,* 85–86.

ation of the noxious humors by means of medications (*pharmaka*) and other methods: emetics and purgatives, diuretics, and bloodletting.[82]

The most notable patients in the Basileios were lepers,[83] who were housed and fed "indefinitely" with treatment of their wounds and bodily care, although they had no hope of recovery.[84] Basil not only exhorted the healthy to treat lepers with dignity as "brothers," but he also cared for them in person as an example to others (Gregory of Nazianzus, *Or.* 43.63). These new healing institutions largely emphasized solidarity with their patients (the lepers, the indigent, and migrants) as the ones who bear God's image and share a common humanity; as in the words of Jerome: "He whom we look down upon, whom we cannot bear to see, the very sight of whom causes us to vomit, is the same as we, formed with us from the selfsame clay, compacted of the same elements. Wherever he suffers we also can suffer (*Quidquid patitur, et nos pati possumus*)" (Jerome, *Epist.* 77.6).[85] These patients are almost certainly the same kind as the destitute (*ptōchoi*) and the leprous (also identified as *ptōchoi*) mentioned in Gregory of Nazianzus's *Oration 14* and Gregory Nyssa's *On the Love of the Poor* (although neither of them mentions the Basileios in their sermons). They are Christ in disguise, and their physical (sick and bedridden) bodies in fact bear Christ's own incarnate body (Gregory of Nazianzus, *Or.* 14.39–40; Gregory of Nyssa, *Paup.* 1). As fellow bearers of God's image, "our kindred" (*synagenēs*), brothers (*adelphoi*), and the very *prosōpon* of Christ, they not only deserve and are entitled to Christian care but also are holy and pious. Lepers, in particular, are compared with Lazarus, the "sacred beggar" in Luke's Gospel, and bear the "public infirmity" (*hē koinē asteveia*; *Or.* 14.8) and the "sacred disease" (*hē hiera nosos*), a designation ancient medical writers attributed to epilepsy[86] (*Or.* 14.6; Gregory of Nyssa, *De anima et*

82. Risse, *Mending Bodies, Saving Souls*, 85.

83. Daniel Caner, "Not a Hospital but a Leprosarium: Basil's Basilias and an Early Byzantine Concept of the Deserving Poor," *DOP* 72 (2018): 25–48, recently argued that the Basileios was not a general hospital but a leprosarium.

84. Crislip, *From Monastery to Hospital*, 113–14.

85. Cf. Anne Elizabeth Merideth, "Illness and Healing in the Early Christian East" (PhD diss., Princeton University, 1999), 113, 151: "There was a change in the perception of and attitudes towards the sick and . . . this shift in perception of the sick was accompanied by the establishment of institutionalized care for the sick"; "no longer marginalized and rejected, left to decay in the streets of the ancient city, the sick now belonged at the center of the new Christian city."

86. Susan R. Holman, *The Hungry Are Dying: Beggars and Bishops in Roman Cappadocia*, OSHT (Oxford: Oxford University Press, 2001), 161.

resurrectione). Due to their poverty or leprosy, then, the sick bodies of the poor, that is, of the patients taken to the hospitals from the streets, are imbued with a certain sanctity and require physical care from the "healthy."[87] However, the sanctity of these patients is the kind of sanctity that does not require their action or active role in the bearing of their illnesses; it was conferred on them by their patrons and the bishops who constructed their image and role.

While the urban poor, the bishops' main clients, were the main beneficiaries of the hospitals under the authority of the bishops, some of their healthy clients served the sick at the bishops' behest. A case in point is the *parabalani*, a corps of stretcher-bearers or ambulance personnel for churches, Christian hostels, sick houses, and leper houses. They are twice mentioned in the Theodosian Code (16.2.42/416 by Theodosius and 16.2.43/418 by Honorius) as an organized body of the poor (*pauperes*) of Alexandria to "occupy themselves, under the supervision of the patriarch, with the care of 'the sick bodies of the weak.'"[88] The code specified a large group of five hundred in the first document and six hundred in the second, even after the horrible reputation they had earned as Patriarch Cyril's intimidation force in the murder of Hypatia in the year 415. As Glen Bowersock notes, the fact that an extra hundred people had to be authorized indicates that the services of this body were sufficiently in demand.[89] As they had to be poor by law, but in good health, their main job was to transport the ailing, including lepers, from the city streets into places of isolation and care.[90] According to George Gask and John Todd, it is possible that in their earlier days, the *parabalani* were partly staffed at *xenodocheia*.[91] The term *parabalani* most likely comes from *paraboloi*, meaning "taking risks," used by Eusebius to mean taking risks in the service of Christ (*Theoph.* 5.2.4); and by the time of Cyril of Alexandria, the *parabalani* served as both the

---

87. See Holman, *The Hungry Are Dying*, 161–62, quoting John Chrysostom, "Sixth Sermon on Lazarus and the Rich Man/On the Earthquake," in *Saint John Chrysostom: On Wealth and Poverty*, ed. and trans. Catherine P. Roth (Crestwood, NY: St. Vladimir's Seminary Press, 1984), 108.

88. Glen W. Bowersock, "*Parabalani*: A Terrorist Charity in Late Antiquity," *Anabases* 12 (2010): 49.

89. Bowersock, "*Parabalani*," 49.

90. Bowersock, "*Parabalani*," 50.

91. George Gask and John Todd, "The Origin of Hospitals," in *Science, Medicine, and History: Essays on the Evolution of Scientific Thought and Medical Practice Written in Honour of Charles Singe*, ed. E. Ashworth Underwood (Oxford: Oxford University Press, 1953), 1:122.

risk-taking minor clerics to perform social works for the poor and the sick and also the patriarch's unruly and terrifying bodyguards.[92]

Ferngren postulates the origin of the *parabalani* as the risk-takers in aiding the sick in the context of the mid-third-century plague, during which they risked remarkable exposure to contagion.[93] As discussed earlier, during the awful plague of 252, Cyprian appealed to and gathered Christians of all social ranks; while the rich donated money, the poor provided their labor and service for the victims of the plague (Pontius, *Vita Cypriani* 10). Around 260, Dionysius of Alexandria also mobilized "very many" Christians, with both clergy and laity tending the overwhelming number of victims of the plague, risking their own lives and perishing in the midst of their care for the afflicted (Eusebius, *Hist. eccl.* 7.22.7–8). Eusebius reports the ravaging plague and famine of 312 in the East, where certain Christians selflessly and diligently performed needful tasks such as burying the dead, assembling the multitude who were famished throughout the city, and distributing bread to them (9.8). While it is unclear whether these groups of Christian caretakers were volunteers or hired as semimedical attendants, it is possible to see them as "a precursor to later organizations of medical attendants like the *parabalani* in Alexandria."[94]

Along with the *parabalani*, another, better known group of numerous caretakers of the sick, the *philoponoi*, emerged in Alexandria.[95] The *philoponoi*, "the lovers of toil," were risk-takers like the *parabalani*, zealous in church services and dedicated to helping the poor and the sick, as mentioned by Zecharias, author of the *Life of Severus* (12).[96] Whereas Christopher Haas associates them with the aristocracy because of their educational level, Sophronius represents them as indigent citizens (*Miracles of Saints Cyrus and John* 6).[97] While they might have drawn their members from a broad social spectrum, it is significant that they were (mainly) known as the poor by their contemporaries. An Egyptian papyrus enumerates a list of *philoponoi* as "among others three farmers, two tailors, an embroiderer, a carpenter, a cantor, and a merchant" (*SB* XX 14105 = P.Lond. III 1071b).[98]

---

92. Bowersock, "*Parabalani*," 52.

93. Ferngren, *Medicine and Health Care*, 120.

94. Ferngren, *Medicine and Health Care*, 120.

95. Bowersock, "*Parabalani*," 52. On the *philoponoi*, see also Timothy S. Miller, *Birth of the Hospital*, 130–32.

96. Bowersock, "*Parabalani*," 52.

97. Christopher Haas, *Alexandria in Late Antiquity: Topography and Social Conflict* (Baltimore: Johns Hopkins University Press, 2006); Bowersock, "*Parabalani*," 52.

98. Serfass, "Wine for Widows," 96.

According to John of Ephesus in the sixth century, the responsibilities of the *philoponoi* included clothing, bathing, anointing, and providing palliative care for the sick.[99] Unlike the *parabalani*, the *philoponoi* were organized as lay confraternities connected with urban churches; but, like the *parabalani*, they also had a reputation for being a "frightening or terrifying" group to non-Christians.[100]

The *philoponoi* had parallel confraternities in other Eastern cities called *spoudaioi*, "the zealous ones" (mainly mentioned from the fourth through the seventh centuries), who comprised both laymen and laywomen who practiced chastity (or continence if married) and were attached to the large churches under the direct supervision of the bishop (like the *philoponoi*).[101] As unskilled, nonprofessional medical attendants, the *spoudaioi* typically searched the streets and alleys by night for the sick, distributed money and food to them, and took them to the baths and the churches for care; these were the typical functions performed by members of the lowest social order.[102]

While the functions of the *parabalani*, *philoponoi*, and *spoudaioi* were centered in the great urban churches and were viewed as an extension of those churches' ministry under episcopal or patriarchal directives,[103] monastics, the new pious poor, were to serve in the hospitals "as one of the standard monastic duties."[104] Starting with the Basileios, monks came to be involved with "nearly every hospital in the Eastern Roman Empire" as main administrators and also caretakers.[105] According to Basil's *Short Rule*, the monks who serve the sick in the hospice (*xenodocheion*) should serve them "with the same disposition we would have towards brothers of the Lord" (*SR* 155).

Monastic Health Care

Even before monastic involvement with hospitals, both lavra monasteries and cenobitic monasteries took care of their own sick not only through

99. Cf. Ferngren, *Medicine and Health Care*, 134.
100. Bowersock, "*Parabalani*," 52.
101. Cf. Ferngren, *Medicine and Health Care*, 131.
102. Ferngren, *Medicine and Health Care*, 133.
103. Ferngren, *Medicine and Health Care*, 133.
104. Crislip, *From Monastery to Hospital*, 116.
105. Ferngren, *Medicine and Health Care*, 133, based on Timothy S. Miller, *Birth of the Hospital*, 130–32.

praying for healing but also by sharing special foods and (sometimes) medical care. In cenobitic monasteries, health care was institutionalized by the father superiors through their infirmary, "an inpatient facility that provided beds for sick monastics, nurses, and doctors, dietary treatment, pharmaceuticals, and surgery."[106] Pachomius, who founded the earliest cenobitic monastery (c. 324) and its infirmary in Egypt, framed the monastic care of the sick as a matter of fearing God and loving their neighbors as themselves (*The Bohairic Life of Pachomius* 48). Horsiesios likewise exhorted his brothers to care for the sick with great compassion, and encouraged those who cooked for the sick to cook "with great care in the fear of the Lord" (*Reg.* 23–24). And Basil's *Short Rule* laid down the kind of disposition in which fellow brothers ought to serve their sick brothers: "As offering our service to the Lord himself who said 'insofar as you did it even to the least of these my brothers, you did it to me' (Matt 25.40)" (*SR* 160).

In the monastic infirmaries of Pachomius, Shenoute, and Augustine, monastic patients were socialized into a new and distinctive "sick role" with set expectations as a special class for benefits and obligations, as shown by Andrew Crislip.[107] Crislip identifies four main components of this "sick role." First, it legitimizes withdrawal from certain monastic obligations such as manual labor, common worship/prayer, and the usual strict dietary regimen.[108] For example, the Pachomian rule says that "the sick are sustained with wonderful care and a great abundance of food," indicating an exemption from the monks' usual *askēsis* and strict diet (5). Horsiesios also mentions that the needs of the sick monks should determine their diet (*Reg.* 24). The anonymous *Rule of the Master* of the early fifth century permits the sick (*infirmi*) to break the fast, quoting Matthew 26:41 ("the spirit is willing but the flesh is weak") (*Reg. Magistri* 28.13–20). According to the rules of Shenoute's fifth-century White Monastery, the sick brothers and sisters are legitimately excused from the following ordinary responsibilities: the common gathering and the Eucharist (233), the four weeks of self-examination (79), flax work (280), making ropes (329), two-day fasts (357), constant meditation (333), and the general ascetic yoke (409). They are given special food and drink and other comforts upon demand: for example, bread (150), "oil to the brined anchovies or to the salted food or the charlock" (187), and a little wine (189).

---

106. Crislip, *From Monastery to Hospital*, 64.
107. Crislip, *From Monastery to Hospital*, 68–99.
108. Crislip, *From Monastery to Hospital*, 70–76.

Second, the monastic sick role exempts the patients from responsibility for their own sickness in terms of its causation and restoration,[109] that is, it severs the tie between sin, sickness, and punishment. Pachomius's heir Theodore instructs his brothers to have compassion for the sick: "As for the brothers who are sick, . . . let them not be discouraged. Let no one among us say, 'no doubt it is because he is a very wicked man that these tribulations have fallen on him!' He who will say that to himself does not perhaps yet deserve remedy himself. And let no one rejoice to have made another greatly suffer, or be glad that another has fallen sick. That is a great wickedness of the enemy" (Theodore, *Instr.* 15).

The monastic rules of Shenoute also validate legitimate illness while distinguishing it from "the illness of the demons" (*accidie*) (32) and pretended illness (156); they confirm the dignity of the truly sick by reproaching, casting doubt on, or scorning another's sickness (33, 127); and they allow the sick monks or nuns to stay in the infirmary as long as necessary (451). Furthermore, the rules also mention permanently disabled monks or nuns without any link to possible sin or punishment: "The sick, the lame, the crippled, or the blind: they shall all live with one another" (438, 450). These instructions and rules depict sickness as a reality of human existence, and by severing the connection between sin, sickness, and punishment, these monastic instructions/rules destigmatize sickness and the sick.

Third, at the same time, the rules impose on the sick a social obligation to get better, to return to the healthy role by highlighting the temporary nature of these sick benefits and the general undesirability of illness.[110] According to Augustine's rule, "when sick people have fully recovered, they should return to their happier ways, which are all the more fitting for God's servants to the extent that they have fewer needs" (*Reg. Praeceptum* 3.5). The Shenoute's rules expect the sick to endure the sickness (153) and imply that the sickness is temporary (127).

Finally, the rules require patients to seek out proper lay and professional medical treatment approved by the monastic triage, the elders, and to comply with prescribed therapies and procedures.[111] Pachomius carefully appointed some capable and faithful brothers as his assistants to look after the food and to care for sick brothers (*The Bohairic Life of Pachomius* 26); his rule also mandated that only the sick monks could be

---

109. Crislip, *From Monastery to Hospital*, 76–81.
110. Crislip, *From Monastery to Hospital*, 81–83.
111. Crislip, *From Monastery to Hospital*, 83–86.

oiled, bathed, or washed according to the "manner established for them" (*Rules* 92–93). While stressing the significance of good monastic hygiene and health, Augustine's rule mandates the consultation of a physician by the member with bodily pain; if a physician orders bathing, even if the sick monk does not want to, he must follow the physician's prescription (*Reg. Praeceptum* 5.5–7). Caesarius of Arles's rule for female monastics in the early sixth century (512 CE) repeats this mandate (*Statuta Sanctarum Virginum* 31.1–3). The Shenoute's rules are rather impressive in their stipulations regarding the medical treatment of the sick. A professional male or female doctor examines, gives prognosis, and prescribes treatment, and the sick are expected to follow instructions (458–460). Nurses are carefully chosen from the members of the community, and a pair of them are to work in shifts (335–337). Ultimately, it is the father superior's (or mother superior's) responsibility to direct and manage the infirmary and oversee the diet of the sick (154–155).

While monastic patients were expected to conform to this institutionally imposed sick role in light of the institutional ideals of ascetic piety, discipline, and health, at least some of them attempted to negotiate and even deviate from this role. These patients sought to appropriate that very ideal of ascetic piety and virtue by "hypervaluating" and spiritualizing illness, as in the case of Melania the Elder and Amma Syncletica, the latter of whom we studied in chapters 2 and 3.[112] Melania the Elder insisted on the steadfast endurance of her illness from her *askēsis* and rejected medical treatment as an ascetic virtue: "Look, I am sixty years old and neither my feet nor my face, nor any of my members except for the tips of my fingers has touched water, although I am afflicted with many ailments and my doctors urge me. I have not yet made concessions to my bodily desires, nor have I used a couch for resting, nor have I ever made a journey on the litter" (Palladius, *HL* 54.2). As mentioned in chapter 2, monastic patients wrestled between the apparently contradictory ideals of ascetic piety and discipline. On the one hand, exceptional health and healing served as a (traditional) symbol of ascetic sanctity and self-control modeled after the life of Antony, along with a corresponding self-understanding of illness as a threat to ascetic aims. On the other hand, the experience and endurance of illness was becoming a new form or mode of asceticism and ascetic progress, as we have

---

112. Crislip, *From Monastery to Hospital*, 92–99. See also Andrew T. Crislip, *Thorns in the Flesh: Illness and Sanctity in Late Ancient Christianity* (Philadelphia: University of Pennsylvania Press, 2012), 100–108.

seen. This new model of asceticism was provided in sources such as the Coptic life of Pachomius, Basil, and Syncletica, accompanied by a corresponding self-understanding of illness as a "training exercise" (*gymnasion*) for souls (Basil, *Quod Deus non est autor malorum* 9).[113] Thus, similar to the ideological and functional dynamics between philanthropic institutions and their patients, on the one hand the monastic infirmaries underscored solidarity with their patients, including the elderly, the disabled, and the chronically infirm as "the brothers of the Lord" (Basil, *SR* 155) and sought to destigmatize them; on the other hand the monastic rules also sought to control their patients' behavior by prescribing a normative monastic ideal of wholeness with a corresponding sick role. However, unlike the sick role in the hospital (philanthropic) context, the monastic sick role carried an inherent theological and existential ambiguity because of the aforementioned tension. This was reflected in the varying degrees to which patients were willing to conform to or transform their roles in light of the new significations of illness. Equally unlike their hospital counterpart, the monastic patient had to act "holy" in illness, either by dismissing, valorizing, or enduring it, as new significations of illness in turn provided a "safe" context for acting "holy."

In order to demonstrate these tensions, we can use cauterization as an example. When it came to cauterization for both internal and external conditions, whether within institutions or at home, patients were expected to submit to this procedure and to endure the pain inflicted by physicians (Theodoret, *Curatio* 1.1–5).[114] Nonetheless, as discussed in the previous chapter, there was a considerable fear of and even hostility to the scalpel and knife used in surgery in the absence of effective anesthetics. Although this pain could be regarded as redemptive in imitation of Christ's own suffering and as a necessary process toward healing, we find that "religiously sanctioned stoicism" was the exception rather than the rule, ascribed only to the saintly ascetic in hagiography.[115] For example, while undergoing a tormenting surgery for an ulcer, a certain Stephan was weaving palm leaves and talking with his ascetic colleagues; he was impervious to pain due to his ascetic valor (Palladius, *HL* 2.24). Be that as it may, Gregory of Nazianzus mentioned loud cries and screams following the placement of red-hot irons on several areas of another patient's body. Basil was all

---

113. For a detailed study, see Crislip, *Thorns in the Flesh*, particularly chaps. 3, 4, and 5.
114. Risse, *Mending Bodies, Saving Souls*, 86.
115. Risse, *Mending Bodies, Saving Souls*, 86.

too aware of the noxious pain of the physician's "cutting, burning, and complete mutilations of the body" (*Quod Deus non est auctor malorum* 3; *LR* 55.3).[116] And Eusebius of Caesarea reported patients' pleas to their physicians to prescribe bitter *pharmaka* rather than to try to heal them with fiery irons and sharp knives.[117]

Private Health Care[118]

Despite occasional denunciations of "secular" or "pagan" medicine,[119] early Christian writers generally expressed positive evaluations of Greco-Roman medicine, and early Christians utilized a variety of medical options as patients.[120] In private contexts, while patients were expected to follow the prescriptions and counsels of their doctors, their social location and economic means also accorded them varying degrees of control or agency in relation to their physicians and health care. Augustine's famous eyewitness account of a surgical scene in Carthage involving Innocentius, a Roman ex-advocate and devout Christian, his various doctors, and holy men, sheds light on the surgical pain and the active role and perspective of the patient in that triangular relationship (*Civ.* 22.8). Innocentius was under treatment for multiple rectal fistulas (*fistulae*) and had already suffered an excruciatingly painful operation, which turned out to be incomplete. When his household doctor (*medicus domesticus*), who had not been allowed by the other doctors (*medicis*) to witness the first operation, advised a second operation, Innocentius was horrified, threw him out, and readmitted him only reluctantly. In response to the patient's pleading against that second surgery, the doctors who had operated on him mocked his domestic doctor's advice and eased his fears with a promise of a cure through nonsurgical treatment. Innocentius then regained hope and confidence by the authority of another reputable, elderly doctor's confirmation of that prog-

116. See also Basil, *Ep.* 151.

117. H. J. Frings, "Medizin und Arzt bei den griechischen Kirchenvätern bis Chrysostomos" (PhD diss., University of Bonn, 1959), 44, 48, 79; see also Risse, *Mending Bodies, Saving Souls*, 86.

118. This section is drawn from Rhee, "Portrayal of Patients," 133–37.

119. E.g., Arnobius, *Ad. nat.* 1.48; 3.23. On Christian suspicions of doctors of pagan medicine, see also Vivian Nutton, "God, Galen and the Depaganization of Ancient Medicine," in *Religion and Medicine in the Middle Ages*, ed. P. Biller and J. Ziegler (York, UK: York Medieval Press, 2001), 22.

120. See Ferngren, *Medicine and Health Care*, chap. 2, in particular.

nosis. However, his condition did not improve at all, even after a considerable time, and Innocentius was informed that the second operation was the only hope for cure by the same, now embarrassed doctors. Aghast, pale, and distraught with extreme terror, Innocentius then ordered them to go away for good. Finally accepting reality, he called in an Alexandrian "surgeon of genius" to perform the operation for him; but the Alexandrian surgeon (*medicus*) was so impressed by the work done by the former doctors that he persuaded Innocentius to let them do the operation. Innocentius then reinstated those doctors and agreed to the second operation, without which the fistula would be incurable, on the next day in the presence of the Alexandrian doctor.

After the doctors' departure, the patient's fear and distress brought the whole household to a state of uncontrollable lament. When a group of holy men, including a bishop, a presbyter, and deacons of the church of Carthage, were paying Innocentius a customary visit that night, he desperately asked for their presence in what he thought would be his funeral under the surgeon's hands the next day. The holy men consoled and exhorted him to trust in God and submit to God's will "like a man" (*viriliter*) (*Civ.* 22.8). They all then prayed, but Innocentius hurled himself to the ground and prayed so passionately with groans, tears, and the shaking of his whole body that Augustine could only utter in his heart, "Lord, what prayers of your people do you hear, if you do not hear these?" (22.8). The bishop then gave him a blessing, and they all left.

The dreaded morning came, and "the servants of God" (*servi Dei*) as well as the doctors arrived (22.8). When the surgeon examined the diseased fistula with his fingers, ready to cut with knife in hand, and inspected it with utmost scrutiny, he found a perfectly solid scar, completely cicatrized! Augustine reports that the praise and thanksgiving to God from every mouth, accompanied by tears of happiness, was beyond description.

Augustine then goes on to recount the healing of Innocentia, a matron and devout Christian in the same city, who had a breast sore (*mamilla cancrum*) that was, according to her physicians (*medicis*), incurable (*Civ.* 22.8). Following the Hippocratic precept, her physician, a competent doctor and family friend, advised her not to attempt any treatment if she wished to prolong her life to some extent. However, Innocentia turned to God for help in prayer; as Easter was approaching, she was instructed in a dream to ask the first newly baptized woman she would meet at the baptistery to sign her affected area with the sign of Christ. This she did, and she was immediately restored to health. When her doctor examined her and found her completely cured, he interrogated her about the remedy she had used, by which the Hippocratic precept could be refuted (22.8). But when what he heard from her story of healing was not what he apparently hoped to hear, the doctor found that healing rather unremarkable, saying, "What was so extraordinary in Christ's healing *cancrum*, when he once raised to life a man four days dead?" (22.8).

While Augustine's accounts ultimately highlight his apologetic point that contemporaneous Christian miracles (healings) stand as proof of God's power, they also serve to illuminate patients' understanding of illness and healing and reveal patients as, to use the words of Ido Israelowich, "actors" in the contemporary medical market,[121] a market in which "*authority* was in the hands of the health care providers, but *power* remained in the hands of the sick, insofar as they chose which of the available treatment options to follow."[122] Examining the patient-healer dynamics during the High Roman Empire, Israelowich has argued that patients played an important role in determining medical authority,[123] expecting and shaping the convergence of religious and rational medicine.[124] As Owsei Temkin pointed out, in earlier centuries, a healing miracle "had been greeted by the cry, 'Great is Asclepius,' in which the doctors had joined," though not without some reservations.[125] Here, if we read Augustine's account literally, on both healing occasions the physicians also joined in saying, "Great is Christ," as it were. At the same time, in the first account, the patient, the holy men (clergy), and Augustine himself accepted the doctors' role without any reservation. Each group did what the patient expected of each—the clergy prayed, the doctors performed operations, and all praised God for the healing.[126]

Given their socioeconomic standing, the patients in Augustine's accounts were in control of their choice of doctors and were free to seek and accept (or not) their counsel and treatments, not only at the outset of their illnesses but also at the conclusion of their healings, as the doc-

---

121. Ido Israelowich, *Patients and Healers in the High Roman Empire* (Baltimore: Johns Hopkins University Press, 2015), 3, 44.

122. Israelowich, *Patients and Healers*, 31, 42–43, 67–69.

123. Israelowich, *Patients and Healers*, 44.

124. Israelowich, *Patients and Healers*, 57–58.

125. Temkin, *Hippocrates*, 225.

126. See also a story of "St. Daniel the Stylite," in *Three Byzantine Saints: Contemporary Biographies of St. Daniel the Stylite, St. Theodore of Sykeon, and St. John the Almsgiver*, trans. Elizabeth Dawes and Norman H. Baynes (Crestwood, NY: St. Vladimir's Seminary Press, 1977), 60–61.

tors confirmed those divine healings, after all. The patients, as Augustine emphasized their piety, were also free to seek divine help. Innocentius, out of desperation and terrifying fear of the doctor's operation, turned to the holy men as a last resort. And Innocentia, also as a last resort, sought God's help in prayer and dreams, and through these means, instructions for healing came. In the first case, divine healing completed the process of healing begun, but not finished, by previous therapeutic procedures, and in the second, it served as an efficacious substitute for a medical procedure. In both cases, the divine healing did not defy or reverse human efforts and medicine, and the doctors and patients themselves saw no dichotomy between religious and "secular" medicine, reflecting the larger culture. Here, patients' devotion or holiness, stressed by Augustine, is what drove them to pursue both religious and "secular" medicine. As Vivian Nutton puts it, "divine and human healing were thus complementary, and how and when each was to be used was a personal decision" by the patient.[127]

Indeed, even in later hagiographical healing accounts where the medical saints were typically portrayed antagonistically to "secular" medicine as inferior competition, divine-human, religious-"secular" complementarity or synergy was still significant, as in the healing cases by Asclepius. Miracle 22 of the seventh-century *Miracles of St. Artemios* tells a healing story of a devout elderly man (who also appears as the victim of a burglary in Miracle 18, recovering his stolen property through Saint Artemios's help). He was suffering with genitals that had sunk to below his knees (in addition to having water in his chest and dropsy). The patient, whose condition was diagnosed as incurable by a chief physician (*archiatros*) of the Christodotēs Hospital (*tou xenōnos tōn Christodotēs*), declined the latter's offer to treat him to lessen his pain and, with lamentation and tears, implored Saint Artemios for healing. In his sleep, the saint made an incision on his right testicle with a scalpel and instructed the chief physician's assistant (*hypothrgos*) in a dream to drain the patient's testicles and to have him constantly shouting, "Lord, have mercy, glory to you, O God." When the chief physician (who had diagnosed the patient's condition as incurable) confirmed the miracle, both the physician and his assistant glorified God. The assistant then "scrupulously did whatever was ordered by the saint, even up to the Holy Epiphany, that is for twelve days; this bed-ridden patient was cured beyond expectation."[128] The assistant and cured patient

---

127. Nutton, *Ancient Medicine*, 286.
128. From Virgil S. Crisafulli and John W. Nesbitt, eds., *The Miracles of St. Ar-*

then confided their visions to each other and glorified God. In this unique, multilayered healing scene, divine healing mediated through a saint entailed the saint's medical performance and sustained care by a medical staff as the patient consciously pursued both kinds of healing, which ultimately came from God.

Similarly, in Sophronius's *Miracles of Saints Cyrus and John*, the saints occasionally performed medical roles by giving medications that reflected late antique pharmacopeias, particularly as evidenced from papyrological documents.[129] These documents consist of recipes for plasters or poultices applied to ulcers, wounds, muscle ruptures, and surgical incisions and fractures; according to Dominic Montserrat, these recipes are made of ingredients like animal lard, wine, and *materia medica*, showing a close proximity to recipes of Galen, Aetius of Amida, and Oribasius.[130] Cyrus and John's prescriptions included similar flesh-based plasters for many kinds of illnesses: salt quail applied to the foot for lameness (Miracle 43); a bath and a poultice of lentils applied to Joannia's poisoned stomach, through which the poison was purged (Miracle 68); and a poultice of sea fish pounded in local wine for a woman's wounded hand (Miracle 9), just to name a few.[131] In each case, the patient showed his or her piety in following through with these prescriptions and was rewarded by being healed. While Sophronius typically presents patients' pilgrimages to the incubation shrine of these saints as a last resort after the repeated failure of "secular" medicine, he also narrates ample cases of healing at the shrine through medical means that accommodate and incorporate traditional medical theories and practices such as purging and emetics.[132]

These accounts illumine yet another aspect of patient-healer relationships in late antiquity. As previously noted, Israelowich, among others, has shown that both physicians and healing deities/temple medicine (e.g.,

---

temios: A Collection of Miracle Stories by an Anonymous Author of Seventh-Century Byzantium (Leiden: Brill, 1997), 136.

129. Dominic Montserrat, "'Carrying on the Work of the Earlier Firm': Doctors, Medicine and Christianity in the Thaumata of Sophronius of Jerusalem," in *Health and Antiquity*, ed. Helen King (London: Routledge, 2005), 230–42, here 235. For a helpful and succinct comparative study on *The Miracles of Saints Cyrus and John*, see Susan R. Holman, "Rich and Poor in Sophronius of Jerusalem's *Miracles of Saints Cyrus and John*," in Holman, *Wealth and Poverty in Early Church and Society*, 103–24.

130. Montserrat, "'Carrying on the Work,'" 235.

131. Montserrat, "'Carrying on the Work,'" 236.

132. Montserrat, "'Carrying on the Work,'" 237.

Asclepius) shared a common medical language and common methods (of Hippocrates and Galen), particularly in their use of terms for illnesses, drugs, and therapeutic practices; it was indeed patients' assumption and expectation that gods could perform in the same way as physicians that encouraged this complementarity or convergence.[133] Furthermore, scholars have noted that Christian saints "carried on the work of the earlier firm" of the pagan healing deities such as Asclepius,[134] though not necessarily in a linear way. The late antique Christian medical market also provided comparable choices for Christian patients. These accounts of patients' roles and understanding in their health care illustrate the dynamic yet critical incorporation and transformation of medicine into the Christian framework in late antiquity.[135]

Christian Therapy of the Soul

Like non-Christian elites, Christian leaders were concerned with not only the therapy and healing of the sick body but also the therapy and healing of the sick soul.

Bishops and Monastic Superiors as Therapists of the Soul

They applied the physician metaphor not only to Christ but also to church leaders, just as the philosophers styled themselves as the physicians of the soul. In the early third century, Origen regarded the apostles Peter and Paul and the prophets as practicing the therapy of the soul, and added bishops to the list of the physicians of souls (*Hom. Ps.* 37.1.1). This was based upon Origen's fundamental analogy between the healing of the body and that of the soul that was rooted in Greco-Roman philosophy. According to David Bostock, Origen saw the work of providence as a work of divine healing of

---

133. Israelowich, *Patients and Healers*, 57–58.

134. Montserrat, "'Carrying on the Work,'" 230–31, quoting E. M. Forster's book review in the *Egyptian Mail* (December 29, 1918). See also Patricia Cox Miller, *Dreams in Late Antiquity: Studies in the Imagination of a Culture* (Princeton: Princeton University Press, 1994), 116–17: "It is one of the ironies of history that, despite virulent attacks on Ascelpian religion by Christian apologists from Tertullian to Eusebius, who viewed Asclepius as a rival to their own healing god, Asclepius lived on in Christianity in the cult of the saints."

135. Nutton, *Ancient Medicine*, 302, 316.

humanity in three ways.[136] First, providence gives humanity "the truth and wisdom of God in proportion to their needs and their spiritual capacity."[137] The Hippocratic Corpus and Galen emphasized that different treatments and diets were necessary for different individuals; likewise, according to Origen, "if ten men with ten different illnesses visit a physician, he does not cure them all by the same method. . . . Similarly the Word of God speaks in accordance with the different sorts and conditions of men" (*Hom. Ezra* 3.8). Second, providence saves humanity from their sins by diverse forms of punishment. Origen was fully aware that "some acts of healing need both the surgeon's knife and cauterization," so "the cancer of our sin requires similar treatment. . . . Physicians cut, cauterize and dose their patients with bitter medicines . . . even so God applies his punishments to sinners with wisdom and with an impressive organization" (5.1). Finally, providence saves humanity from their sins by the use of contraries. Just as the Hippocratic Corpus and Galen taught removal of disease by the use of contraries, according to Origen, "at the heart of this healing by contraries was the Cross on which the humility of Christ confronted the pride of Satan and the forces of life confronted the powers of death."[138] As a result, "we have all been purified by the death of Christ, which was given like a medicine (*tropon pharmakou*) to the opposing forces" (*Comm. Jn.* 1.32).

Cyprian also expounded on the duty of the bishop to provide "the remedies needed for salvation" (*Laps.* 14). Facing the thorny situation of mass apostasies during the Decian persecution, Cyprian urged his clergy to apply beneficial remedies of rigorous penitence in the case of the lapsed Christians. They should not act like untrained doctors who treat "the swelling, festering wound" of apostasy with gentle hand, allowing the poison to remain buried deep in the body; rather, they should cut open the wound and eliminate it "with stringent remedies," that is, by imposing rigorous penance. Cyprian says, "let the patient shout and cry never so much, let him protest in exasperation at the pain—later he will be grateful, when he feels his health restored" (14). The physician does "not give to the sick what is appropriate for healthy bodies," since unseasonable food may aggravate the illness instead of mitigating it (*Ep.* 31.6.4). Likewise, the clergy should

136. See David G. Bostock, "Medical Theory and Theology in Origen," in *Origeniana Tertia: The Third International Colloquium for Origen Studies*, ed. Richard Hanson and Henri Crouzel (Manchester: University of Manchester Press, 1981), 193–97.

137. Bostock, "Medical Theory and Theology," 194.

138. Bostock, "Medical Theory and Theology," 197.

let those who are ailing through their apostasy "struggle with their suffering" in patience; if they apply the remedy of forgiveness too hastily without allowing the slow passing of time for genuine penitence, "the healed wound splits open at the slightest mishap" (31.7.2; cf. 30.3.3). The bishop should regard "those who [he sees] were injured in the fatal persecution" not as dead but as "lying unconscious," and he ought to provide his assistance and healing remedies as part of his healing art (55.16.3).

The *Didascalia Apostolorum*, in its rather comprehensive portrait of institutional church life in third-century Syria, likens the bishop to "a compassionate physician" who heals "all those who sin," and those who sin are likened to "the sick" (10). The bishop is to "distribute with all skill, and offer healing for the remedy of their lives" using "the word of bandages and the admonitions of fomentations and the compresses of intercession." However, if the "ulcer" of sin "goes deep and decreases" the sinner's flesh, the bishop should "nourish it and counter-act it with healing medicine." If there is "filth" in the flesh, the bishop must "cleanse it with sharp medicine, that is with a word of reproof." If the flesh is "over-swollen," he should "reduce it and counter-act it with strong medicine, that is with the threat of judgement." The author then exhorts the bishop to "cut away and clear out the foulness of the ulcer" "with incisions of much fasting," so as to avoid gangrene in the flesh. If the gangrene prevails "even over the burning," the bishop, with advice and consultation with "other physicians [bishops]," should "cut off that decayed member, that it may not corrupt the whole body." However, amputation should not be done hastily, and the bishop should "use first the scalpel and cut the ulcer, that it may be clearly seen, and that it may be known what is the cause of the pain that is hidden inside, so that the whole body may be kept uninjured." Only when the bishop sees the person unrepentant, without hope of restoration, should he "with grief and mourning cut him off and cast him out of the church" (10). It is noteworthy that the author of the *Didascalia* is clearly familiar with the process of diagnosis, medical treatment, and surgery,[139] and medicalizes the spiritual problem and solution in a comprehensive way with an apparent expectation that his audience will readily understand them.

A century later, Gregory of Nazianzus compares physicians and pastors extensively in his *Oration* 2 in the context of explaining his retreat to Nazianzus after his ordination. The church is the body of Christ, and

---

139. *The Didascalia Apostolorum in Syriac I: Chapters I–X*, trans. Arthur Vööbus, CSCO 402 (Leuven: Secrétariat du CorpusSCO, 1979), 115n32.

just as a body is ruled by one member, the local body of Christ is to be presided over by its pastor (bishop), "performing the function of the soul in the body" (2.3). As this body suffers from sickness and needs healing, it is subject to pastoral leadership and care, whose goal is "the perfecting of the church" (2.3). In this pastoral ministry, "any one may recognize this, by comparing the work of the physician of souls with the treatment of the body; and noticing that, laborious as the latter is, ours is more laborious" (2.13); for the one labors about bodies, which are perishable, whereas "the other is concerned with the soul, which comes from God and is divine" (2.16, 17). A physician scrutinizes a patient, prescribes medicines and diet, overrules the wishes of a patient who might work against his own healing, and even uses "the cautery, the knife, or the severer remedies" (2.18). However, according to Gregory, "none of these, laborious and hard as they may seem, is so difficult as the diagnosis and cure of our habits, passions, lives, [and] wills" (2.18). We should not hide away our sin or cover it deep in the soul "like some festering and malignant disease" (2.20); nor should we reject the treatment with medicines of wisdom by which the spiritual sickness is healed (2.20). Therefore, Gregory states that "our office as physicians far exceeds in toilsomeness, and consequently in worth, that which is confined to the body" (2.21). While the bodily medicine "is mainly concerned with the surface, and only in a slight degree investigates the causes which are deeply hidden," the careful pastor's holistic treatment is concerned with remedying what's hidden away in the soul (2.21), so that the pastor will equip "the soul with wings . . . rescue it from the world and give it to God" so that it advances toward the state of *theosis* (2.22). In this task of providing a therapy for the soul, the pastor is to heal and cleanse not only the souls of his congregation but also his own soul from passion and sickness (2.26). As the physician does not administer the same medications (*pharmakeia*) and the same food to all bodies, but to some one kind and to others a different kind, according to their degree of health or illness, so too the pastor must treat souls according to different words and methods in order to achieve the state of *theosis* (2.30). "The patients themselves bear witness to the cure: the word (*logos*) guides some; others are led by example (*paradeigma*); some require incentives; others restraints" (2.30).[140] Gregory then lays out the familiar paths of Platonic

---

140. This translation comes from Susanna Elm, *Sons of Hellenism, Fathers of the Church: Emperor Julian, Gregory of Nazianzus, and the Vision of Rome*, Transformation of the Classical Heritage 49 (Berkeley: University of California Press, 2012), 170.

and Aristotelian models of "finding the right words to cure each soul of its false beliefs and imprint it with the right ones" (2.31–33):[141] some must be praised, others exhorted; some must be encouraged, others rebuked; and some publicly blamed, and others privately corrected; in all these cases the pastor must speak at the right time (*kairos*) so as to avoid injuries (2.31). Following the classical example of discerning the various needs of patients and responding appropriately to each with the psychagogic mixed method of exhortation, Gregory indeed stresses that no single cure works "at all times and in all cases . . . as time and circumstance and the disposition of the patient admit" (2.33).

Furthermore, the same Gregory of Nazianzus, in his funeral oration for Basil the Great, characterizes the latter as a physician who, again, as he relieves their wounds in his "New City," combines tenderness and severity in his pastoral ministry and care for the sick (*Or.* 43.40, 63, 81). In fact, Basil himself utilizes medical imageries extensively, referring to the superior of the monastic community as a "spiritual physician" and the disobedient member as a "diseased member" (*LR* 28.1). The superior should "endeavor to cure his ailment through private admonitions"; but if the disobedient member persists in disobedience, the superior "must rebuke him sharply before all the community and apply remedies" such as no food for a time or withdrawal of a blessing (*LR* 28.1; *SR* 44, 122). When correcting a sinner's fault, the superior is to "apply remedies to those trapped in passions (*tois empathesi*) in the manner of a doctor, not being angry at the infirm but doing battle with the disease";[142] he is to "set up a counter-irritant to the passions and heal the soul's debility by means of a more drastic regimen if necessary, as for example, countering vainglory by means of practices of humility" (*LR* 51). Indeed, the superior is the healer of the sickness of the community, and as such should have the disposition of a servant; for he "tends many wounded, wiping away the discharge from each wound and applying remedies according to the special need of each injury as it presents itself" (*LR* 30; cf. 43.2). Just as in bodily healing only the seasoned doctor with experience and expertise should use the knife on the sick, so, in spiritual healing, only the best qualified should dispense spiritual food judiciously and carefully to those who ask for it (45.2). Likewise, "if rebuke is a doctoring of the soul (*iatreia psychēs estin hē epitimēsis*), it is not for everyone to rebuke, anymore

---

141. Elm, *Sons of Hellenism*, 171.

142. Note the Hippocratic understanding of medical treatment as "combating" the disease.

than it is to practice as a doctor, but only the duty of him to whom" the superior "has entrusted it after thorough examination" (53).

Augustine also compared the task of pastoral minister to that of a physician, as early as 388 (three years before his forced ordination to priesthood): bishops, priests, and deacons have "in their care, not only the healthy, but those in need of cure, and the vices of the throng must be borne with in order that they may be cured, just as the plague must be tolerated until it is quelled" (*Mor.* 1.32.69). While Augustine would spend the rest of his life as the physician of souls in the midst of a "plague" among the diseased, unlike Gregory of Nazianzus, he was very reticent to present himself as the physician "seeking to cure the soul with philosophical knowledge and rhetorical skill."[143] Rather, Christ is the Physician (*medicus*)[144] who provided "many medicines from the holy scriptures" to "heal every ailment of the soul" (*Serm.* 32.1): "The man who diagnoses (*legit*) your illness for you from Hippocrates is not more to be relied on by you than the one who demonstrates [Christ] to you from the divine scripture how you are internally ill. So listen to the scripture saying, 'The beginning of every sin is pride'" (*Serm.* 360B.17). For this sin of pride, the terrible sickness of soul and the source of all sins, Christ the Physician becomes the remedy of the opposite, offering himself in humility: "Christ's humility is the remedy (*medicamentum*) for your pride" (*Serm.* 360B.17; 341.4; *Enarrat. Ps.* 18.2.15). Thus, Augustine exhorts his congregation to "listen to the doctor as he hangs on the cross" (*Serm.* 360B.18). As we can see, for Augustine the sickness of the soul basically corresponds to sin: "the soul wrestling with its diseases" is the one that has fallen into great harm through sin (*Enarrat. Ps.* 6.4).[145] This is fundamentally different from the Stoic understanding of "sickness of the soul," which refers to passions (*pathē*), that is, erroneous

---

143. Paul R. Kolbet, *Augustine and the Cure of Souls: Revising a Classical Ideal* (Notre Dame: University of Notre Dame Press, 2010), 184.

144. For the idea of Christ as the physician in Augustine, see the following articles: Susan B. Griffith, "Medical Imagery in the 'New' Sermons of Augustine," *Studia Patristica* 63 (2006): 107-12; Rudolph Arbresmann, "Christ the *Medicus Humilis* in St. Augustine," in *Augustinus Magister, Congrès International Augustinien, Paris, 21-24 September 1954* (Paris, 1954), 2:623-29; Rudolph Arbresmann, "The Concept of 'Christus Medicus' in St. Augustine," *Traditio* 10 (1954): 1-28; Thomas F. Martin, "Paul the Patient: *Christus Medicus* and the '*Stimulus Carnis*' (2 Cor. 12:7): A Consideration of Augustine's Medicinal Christology," *AugStud* 32, no. 2 (2001): 219-56.

145. See *Agon.* 11.12 and also *Serm.* 87.10.13: "The human race suffers not from bodily diseases but from sins."

judgments.[146] However, Augustine refuses to equate passion (or affective pain) with an error of judgment, just as he refuses to consider all passions as illnesses of the soul. In chapter 3, we saw how Augustine, in the *City of God*, uses Scripture to affirm fear, desire, grief, sadness, or joy as necessary parts of life in this world (14.9). Since the sickness of the soul is primarily sin, Christians should find a remedy in the Christian word, Scripture, particularly in the Psalms: "every illness of the soul finds its medicine in the scriptures, so any of us who are suffering from the malady . . . will do well to drink a potion from this Psalm" (*Enarrat. Ps.* 36.1.3). Therefore, Augustine instructed his congregants to receive the Word of God as "our daily food on this earth" with "a regular discipline of listening to the word of God," so that they could be formed by it (*Serm.* 22.7; 56.10). In fact, listening to Scripture was a part of the whole liturgical practice involving the Christian mysteries, and as such they were intended to be a holistic curative experience:[147]

> Brothers and sisters, what calls for all our efforts in this life is the healing of the eyes of our hearts, with which God is to be seen. It is for this that the holy mysteries are celebrated, for this that the word of God is preached, to this that the Church's moral exhortations are directed, those, that is, that are concerned with the correction of our carnal desires, the improvement of our habits, the renunciation of the world, not only in words but in a change of life. Whatever points God's holy scriptures make, this is their ultimate point, to help us purge that inner faculty of ours from that thing that prevents us from beholding God. (*Serm.* 88.5)

Thus, regular participation in liturgy forms a therapeutic process in which the Holy Spirit heals those who are sick in the soul by the reordering of their hearts, minds, and lives, and leads them to the vision of God.

We now come to Augustine's contemporary John Chrysostom on the therapy for the soul, and on this topic there is a plethora of recent publications, led by those of Wendy Mayer.[148] According to these studies and in

---

146. Isabelle Bochet, "Maladie de L'Âme et Thérapeutique Scripturaire selon Augustin," in *Les Pères de L'Église Face à la Science Médicale de Leur Temps*, ed. Véronique Boudon-Millot and Bernard Pouderon (Paris: Beauchesne, 2005), 381.

147. Kolbet, *Augustine and the Cure of Souls*, 190.

148. See, for example, Robert G. T. Edwards, "Healing Despondency with Biblical Narrative in John Chrysostom's *Letters to Olympias*," *JECS* 28, no. 2 (2020): 203–31; James Cook, "'Hear and Shudder!': John Chrysostom's Therapy of the

continuity with the previously mentioned church fathers, Chrysostom saw himself as a medico-philosophical therapist of souls, and his pastoral ministry, particularly through his homilies and writings, as the therapy of souls. Unlike Augustine, Chrysostom does identify passions as the sickness of the soul: greed (*philargyria*), vainglory (*kenodoxia*), and usury are the diseases (*nosēma*) of the soul; and greed in particular causes spiritual blindness just as an evil humor (*chumos*) causes physical blindness (*Hom. Matt.* 20.2, 4, 5; 19.1). A host of other passions constitute the diseases of the soul as well: desire (*epithymia*), luxury (*tryphē*), anxiety, pleasure, envy, and vanity (*Hom. Matt.* 89.3). To top the list, pride (*hyperēphania*), which caused the devil to fall, is wretched and miserable and "the most grievous disease (*nosos*) of the soul" (*Hom. John* 16.4). What causes these diseases of the

[handwritten margin note: Vices (corrupt) (sickness) & Virtues]

Soul," in *Revisioning John Chrysostom: New Perspective, Theories, and Approaches*, ed. Chris L. de Wet and Wendy Mayer (Leiden: Brill, 2019), 247–75; Chris L. de Wet, "The Preacher's Diet: Gluttony, Regimen, and Psycho-Somatic Health in the Thought of John Chrysostom," in Wet and Mayer, *Revisioning John Chrysostom*, 410–63; Wendy Mayer, "John Chrysostom: Moral Philosopher and Physician of the Soul," in *John Chrysostom: Past, Present, Future*, ed. Doru Costache and Mario Baghos (Sydney: AIOCS Press, 2017), 193–216; Wendy Mayer, "Madness in the Works of John Chrysostom: A Snapshot from Late Antiquity," in *Concept of Madness from Homer to Byzantium: Manifestations and Aspects of Mental Illness and Disorder*, ed. Hélène Perdicoyianni-Paléologou, Byzantinische Forschungen 32 (Amsterdam: Adolf M. Hakkert, 2016), 349–73; Wendy Mayer, "Medicine in Transition: Christian Adaptation in the Later Fourth-Century East," in *Shifting Genres in Late Antiquity*, ed. Geoffrey Greatrex and Hugh Elton (Farnham, UK: Ashgate, 2015), 11–26; Wendy Mayer, "Persistence in Late Antiquity of Medico-Philosophical Psychic Therapy," *Journal of Late Antiquity* 8, no. 2 (2015): 337–51; Wendy Mayer, "Shaping the Sick Soul: Reshaping the Identity of John Chrysostom," in *Christians Shaping Identity from the Roman Empire to Byzantium: Studies Inspired by Pauline Allen*, ed. Geoffrey Dunn and Wendy Mayer, Supplements to Vigiliae Christianae 132 (Leiden: Brill, 2015), 140–64. Furthermore, see also Blake Leyerle, "The Etiology of Sorrow and Its Therapeutic Benefits in the Preaching of John Chrysostom," *Journal of Late Antiquity* 8, no. 2 (2015): 368–85; Raymond Laird, *Mindset, Moral Choice, and Sin in the Anthropology of John Chrysostom*, Early Christian Studies 15 (Strathfield, Australia: St. Paul's Publications, 2012); Peter Moore, "Chrysostom's Concept of γνώμη: How 'Chosen Life's Orientation' Undergirds Chrysostom's Strategy in Preaching," *Studia Patristica* 67 (2013): 351–58; David Rylaarsdam, *John Chrysostom on Divine Pedagogy: The Coherence of His Theology and Preaching* (Oxford: Oxford University Press, 2014); Jessica Wright, "Between Despondency and the Demon: Diagnosing and Treating Spiritual Disorder in John Chrysostom's Letter to Stageirios," *Journal of Late Antiquity* 8, no. 2 (2015): 352–67; Junghun Bae, "John Chrysostom on Almsgiving and the Therapy of the Soul" (PhD diss., Australian Catholic University, 2018).

soul? According to Raymond Laird, the sick soul is caused by the disorder of the mind-set (*gnōmē*), as the mind-set controls both the soul (attitudes, desire, will, and choice) and the body as the critical faculty of the soul; therefore, the right mind-set guards the health of the soul.[149] According to Peter Moore, then, Chrysostom's preaching methods sought to transform the mind-set of diseased souls with biblical exemplars such as Christ and Paul, as Chrysostom believed that the consistent exposure to the models of a healthy mind-set transformed the disordered mind-set.[150] The mind-set's "orientations embodied in Christ's sacrifice, Scripture, and the lives of biblical characters and saints, portrayed to his congregation through preaching, [had] a transformative power upon" their mind-set's orientation and would lead to its reorientation at a profound level.[151] Scholars further suggest that Chrysostom used mind-set (*gnōmē*), mind (*nous/dianoia*), and choice/intention (*proairesis*) synonymously, and that he stressed the role of rational faculty (mind, reason, intention) in the health of souls.[152] And we see Chrysostom encouraging his listeners who are in despair to come continually to the church as a house of healing, to listen to the law of God, and to stamp it in their mind (*dianoia*) (*Hom. John* 3.1; cf. *Hom. Matt.* 32.6; 61.3). Since the confusions or corruptions in our souls are not due to "the nature of circumstances" but to "the infirmity of our minds (*dianoias*)," the remedy of consistent meditation on his preaching will bring about healing, which will not be attacked by the devil (*Hom. John* 3.1–2; cf. *Hom. Matt.* 20.3–4). Hence he writes:

> Doctors who treat the human body have discovered a multiplicity of drugs, various designs of instruments, and appropriate forms of diet for the sick, and the character of the climate is often sufficient by itself to restore the patient's health. Sometimes a timely bout of sleep relieves the doctor of all trouble. But in the present case there is nothing like this to rely on. When all is said and done, there is only one means and only one method of treatment (*therapeias*) available, and that is teaching by word (*logou*). This is the best instrument, the best diet, and the best climate. It takes the place of medicine and cautery and surgery. When we need to

149. Laird, *Mindset, Moral Choice*, 26–112, 221–56.
150. Moore, "Chrysostom's Concept of γνώμη," 356–58.
151. Moore, "Chrysostom's Concept of γνώμη," 358.
152. Samantha L. Miller, *Chrysostom's Devil: Demons, the Will, and Virtue in Patristic Soteriology* (Downers Grove, IL: IVP Academic, 2020), 97–110; Bae, "John Chrysostom on Almsgiving," 60.

cauterize or cut, we must use this. Without it all else is useless. By it we rouse the lethargy of the soul, reduce its inflammation, remove excrescences, and supply defects, and in short, do everything which contributes to its health. (*Sac.* 4.3)[153]

Indeed, the priest must convince the spiritually ill to follow this therapy of the word appropriate for the illness (*Sac.* 2.3–4).

In the preface of *On the Providence of God* (*Ad eos qui scandalizati sunt* [*Scand.*]), one of his last treatises, Chrysostom, as Wendy Mayer has shown,[154] puts forth a similar comparison between the work of the physician of the body and that of the physician of the soul:

When doctors intend to treat those with a fever or those sick with some other illness, they first seek to see the afflicted ones themselves, since from far away they are unable to provide for these illnesses. But we, who are earnest to treat not one or two but all in the world who have been scandalized ("fallen into moral error"),[155] do not require anything like that. . . . But while staying in one place, without instruments or drugs or food or drink or money or a long journey, we drive out this sickness. How and in what way? By preparing the medicine (*pharmakon*) of the *logos*, which for the sick becomes all these things and better than all we have mentioned. It is more nourishing than bread, it restores to health better than medicine, and it cauterizes more vigorously than fire without causing any pain at all. It restrains the foul-smelling streams of wicked thoughts; sharper than iron, it cuts out without pain that which is rotten. And it does this without causing any money to be spent and without increasing poverty. Thus, having prepared this medicine, we are sending it on to everyone, and I know that everyone will benefit from the treatment, provided they pay heed with exactitude and right-mindedness (*eugnōmosynē*) to what is said. (*Scand.* 1)

Like Gregory of Nazianzus, Chrysostom here highlights the superior work of the physician of the soul with the superior medicine, the *logos*, capable of treating many patients of the diseased soul ("those fallen into moral

---

153. John Chrysostom, *Six Books on the Priesthood*, trans. Graham Neville (Crestwood, NY: St. Vladimir's Seminary Press, 1984), 114–15, slightly altered.

154. See Mayer, "Persistence in Late Antiquity," 337–45.

155. This is Wendy Mayer's translation in "Persistence in Late Antiquity," 340.

error") powerfully, simultaneously, and freely. Since the treatment should correspond to the cause, he probes the cause of this disease of the soul, which is "an inquisitive and curious mindset" (*hē polypragmōn kai periergos gnōmē*), "wanting to know all the causes of everything that happens" and "questioning the incomprehensible and unspeakable providence of God" (2.1). If the patients understand the cause "and if they are willing to pay heed to it scrupulously, they will be delivered not only from this sickness but from many others, and not only for the present but perpetually" (1.1–2). For the nature of this medicine is such that "it both cures the illness at hand and acts as a preventative against other diseases" (1.2). Again, the wrong, disordered mind-set is the cause of the disease of the soul; and we should note the importance of the right mind-set to understand the cause and apply rational thoughts to this treatise (*logos*), which incorporates Scripture. In the context of this treatise, only with the willing right mind-set replacing its false belief, this therapy of *logos* becomes a panacea and prophylactic for Chrysostom's many supporters who have questioned God's providence in light of his exile and suffering (1.5).

Medico-Philosophical Treatises

In this last section, we will dig deeper into the Christian application of the therapy of the soul and study two more specific Christian texts. By focusing on otherness and difference in the context of the Christianized Roman East, we will draw out particular kinds of Christian identity. One is a rather infamous heresiology, Epiphanius of Salamis's *Panarion* (c. 375–378), and the other is Theodoret of Cyrrhus's apology, *A Cure for Greek Maladies* (*Graecarum affectionum curatio*) (c. 441–449). As exemplified by other bishops in the previous section, both Epiphanius and Theodoret adapt the fundamental premises of medical-philosophical treatises: (1) the interconnectedness of the psychic and somatic parts of a human being; (2) the medicalization of both problems (e.g., sickness of heresy or unbelief) and solutions addressed (e.g., medicine of the *logos*); and finally, (3) the construction of both a healthy belief-set/mind-set and ethical practice, as the main part of therapy. As mentioned in the previous chapter, according to Christopher Gill, there is a threefold distinction in the writings of ancient philosophical therapy: protreptic, therapy, and advice.[156] Having interrelated functions, they form a well-designed therapeutic

156. Christopher Gill, "Philosophical Therapy as Preventive Psychological Med-

process toward health.[157] "Protreptic offers encouragement to undergo therapy"[158] by "demonstrating value and benefit of a particular philosophy and defending objections of the adversaries in refutation" (especially in the literary and Christian apologetic tradition).[159] Therapy "removes false beliefs that produce psychological sicknesses" and is strongly tied to medicine. Finally, "advice replaces the false beliefs with true, or at least better-grounded, ones."[160]

At the outset, Epiphanius frames the purpose and goal of his *Panarion* as follows:

> Since I am going to tell you the names of the sects (*haireseōn*) and expose their unlawful deeds like poisons and toxic substances (*ious kai dēlētēria*), and at the same time match the antidotes (*antidotous*) with them as cures (*alexētēria*) for those already bitten and preventatives (*prokatalēptika*) for those who will have this misfortune, I am drafting this Preface for the scholarly to explain the "Panarion," or chest of remedies (*kibōtion iatrikon*) for those whom savage beasts have bitten. It is composed in three Books containing eighty Sects, symbolically represented by wild beasts or snakes. But "one after the eighty" is at once the foundation, teaching and saving treatment (*sotērios pragmateia*) of the truth and Christ's holy bride, the church. ("Proem I" 1.1–3)

Epiphanius here reflects his experience "as ascetic and bishop within the divided fourth century church" within imperial Christianity and sees his work as fundamentally medical in nature.[161] He aims not only to expose the "poisons and toxic substances" of the heresies ("sects"), represented as "wild beasts or snakes," but also to counter them with the *pharmaka* of orthodoxy, which works as a cure for those already affected, and a prophylactic for those yet to be bitten. Equating heretics with "deadly and harmful beasts that threaten innocent Christians with venomous doctrines and practices," he develops "a new taxonomy to stigmatize the heretical 'other'

---

icine," in *Mental Disorders in the Classical World*, ed. William V. Harris, Columbia Studies in the Classical Tradition 38 (Leiden: Brill, 2013), 342.

157. Gill, "Philosophical Therapy," 342.

158. Gill, "Philosophical Therapy," 342.

159. Rhee, *Early Christian Literature*, 24.

160. Gill, "Philosophical Therapy," 342–43.

161. Rebecca J. Lyman, "The Making of a Heretic: The Life of Origen in Epiphanius' *Panarion* 64," *Studia Patristica* 31 (1997): 445.

with these known and feared creatures."[162] Epiphanius thus heightens the danger of real or perceived heresies among readers, and urges them to be armed and therefore healed by "the foundation, teaching and saving treatment of the truth," and the church. In this way, Epiphanius uses protreptic to encourage his readers to undertake a removal of any heretical beliefs (therapy) that infiltrate the body of Christ and thus sicken it.

He then utilizes both therapy (*therapeia*) and advice throughout his work. In his second preface, Epiphanius establishes a connection between his work and the works of Nicander of Colophon[163] and others who catalogued the nature of beasts and reptiles and described the qualities of roots and plants ("Proem II" 3.1). "In the same way," he, "in trying to reveal the roots and beliefs of the sects," is making "a diligent effort, not to point evil out, but to frighten people and ensure their safety" (3.2). Integral to his therapy here is exposing the heresies' false beliefs and wrong behaviors so that his readers "would recognize the dreadful, dangerous beasts and be safe and escape them by God's power, by taking care not to engage with such deadly creatures, if they encountered them, and were menaced by their breath or bite, or by the sight of them" (3.3). Thus, he sees his work as "a defense" against "these snake-like teachings of the sects" (3.4).

For example, Epiphanius exposes "horrid, snake-like wickedness" of Cerinthians (also known as Merinthians) (*Pan.* 28.8.3), a type of Jewish group derived from Cerinthus and Merinthus, as the ones who "boast of circumcision, but say that the world was made by angels and that Jesus was named Christ as an advancement" ("Anacephalaeosis II" 28). He likens them to the "rot viper" with two heads whose bite harms "those who happen on it. For it ruins its adherents, sometimes by destroying the New Testament's teachings with material from the old religion [i.e., Judaism], and sometimes by circulating their false charges against the apostles who had come from circumcision to faith in Christ, with lying words as though from the New Testament" (28.8.4). His antidote (advice) to their false teaching is the truth gained from a proper exegesis of Scripture: first, the Holy

---

162. Young Richard Kim, *Epiphanius of Cyprus: Imagining an Orthodox World* (Ann Arbor: University of Michigan Press, 2015), 174. On Epiphanius's discursive violence, see 201–3.

163. He was a Hellenistic (third- or second-century BCE) poet who wrote didactic poetry, including the *Theriaca*, on poisonous animals and natural remedies. According to Richard Flower, Epiphanius may have engaged directly with Nicander's writings: Richard Flower, "Medicalizing Heresy: Doctors and Patients in Epiphanius of Salamis," *Journal of Late Antiquity* 11, no. 2 (2018): 254–59.

Spirit fell upon the uncircumcised Cornelius in Acts 10, which is proof of the salvation of the gentiles (28.3.1–5); second, the apostle Paul forbade the gentiles to get circumcised in Galatians 2 (28.4.1–3); third, Christ was conceived by the Holy Spirit and born of the Virgin (incarnation) and indeed suffered and was raised from the dead (resurrection) (28.6.7–28.7.7). Therefore, Epiphanius "strikes and thrashes" the viper's "rot, poison and fangs with the cudgel of the truth" and effects the cure of its bites, and serves as a remedy and preventative for its poison (28.8.5; 28.7.7).

Another example of Epiphanius's utilization of therapy and advice is his revelation of Montanists (Pepuzians), who "accept the Old and the New Testament but, by boasting of a Montanus and a Priscilla, introduce other prophets after the [canonical] prophets" ("Anacephalaeosis IV" 48). This sect is "like the viper of hemorrhage, whose mischief is to drain the blood from its victims' entire bodies and so cause their deaths" (48.15.6). They "stab the body of an innocent child and get its blood to drink, and delude their victims by [pretending], if you please, that this is initiation in the name of Christ" (48.15.7). In addition, Montanus glorifies himself—this is proof that "he is not Christ, was not sent by Christ, and has received nothing from Christ" (48.11.8). In contrast, the truth of Scripture is that "the holy apostles glorified the Lord after receiving the Paraclete Spirit" (48.11.6). Again, Epiphanius's point is clear:

> I give all the facts, as I said, with accuracy, . . . and make these shocking disclosures for the readers' correction. And I prepare a sort of medicine made of refutation from the words of sacred scripture (*theiōn logōn*) and right reasonings (*logismōn*), and compound [it] in the Lord for two purposes: for the recovery of the sufferers from their illness (*nosou*) and great pain (*algēdonos*), but for a prophylactic (*prokatalēptikon*), as it were, for those who have never contracted the disease. Thus may I too be called a disciple of the Lord's disciples for imparting the medicine (*pharmakou*) of the truth to the wise, and a disciple of the Savior himself, the help of bodies and souls. . . . I have crushed [the viper's] poison, and the venom on its hooked fangs, with the cudgel of the truth of the cross. (48.15.3–5, 6)

The last example is regarding Paulianists, derived from Paul the Samosatian, the adoptionist bishop of the metropolis of Antioch. Paul the Samosatian insisted that Christ did not preexist, but came to existence in Mary's time through the incarnation ("Anacephalaeosis V" 65). Epiphanius likens Paul to a viper called the dryinas, whose "camouflaging of itself among the fallen leaves with the color of each leaf" is well known (65.9.4). This animal's bite is not particularly painful, but if it remains undetected, it can cause death (65.9.5). "In the same way," Epiphanius believes, "this man, with his sect, pretends to belong to the faithful by bearing Christ's name while adopting Jewish doctrine. He confesses that Christ is the Word but does not believe that he is; and he is not ashamed to make a parade of himself in many ways" (65.9.6). Epiphanius's antidote again comes from Scripture (John 1:1, 3, 14): "In the beginning was the Word, and the Word was with God, and the Word was God. . . . All things came into being through him, and without him not one thing came into being. . . . The Word became flesh and lived among us" (see 65.3.5, 8). These scriptural verses make clear that the Word's existence is "not just as an utterance but as an entity" and that the Word was God, "not the Word was in God" (65.3.6, 9). Therefore, by preaching this "doctrine of the truth," Epiphanius has pointed out the deadliness of Paul's "thorns" (65.9.2); and "with the healing scalpel of the Gospel," Epiphanius has "drawn the poison out of them" (65.9.7, 2).

Epiphanius's therapy and advice presuppose rational capacities and willing agencies of the affected sick souls. Indeed, Gill identifies personal agency as fundamental to the first of four key elements in the therapeutic strategies of philosophical theories.[164] At the end of his exposure of Simonians, Epiphanius presents his therapy as "an occasion of truth and healing," which his readers will voluntarily choose (*Pan.* 21.7.1). At the end of his revelation of Origenists' false beliefs, Epiphanius offers his advice as "a medicine, like an antidote" to "those who are willing to read attentively for exercise in truth" (63.72.3). Furthermore, after unveiling Manichaeans' errors, Epiphanius expresses his expectation that "the sensible can understand the meaning in all the words of the truth, and in those of this so-called Mani's falsehood" (66.87.8). Finally, he attributes the success of his therapy and advice concerning the Arians to "the right reasoning in [his] mind" and expects his readers, who share the same reasoning, to follow through with "the cure of those who have been bitten, and the correction of those who have already joined the ranks of the evil" (69.81.2, 7).

At the conclusion of the *Panarion*, Epiphanius adds *De fide*, a short, "accurate account of the faith of the Catholic and Apostolic Church," as the culmination of his therapy and advice. Against the "various, multiform,

---

164. Gill, "Philosophical Therapy," 349–50. Mayer shows the importance of personal agency in John Chrysostom's psychology in "The Persistence in Late Antiquity," 343–45.

much divided, rash teachings of the crooked counsels of our opponents," distinguished by species and genus, he presents the "one faith" of the catholic church born with the help of the Holy Spirit (1.1; 6.1).[165] It confesses the Trinity and its coeternal and everlasting nature in unity. Epiphanius then puts forth the biblical accounts of the creation, incarnation, Christ's baptism, earthly ministry, suffering, death, resurrection, and ascension, and the judgment, all of which constitute orthodox cognitive therapy in addition to the specific scriptural antidotes to the specific heresies (14.1–18.3). As for behavioral therapy, Epiphanius highlights, first, the ascetic disciplines of the monks (22.7; 23.4, 6–7); second, "the custom of hospitality, kindness, and almsgiving to all," which "has been prescribed for all members of [the] holy catholic and apostolic church" (24.1); finally, many forbidden activities such as fellowship with any sect, fornication, magic, charms and amulets, and theatrical shows. These ordinances, drawn from "the Law, the Prophets, the Apostles and the Evangelists," function "like a good antidote compounded of many perfumes for the health of its users" (25.1). In this way, Epiphanius sees his *therapeia* effective for drawing a clear-cut Nicene orthodox Christian identity in the Christian Roman Empire of late antiquity.

If Epiphanius's *Panarion* employed a medico-philosophical therapy for his polemical aim of otherizing heretical sects from orthodox Christianity, Theodoret of Cyrrhus's *A Cure for Greek Maladies* also utilized a medico-philosophical therapy whose use of medical language and analogy is central to his apologetic aim.[166] Theodoret's goal in this treatise is to cure (i.e., persuade) his diseased audience (*nosountōn*), steeped in the "stain of unbelief" (*hē apistis tēn lōbēn*), that is, Greek *paideia* and religion, with the holistic *therapeia* of Christian *paideia*. At the outset, a quick note about the genre of this treatise: Yannis Papadogiannakis regards it as an apology using a traditional philosophical concept of *therapeia*,[167] and Wendy Mayer

---

165. This one (truth) versus many (false) argument was a typical feature of early Christian apologetics in presenting the superiority of Christianity's monotheism over Greek polytheism. For examples, Athenagoras, *Leg.* 14.1–2; Theophilus, *Autol.* 1.1, 10; 2.3.

166. Chris de Wet regards the discursive operations and aims of Epiphanius's *Panarion* and Theodoret's *Curatio* the same: "They classify pathological ideologies, and aim to correct or anathematize the groups associated with the ideologies." Chris L. de Wet, "Medical Discourse, Identity Formation, and Otherness in Late Ancient Christianity" (paper presented at the International Patristic Conference at Oxford, August 2019), 4.

167. Yannis Papadogiannakis, *Christianity and Hellenism in the Fifth-Century Greek*

takes it as a "therapeutic medico-philosophical treatise for the sick soul."[168] I take a blended approach, to view the treatise as a "medico-philosophical therapy" with an apologetic aim. The treatise with its *topos* stands in the long tradition of Christian apologetics that has continued from the second century, and at the same time utilizes the form and methodology of medico-philosophical therapy to achieve that aim.

In the prologue and throughout the treatise, Theodoret aims to present counterarguments to Greek objections to Christian faith, and to put forth "the knowledge (*epignōsis*) of Gospel truth based on Greek philosophy" so as to cure the ill and to assist the healthy (Prologue, 16). Thus, like Epiphanius, Theodoret uses protreptic as an overall framework of the treatise and employs both *therapeia* and advice throughout the twelve discourses. For therapy in particular, Theodoret in his first discourse on faith establishes a parallel between physical and spiritual illness and treatment using a medical analogy. God has assigned a medical treatment (*therapeia*) for bodily ailments, which are involuntary, and a psychic *therapeia* for spiritual ailments, which are voluntary (*Curatio* 1.1–2).[169] Physicians administer to the bodily ill, who desire to return to good health, treatments ranging from gentle to bitter *pharmaka*, to strict diet, and even cautery (1.3).[170] On the other hand, the spiritually ill, who have contracted the "stain of unbelief" (*hē apistis tēn lōbēn*), not only fail to perceive the seriousness of their sickness but even imagine they are enjoying the height of health. Thus, if a skilled physician of the soul wishes to apply *pharmaka* for the disease, the sick retreat immediately, "like those afflicted with phrenitis" (*hoi phrenitidi katechomenoi nosōi*); and they push aside the proffered treatment (*therapeian*) with boxing and kicking and running from the cure (*tēn iatreian*) (1.4, 5). John Chrysostom described the symptom of phrenitis in a similar way: "When their brain becomes inflamed and they kick and bite those who wish to deliver them from their infirmity, then their disease becomes incurable" (*Pan. Bab.* 55). So has Augustine: "[A doctor] cuts to heal; but when he cuts the sick patient, the latter feels pain, and screams and struggles; if he has lost his mind with fever, he even strikes the doctor" (*Enarrat. Ps.* 34.2.13). Then physicians not only restrain those with phrenitis and

---

East: Theodoret's Apologetics against the Greeks in Context (Cambridge and London: Center for Hellenic Studies, 2012), 31.

168. Mayer, "Medicine in Transition," 24n90; Mayer, "Persistence in Late Antiquity," 338; Mayer, "Shaping the Sick Soul," 153.

169. Clement of Alexandria, *Paed.* 1.1.3; cf. Theodoret, *Prov.* 4.28.

170. Cf. Clement of Alexandria, *Strom.* 1.27.171; *Protr.* 10.109.

forcibly apply fomentations, but also devise every means to drive out the sickness (*pathos*) and "restore the former equilibrium (*harmonia*) of the parts to the whole" (*Curatio* 1.5). The physicians of the soul must likewise seek every means to "dissipate the mist settled upon them and show the splendor of the spiritual light" (1.6).

The proper diagnosis of internal disorder as a major reason for psychic illness was a characteristic feature of philosophical therapy.[171] Theodoret here diagnoses unbelief (*apistia*), that is, a rejection of Christian doctrine and way of life, as a spiritual disease that manifests as madness with delirious responses. "Just as insanity involves fundamental changes in perception and the physical state," unbelief, construed as a psychosomatic phenomenon, "affects deeply those who suffer from it."[172] Theodoret further defines unbelief as "the disease of smug self-conceit" (*to tēs oiēseōs pathos*; 1.9; 2.5, 21).[173] This disease prevents those who espouse Greek *paideia* from dispersing the mist (*achlyn aposkedasai*) from their eyes (2.5). "Self-conceit" (*oiēsis*) is their deep-seated pretentious feeling of intellectual and cultural superiority, and thus they despise the Christian Scripture in its style (1.9; 2.5). It is also an illusory and blind knowledge that is based on mere conjecture, not true knowledge (2.21). Indeed, Clement of Alexandria lamented that pride and *oiēsis* (self-conceit) had brought philosophy into evil repute, and so had false knowledge (*gnōsis*) tainted true knowledge (*Strom.* 2.11). In this sense, self-conceit for Theodoret is a complex *pathos* with cognitive content (in this case, false belief/knowledge) that leads the sick to resist Christian *paideia*.[174] Indeed, *apistia*, literally a lack of *pistis*, arises from a false belief, as the opposite of *pistis*, defined as "a voluntary assent of the soul" and "the contemplation of an invisible object" that is accompanied by true knowledge (*Curatio* 1.91–92, 108).

So how does Theodoret proceed in his *therapeia*? Theodoret focuses on a cognitive therapy that identifies false beliefs as the main cause of unbelief (*apistia*) and self-conceit (*oiēsis*) by showing that

> Just as physicians prepare effective remedies from poisonous beasts, even from vipers, they first throw away some parts and boil the rest, and with this they expel many maladies, we also, who have taken in hand the works

---

171. Gill, "Philosophical Therapy," 348–52.
172. Papadogiannakis, *Christianity and Hellenism*, 34.
173. Cf. Clement of Alexandria, *Strom.* 7.16.95; 7.16.98, 99; 2.11.
174. Papadogiannakis, *Christianity and Hellenism*, 36.

of your poets, your historians, your philosophers, we first lay on one side what is injurious and manipulate the rest for the knowledge of teaching. We apply to you this remedy that acts as an antidote. And even those whom you regard as our enemies we show you that they champion our teachings and we make you see that even they teach you the faith. (1.127)

Theodoret has already applied this treatment when he "proved" the need for faith with the selected opinions (*doxai*) of the Greek poets and philosophers themselves (e.g., Antithenes, Euripides, and Aristotle; 1.91–92, 108). In proving the superiority of Christian monotheism and the Trinity (in the second discourse), like his apologetic predecessors before him, Theodoret employs "testimonies" (*martyriai*) from Greek poets and philosophers (especially Plato) to expose their conflicting views on the first principle (2.1–11) and their dependence on the Egyptians and the Hebrews (2.21–50).[175] In fact, Porphyry, "our worst enemy" (12.97), testifies to the antiquity of Moses prior to Homer and Hesiod (2.50), and Moses's priority and divine inspiration in Scripture make Christian revelation on the Trinity the only harmonious, therefore certain, teaching (2.51–94).[176] Monotheism and the Trinity in the writings of Plato, Plotinus, Plutarch, Numenius, and others confirm the truth of Christian doctrine, but also show that they have plagiarized the biblical teaching (another familiar theme in Christian apologetics); and others mixed falsehood with the pure scriptural teaching, and with Greek myths and theogonies (2.70–89, 94–114). Theodoret positions himself as a skilled physician of the soul in applying, with calculated selectivity and psychological precision, his *therapeia* to the opinions and testimonies of the who's who of Greek *paideia*. In Theodoret's hand, they function as self-exposing, self-correcting, and thus self-healing therapy and advice for suffering souls that help them transform their false beliefs.

Theodoret then makes an explicit correlation between health and true theology on the one hand and deformity and false theology (i.e., atheism and polytheism) on the other: "Regard as 'wholesome' and 'sound' (*hygieis men kai artious*) those bodies which have embraced the true theology (*alēthē theologian*)—what was given by nature at the beginning of the time and confirmed later by the divine oracles—but label as 'deformed'

---

175. Cf. Clement of Alexandria, *Strom.* 1.21, 101.

176. On the priority of Moses to Homer, who was after the Trojan War, see Clement of Alexandria, *Strom.* 1.21.117; Theophilus, *Autol.* 3.20, 23; Tatian, *Or. Graec.* 31.1–4; Cyril of Alexandria, *Juln.* 1.4–5.

(*anapērous*), on the other hand, not merely those that deny the existence of God but also those that have promoted a multiplicity of gods[177] and that have placed the Creator and His creation on the same footing" (3.3).

Both Hippocratic and Galenic medicine regarded health as the state according to nature (*kata physin*), and illness or deformity as the state against nature (*para physin*), as they conceived of nature as "a principle of order, regularity, and normalcy."[178] Theodoret applies this prevalent medical understanding of health and sickness to true and false theology; a belief in Christian doctrine results in physical and psychic health, and a belief in Greek atheism and polytheism causes unbelief and self-conceit, the psychosomatic illness that prevents one from recognizing the true religion. Again, like his apologetic predecessors, Theodoret identifies demons behind atheism and various forms of polytheism such as worshiping creation, and the deification of heroes, emperors, passions (*pathē*), and malevolent daemons (3.1–69). While testimonies by Porphyry, Plato, and Zeno as *pharmaka* expose the vanity of idols (3.70–74), Scripture teaches the true nature of angels as the model of Christian ascetics and the governing and sanctifying work of the Holy Spirit (3.87–99, 105–108).

Like Epiphanius, essential to Theodoret's *therapeia* is the idea that humans are rational and free agents (5.4, 6); since *apistia* and *oiēsis* are a voluntarily contracted disease, their cure involves, or rather must involve, the agency of the sick themselves. Theodoret's understanding of cure as persuasion and his juxtaposition of false and true theology with the exhortation to choose the truth underscore the importance of personal agency. Indeed, arguing for scriptural affirmation of human free will and accepting the Pythagorean and Platonic division of the soul (in discourse 5) (4–8, 19, 28–32), Theodoret presents the following "truth of the divine dogmas": "the body's formation by God, the immortal nature of the soul, the reasonable part of which controls the passions (*to tautēs logikon hēgoumenon tōn pathōn*), which also have a necessary and useful function to play in human nature" (5.76). The harmony of human nature, that is, reason (*logismos*) ruling over *thymos* and *epithymia*, produces a balance that constitutes health and virtue (*aretē*); but *pathē* becomes disease when it rebels against reason and throws the soul into violent motion that impedes proper judgment (5.77–79).[179] Theodoret's point here is that one can maintain equilibrium or return to equilibrium by

177. See n. 165.
178. Temkin, *Hippocrates*, 190.
179. Cf. Papadogiannakis, *Christianity and Hellenism*, 42.

choosing Christian *paideia*; in fact, this Christian *therapeia* has a democratizing effect— in contrast to the elitism of Greek philosophers, its deliberate, rational pursuit of virtue defies distinctions of gender or social status (5.68–69);[180] and since human action is grounded in human nature and free will, it is liable to God's reward and punishment (5.80).

This leads to the behavioral therapy part of Christian *therapeia*, without which the latter is incomplete. In discourse 12, Theodoret reminds his readers of the object of philosophy, which is the knowledge of the divine and the corresponding practical virtue or behavior,[181] which can be summed up in the following way: "He imitates, as far as possible, the God of the universe. He desires what God desires, and likewise hates what his Master of the universe hates. How what pleases and displeases God is taught in clear terms in the divine laws" (12.8). Like Epiphanius, Theodoret exhorts readers to the lifestyle of Christian ascetics as the embodiment of this ideal (12.4–6, 27).[182] "Having installed themselves on mountain tops, . . . they have chosen a life spent in contemplating divine things and their chosen lot in life is in harmonizing themselves with this contemplation, with no care for wives, children, and material possessions, but directing their souls in accordance with the canon of divine laws, and like the best artists, they paint their spiritual image after the best models of virtue" (12.27).

Just as he did for cognitive therapy, he compiles testimonies from Greek philosophers that both support and come short of the divine law of virtue in the Scripture (from Theodoret's perspective). Plato, drawing from the Old Testament, but in contrast to immoral Socrates (12.26–32), again comes closest to Christian virtue with his teaching on imitating or assimilating to God the Creator through justice, wisdom, and the perfection of virtue (12.19–26). In this teaching, Plato is in fact depicting "the mode of existence of our philosophers [i.e., ascetics] because he certainly did not find such types among the Greeks" (12.26.1–2).[183] Porphyry himself details

180. Cf. Clement of Alexandria, *Strom.* 4.19.122.

181. Cf. P. Hadot, *Philosophy as a Way of Life: Spiritual Exercises from Socrates to Foucault*, edited and introduced by A. I. Davidson, trans. M. Chase (Oxford: Blackwell, 1995), 82–83.

182. On the correlation between ancient medical and philosophical knowledge and early Christian ascetic discourse, see Mayer, "Medicine in Transition," 16; Bae, "John Chrysostom on Almsgiving," 40n176.

183. Theodoret goes on to characterize the Christian ascetics as the fulfillment of Platonic prefiguration of a philosophical way of life: "But among the Greeks who cultivate philosophy no one of them has built a mountain shack and occupied it; a suf-

the immoral proclivities of Socrates (12.57–69), and even Plato falls short with his love of tyranny and the luxury of Sicily (12.70–72). While some Greek women were morally admirable in their modesty (12.73), the biblical teaching on chastity as leading a life of freedom from care and marriage as a means to multiply the human race in chastity is superior (12.74–79). Theodoret highlights that the superiority of a Christian way of life (*philosophia*) rests in its rationality and moral uprightness in contrast to the absurdity of a Greek way of life (12.77–94). Therefore, even Porphyry acknowledges that once Christian faith was established, Jesus replaced Asclepius and other false gods as the true healer and light (12.95–97).

Drawing on the medico-philosophical literature of the Greek *paideia*, Theodoret likens the symptoms of Greek unbelief to those of madness manifest in psychological sickness and moral and rational failings. Then he constructs a holistic therapy of the Christian way of life by selectively drawing from the same kind of literature as the Greek *paideia* that he calls his audience to reject. This therapy of Christian *paideia* consisting of a Christian belief set (Christian monotheism and Trinity) and practical ethics (virtues—asceticism) seeks to move his sick audience away from the wrong belief (Greek unbelief) that causes psychosomatic sickness, and to restore their moral and intellectual faculties to accept the truth toward well-being and wholeness. Theodoret's construction of Christian *paideia* presents an alternative *paideia* to those who might freely embrace the medicine of Christianity still very much taken from the traditional Greek therapy. In this way, Christians in the Christian Roman Empire sought to form a distinctive identity in relation to Greco-Roman *paideia*, but still within its cultural framework.

Conclusion

Having adopted and appropriated Greco-Roman medicines and the philosophical care of the soul, Christians in late antiquity developed unique forms of comprehensive health care, including miraculous healings, the tangible care of the sick, medical charity through hospitals, monastic health care, private health care involving religious and rational medicine, and the therapies of the soul developed by bishops. By the fourth century, Christianity emerged as "the religion of healing" par excellence, and holy men's

---

ficient proof of that is provided by the writings of antiquity, and this is corroborated by you in your hostility to those who opt for such a life" (12.29).

spiritual prowess primarily manifested in healings—even through their relics. Christians sacrificially took care of their own sick people and plague victims in the earlier centuries; by the late fourth century, they established an expansive network of hospitals, especially in the Greek East, for providing public, collective, and holistic care for the poor for the long haul; their hospitals were there to stay, and are the prototypes for the hospital as we understand it today. Christians also made use of rational medicine along with prayers and amulets for healings, fully utilizing their options as patients in the late antique "open medical marketplace." Finally, bishops and monastic superiors took over the doctor-philosopher's role in healing the souls of their congregants and monks through the *logos* of the Scripture and their homilies and writings. Like the larger Greco-Roman health care, these forms of Christian health care were not mutually exclusive, but rather synergistic, convergent, and comprehensive. In incorporating and developing these various elements and forms of health care, Christians were deliberate in both drawing boundaries and blurring boundaries in their identity formation, and in doing so, shaped late antique *Christian* culture and society while constructing an alternative *paideia*, using the building blocks of the Greco-Roman *paideia*.

CONCLUSION

We have charted Christian understandings and practices concerning health, disease/illness, pain, and health care in the larger Greco-Roman context from the second through the fifth centuries. What emerged out of them are Christian constructions of health, illness narratives, pain narratives and pedagogies, and a health-care system as they related to Christian identity formation and community ethos. This brief conclusion presents some summative statements on each area in light of the thesis of this study.

Constructing Christian Visions of Health

First of all, Christians constructed their visions of health fundamentally coming from God the Creator, but they also incorporated the well-trodden concepts of health in Greco-Roman medicine and philosophy into their faith's framework of creation and providence. Therefore, they affirmed health as a balance of bodily elementary qualities and humors in accord with nature and, more importantly, adopted the Aristotelian and Galenic understanding of teleological anatomy. Whether it was Lactantius, Gregory of Nyssa, John Chrysostom, Nemesius, Augustine, or Theodoret, Christian leaders regarded the intricate composition, upright stature, and impeccable functionality of the human body and mind as the prime validation of their doctrines of creation and providence, which had to be defended well into the fifth century. Thus, they found Greco-Roman medicine's naturalism and functional notions of health compatible with their theological anthropology. Furthermore, the influential *Life of Antony* by Athanasius set forth an ideal of prelapsarian ascetic health through strict *askēsis*, which would later be negotiated and reconfigured. CHRISTIAN ETIOLOGIES OF ILLNESS

Constructing Christian Illness Narratives

Second, Christians accepted in general the standard medico-philosophical theory of disease as an imbalance of elementary qualities and humors against nature, and a natural etiology of disease. However, they also conferred moral and symbolic meanings and religious significance on their illness by constructing their illness narratives around multiple etiologies as their non-Christian authors did in their religious contexts. Following the Hebrew Bible and like their non-Christian neighbors, they first understood illness as God's punishment or displeasure for sin. It was a way to make meaning out of their illness in relation to the fallen world as Irenaeus directly linked the primordial sin of disobedience to the illnesses tormenting humanity (*Haer.* 5.15.2). On the basis of that fundamental reality, Christians readily made a specific sin-illness connection not only for themselves but also for their "enemies" (persecutors). This sin-illness paradigm, however, also anticipated the hope of the repentance-healing template replete in the Hebrew Bible and ascetic literature. The second way of Christians constructing meaning out of illness was to see God's pedagogy and providence behind their illness for the eventual good of the sufferer. This good was particularly interpreted as humility and moderation against pride, the necessary virtues to experience God's power in weakness. The third etiology relevant to and meaningful for Christians was demonic causes of illness, which began with the Second Temple Jewish literature and continued in the Christian Bible (NT). It seems that this demon-illness paradigm particularly took roots in "popular" Christianity, as evidenced in the apocryphal Acts in the second and third centuries and also in Christian magical amulets in the fourth and fifth centuries. In the fourth and fifth centuries, however, this etiology was also shared by Christian clergy and ascetics alike in light of the changing contexts of the Christianization of the Roman Empire and prominence of ascetic movements. Finally, the ascetic etiology presented both theological challenges and opportunities to holy men and women whose illnesses came as a result of stringent *askēsis*. The ideal of prelapsarian ascetic health by Antony's *askēsis* was fundamentally based on the Deuteronomic notion of the obedience-prosperity (health) and disobedience-poverty (illness) paradigms. However, this notion no longer corresponded to the reality of ascetics, as their health was often stricken rather than strengthened by their *askēsis*. Therefore, a new theology had to accommodate and reflect this reality as to how to explain ascetic sanctity in light of ascetic illness: illness was no longer a threat to *askēsis* but a new mode of *askēsis*; thus, this new mode of *askēsis*—that is, faithful endurance of illness—became a paradoxical way to achieve ascetic holiness. These multiple etiologies of illness show the common but unique ways that early Christians constructed meanings and significance for their illnesses, and also attempted to understand and negotiate their own sick roles in their respective communities vis-à-vis the Greco-Roman world.

Constructing Christian Narratives and Pedagogies of Pain

Third, Christians constructed pain narratives and pedagogies similar to illness narratives as they wrestled to understand and structure various instances of pain and suffering that they perceived, witnessed, and experienced. In the Greco-Roman medical, literary, philosophical discourses on pain (and suffering), the notions of pain's subjectivity, inexpressibility, and inshareability were prominent, while literary and philosophical discourses, in particular, offered paradoxical ideas about the expressibility and shareability of pain through their own pain narratives and pedagogies. In that larger context, the Christian pain narrative par excellence was the passion narrative of Jesus in the Gospels; and the subsequent interpretations of Jesus's pain narratives provided key perspectives on grounding Christian experiences of pain and suffering in mimetic identification with Christ. The Christians then developed their own pain narratives and thus pedagogies of pain in the forms of martyr narratives, an ascetic narrative, and theological treatises. Each of them evaluated the social agency of pain within the context of their ultimate Christian identities as martyrs, ascetics, and leaders in their respective communities; and thus, each one of them transformed pain into something communal, expressible, and shareable, to the extent that they shared and joined the same symbolic and spiritual communities. In this sense, their pain narratives and their attendant pedagogies of pain, both of which were based on their experiences of torture, martyrdom, *askēsis* (both discipline and bearing illness), longing for the divine, caring for the suffering, showing compassion, and sharing affective pain, shaped their communal identity as a community of cosufferers in mimetic identification with Christ.

Constructing a Christian System of Health Care

Lastly, Christians developed their multifaceted health care with the building blocks of the Greco-Roman health-care system. The Greco-Roman health-care system consisted of various parts, such as religious healing (temple medicine), rational medicine, targeted hospitals, popular medicine, and medico-philosophical therapies of the soul; but the divisions among them were often blurred, and each part, in reality, tended to complement the others. We see the mirroring of Greco-Roman health care in the Christian health-care system. Christians first Christianized religious healing, replacing the pagan gods, especially Asclepius and Apollonius of Tyana, with first the apostles and then ascetic holy men and saints (martyrs) as the main agents of healing par excellence in the second through the fifth centuries. Typically, Christian healings involved not only physical cures but also conversion (religious restoration), behavioral and moral rectitude through repentance, and, at times, the renunciation of "secular" physicians; thus, they drew identity boundaries vis-à-vis non-Christian healings. The charismatic healings also became institutionalized into the church's liturgy (in Egypt) and found their wishes for healing fulfilled via Christian magical amulets, as well.

However, Christians also incorporated and utilized rational medicine as a matter of fact, especially from the fourth century on. In addition to palliative and spiritual care, Christian medical care was offered, in varying degrees, in various hospitals, hospices, and hostels as the early church developed extensive networks of medical charity for the sick poor throughout the Eastern Roman Empire, in particular. As Julian "the Apostate" reticently noted, it was the Christian *philanthropia* through their hospitals that distinguished them from the "Greeks" (non-Christians). In terms of monastic health care, the monastic infirmaries provided medical care for sick or elderly monks or nuns, who were expected to play appropriate sick roles with benefits and obligations while still navigating the ambiguities of ascetic illness. Furthermore, just like their non-Christian neighbors, individual Christians also chose rational medicine as one of their options for healing; for Christians, this option existed alongside religious healing, as seen in the patients featured in Augustine's *City of God*. This divine-human, religious-"secular" medicine's complementarity and synergy, continued even in the later hagiography, show the continuation of earlier complementarity between temple medicine and rational medicine on the one hand, and the critical incorporation and transformation of medicine in late antique Christian frameworks, on the other.

Finally, another dynamic way early Christians transposed and incorporated medicine into the Christian framework was by Christianizing medico-philosophical therapies of the soul. From the third century on, the Christian bishops styled themselves as physicians of the soul engaged in spiritual therapy with their congregations. Bishops persistently described the remedy of *logos*—chiefly the Scripture as delivered in their homilies and treatises—delivered in the right way, and at the right time; Christian leaders also noted the importance of considering the different states of souls when engaged in these kinds of healing practices. This would help the bishops construct a common language of therapy for their ministries and impart a common identity to their congregants as those undergoing spiritual therapy in the midst of an ongoing spiritual convalescence. During the process of Christianization in the Roman East, bishops such as Epiphanius and Theodoret constructed a common Christian identity and *paideia* against "others"—heretics and the "Greeks"—and with those whose therapies espoused right belief (orthodoxy) and right practices (orthopraxy). In this way, Christian health care not only took the diverse forms of their non-Christian counterparts in the larger common context, but also contributed to the reordering and transformation of the existing health-care system while utilizing the same therapeutic resources.

To return to my point on the construction of Christian identity(-ies) in the introduction, this study then has shown the validity of Kathryn Tanner's argument that "Christian distinctiveness is something that emerges in the very cultural processes occurring *at* the boundary, processes that construct a distinctive identity for Christian social practices through the distinctive use of cultural materials *shared with others*."[1]

---

1. Kathryn Tanner, *Theories of Culture: A New Agenda for Theology* (Minneapolis: Fortress, 1997), 115 (emphasis added).

Bibliography

Primary Sources

Acta Apostolorum Apocrypha. Edited by R. A. Lipsius and M. Bonnet. 3 vols. Leipzig: Mendelssohn, 1891–1903.

Aelius Aristides. *P. Aelius Aristides: The Complete Works*. Translated by C. A. Behr. 2 vols. Leiden: Brill, 1981–1986.

———. *Sacred Tales*. Edited by C. A. Behr. In *P. Aelii Aristidis Opera Quae Extant Omnia*. Leiden: Brill, 1976.

The Ante-Nicene Fathers. Edited by Alexander Roberts and James Donaldson. Vols. 1–9. 1885–1887. Reprint, Peabody, MA: Hendrickson, 1994.

Antony. *The Letters of St. Antony: Origenist Theology, Monastic Tradition, and the Making of a Saint*. Translated by S. Rubenson. Reprint, Harrisburg, PA: Trinity Press International, 1995.

Apophthegmata partum. PG 65:71–440. English: *The Sayings of the Desert Fathers*. Translated by B. Ward. Cistercian Studies 59. Kalamazoo, MI: Cistercian Publications, 1975.

Apostolic Fathers. Translated by Bart D. Ehrman. 2 vols. LCL. Cambridge, MA: Harvard University Press, 2003.

Apostolic Fathers: Greek Texts and English Translations. Translated by Michael W. Holmes. 3rd ed. Grand Rapids: Baker Academic, 2007.

Aristotle. *Parts of Animals. Movement of Animals. Progression of Animals*. Translated by A. L. Peck and E. S. Forster. LCL. Cambridge, MA: Harvard University Press, 1937.

Athanasius. *Athanase d'Alexandrie, Vie d'Antoine*. Edited by G. J. M. Bartelink. SC 400. Paris: Cerf, 1994.

———. *Contra gentes*. Edited by R. W. Thompson. In *Contra gentes and De incarnatione*. Oxford: Oxford University Press, 1971.

———. *Life of Antony*. PG 26:835–976.

———. *The Life of Antony and the Letter to Marcellinus*. Translated by R. C. Gregg. CWS. Mahwah, NJ: Paulist, 1980.

Augustine. *The Catholic and Manichaean Ways of Life*. Translated by Donald A. Gallagher and Idella Gallagher. Washington, DC: Catholic University of America Press, 1966.

———. *City of God*. Translated by H. Bettenson, with introduction by J. J. O'Meara. New York and Harmondsworth, UK: Penguin Books, 1984.

———. *De Civitate Dei: Libri I–X*. Edited by B. Dombart and A. Karb. CCSL 47. Turnholt: Brepols, 1955.

———. *De Civitate Dei: Libri XI–XXII*. Edited by B. Dombart and A. Karb. CCSL 48. Turnholt: Brepols, 1955.

———. *De doctrina christiana*. CCSL 32. Turnholt: Brepols, 1962.

———. *Enarrationes in Psalmos. Pars X.1: Sancti Aurelii Augustini*. CCSL 38. Turnholt: Brepols, 1990.

———. *Epistolae. Sancti Aurelii Augustini Opera Omnia*. PL 33. Paris, 1865.

———. *Expositions of the Psalms* III/16 (33–50). Translated and notes by M. Boulding. Hyde Park, NY: New City Press, 2000.

———. *In Iohannis evangelium tractates. Pars VIII: Sancti Aurelii Augustini*. CCSL 36. Turnholt: Brepols, 1954.

———. *Sermones. Sancti Aurelii Augustini Opera Omnia*. PL 38. Paris, 1861.

———. *Sermons* III/8 (273–305A): *On the Saints*. Translated and notes by E. Hill. Hyde Park, NY: New City Press, 1994.

———. *Sermons* III/8 (306–340A): *On the Saints*. Translated and notes by E. Hill. Hyde Park, NY: New City Press, 1994.

Aulus Gellius. *Attic Nights*. Translated by J. C. Rolfe. LCL. Cambridge, MA: Harvard University Press, 1927.

Basil of Caesarea. *Quod Deus non est auctor malorum*. PG 31:329–54.

———. *Regulae brevius tractatae*. PG 31:1080–1305. In *Basil of Caesarea: Ascetical Works*, translated by M. M. Wagner. FC 9. Washington, DC: Catholic University of America Press, 1950.

———. *Regulae fusius tractatae*. PG 31:889–1052. In *Basil of Caesarea: Ascetical Works*, translated by M. M. Wagner. FC 9. Washington, DC: Catholic University of America Press, 1950.

———. *Saint Basil: The Letters*. Edited and translated by R. J. Deferrari. 4 vols. LCL. Cambridge, MA: Harvard University Press, 1962.

———. *St. Basil the Great: On the Human Condition*. Popular Patristics Series. Crestwood, NY: St. Vladimir's Seminary Press, 2005.

Bongie, Elizabeth Bryson, trans. *The Life & Regimen of the Blessed & Holy Syncletica by Pseudo-Athanasius.* Toronto: Peregrina, 1999.

Bradshaw, Paul, ed. *The Canons of Hippolytus.* Translated by Carol Bebawi. Nottingham: Grove Books, 2010.

Caelius Aurelianus. *On Acute Diseases and on Chronic Diseases.* Edited and translated by I. E. Drabkin. Chicago: University of Chicago Press, 1950.

Castelli, Elizabeth A., trans. "The Life and Activity of the Holy and Blessed Teacher Syncletica." In *Ascetic Behavior in Greco-Roman Antiquity: A Sourcebook,* edited by V. Wimbush, 265–311. Minneapolis: Fortress, 1990.

Celsus. *De Medicina.* Translated by W. G. Spencer. 3 vols. LCL. Cambridge, MA: Harvard University Press, 1938–1953.

Cicero. *De finibus bonorum et malorum.* Translated by H. Rackham. LCL. Cambridge, MA: Harvard University Press, 1951.

———. *Tusculan Disputations.* Translated by J. E. King. LCL. Cambridge, MA: Harvard University Press, 1927.

Clement of Alexandria. *Christ the Educator.* Edited by M. Harl. In *Clement d'Alexandrie: Le Pedagogue.* SC 10, 108, 158. Paris: Cerf, 1960–1970.

———. *Clement of Alexandria: Christ the Educator.* Translated by S. P. Wood. FC 23. Washington, DC: Catholic University of America Press, 1954.

Columella. *On Agriculture (De re rustica).* Translated by H. Boyd Ash. LCL 3. Cambridge, MA: Harvard University Press, 1960.

Cotter, Wendy. *Miracles in Greco-Roman Antiquity: A Sourcebook for the Study of New Testament Miracle Stories.* London: Routledge, 1999.

Cyprian. *De Lapsis and De Ecclesiae Catholicae Unitate.* Text and translation by M. Bevenot. Oxford: Clarendon, 1971.

Cyril of Alexandria. *Select Letters.* Translated and edited by Lionel R. Wickham. Oxford: Clarendon, 1983.

The Didascalia Apostolorum in Syriac. Edited by Arthur Vööbus. 4 vols. CSCO 401, 402, 407, 408. Leuven: Secrétariat du CorpusSCO, 1979 (includes Syriac text and English translation).

Edelstein, Emma J., and Ludwig Edelstein. *Asclepius: Collection and Interpretation of the Testimonies.* 2 vols. Baltimore: Johns Hopkins University Press, 1945.

Elliott, James K. *The Apocryphal New Testament.* Oxford: Clarendon, 1993.

Epictetus. *The Discourses as Reported by Arrian, the Manual and Fragments.* Translated by W. A. Oldfather. 2 vols. LCL. Cambridge, MA: Harvard University Press, 1925 (reprint 1961).

Epiphanius of Salamis. *The Panarion of Epiphanius of Salamis: Book I (Sects 1–46).* Translated by Frank Williams. 2nd ed. Leiden: Brill, 2009.

————. *The* Panarion *of Epiphanius of Salamis: Books II and III. De Fide.* Translated by Frank Williams. 2nd ed. Leiden: Brill, 2013.

Eusebius. *Ecclesiastical History.* Translated by K. Lake. 2 vols. LCL. Cambridge, MA: Harvard University Press, 1926.

————. *The Ecclesiastical History and the Martyrs of Palestine.* Translated by H. J. Lawlor and J. E. L. Oulton. London: SPCK, 1927.

Evagrius of Pontus. *Evagrius of Pontus: The Greek Ascetic Corpus.* Translated by R. E. Sinkewicz. OECS. Oxford: Oxford University Press, 2003.

Fronto, Marcus Cornelius. *The Correspondence of Marcus Cornelius Fronto.* Translated by C. R. Haines. 2 vols. LCL. New York: Putnam, 1919–1920.

Galen. *Galen: Hygiene; Books 1–4.* Vol. 1. Edited and translated by Ian Johnston. LCL. Cambridge, MA: Harvard University Press, 2018.

————. *Galen: Hygiene; Books 5–9.* Vol. 2, *Thrasybulus; On Exercise with a Small Ball.* Edited and translated by Ian Johnston. LCL. Cambridge, MA: Harvard University Press, 2018.

————. *Galen: Method of Medicine.* Edited and translated by Ian Johnston and G. H. R. Horsley. 3 vols. LCL. Cambridge, MA: Harvard University Press, 2011.

————. *Galen: On Diseases and Symptoms.* Translated and introduced by Ian Johnston. LCL. New York: Cambridge University Press, 2006.

————. *Galen: On the Constitution of the Art of Medicine; The Art of Medicine; A Method of Medicine to Glaucon.* Edited and translated by Ian Johnston. LCL. Cambridge, MA: Harvard University Press, 2016.

————. *Galen: Psychological Writings.* Edited by P. N. Singer. Cambridge: Cambridge University Press, 2013.

————. *Galen: Selected Works.* Translated by P. N. Singer. Oxford: Oxford University Press, 2002.

————. *Galeni Opera Omnia.* Edited by D. C. G. Kühn. Greek and Latin text. 22 vols. Leipzig: C. Knobloch: 1821–1833.

————. *Galen on Bloodletting: A Study of the Origins, Development, and Validity of His Opinions, with a Translation of the Three Works.* Translated by Peter Brain. Cambridge: Cambridge University Press, 1986.

————. *Galen on the Affected Parts.* Translated from the Greek text with explanatory notes by Rudolph E. Siegel. Basel: S. Karger, 1976.

————. *Galen on the Doctrines of Hippocrates and Plato.* Translated by Phillip De Lacy. Berlin: Akademie Verlag, 1984.

Gregory of Nazianzus. *Faith Gives Fullness to Reasoning: The Five Theological Orations of Gregory Nazianzen.* Introduction and commentary by Frederick W. Norris. Translated by Lionel Wickham and Frederick Williams. Supplements to Vigiliae Christianae 13. Leiden: Brill, 1991.

——. *On God and Man: The Theological Poetry of St. Gregory of Nazianzus.* Crestwood, NY: St. Vladimir's Seminary Press, 2001.

——. *Select Orations.* Translated by M. Vinson. FC 107. Washington, DC: Catholic University of America Press, 2003.

——. *Three Poems.* Translated by D. M. Meehan. FC 75. Washington, DC: Catholic University of America Press, 1987.

Herzog, Rudolf. *Die Wunderheilungen von Epidauros: Ein Beitrag zur Geschichte der Medizin und der Religion.* Philologus, supplement 22. Vol. 3. Leipzig: Dietrich, 1931.

Hilary of Poitiers. *The Trinity.* Translated by S. McKenna. FC 25. New York: Fathers of the Church, 1954.

Hippocrates. *Hippocrates.* Vol. 1, *Ancient Medicine; Airs, Waters, Places; Epidemics 1 and 3; The Oath; Precepts; Nutriment.* Translated by W. H. S. Jones. LCL. Cambridge, MA: Harvard University Press, 1923.

——. *Hippocrates.* Vol. 2, *Prognostic; Regimen in Acute Diseases; The Sacred Disease; The Art; Breaths; Law; Decorum; Physician (Ch. 1); Dentition.* Translated by W. H. S. Jones. LCL. Cambridge, MA: Harvard University Press, 1923.

——. *Hippocrates.* Vol. 4, *Nature of Man; Regimen in Health; Humours; Aphorisms; Regimen I-III; Dreams; Heracleitus.* Translated by W. H. S. Jones. LCL. Cambridge, MA: Harvard University Press, 1931.

——. *Hippocrates.* Vol. 5, *Affections; Diseases I-II.* Translated by P. Potter. LCL. Cambridge, MA: Harvard University Press, 1988.

——. *Hippocrates.* Vol. 6, *Diseases III; Internal Affections; Regimen in Acute Diseases (Appendix); Weights and Measures; Index of Symptoms and Diseases; Greek Names of Symptoms and Diseases; Index of Foods and Drugs; Greek Names of Foods and Drugs.* Translated by P. Potter. LCL. Cambridge, MA: Harvard University Press, 1988.

——. *Hippocrates.* Vol. 7, *Epidemics 2, 4–7.* Translated by W. D. Smith. LCL. Cambridge, MA: Harvard University Press, 1994.

——. *Hippocrates.* Vol. 8, *Places in Man; Glands; Fleshes; Prorrhetic I-Ii; Physician; Use of Liquids; Ulcers; Haemorrhoids; Fistulas.* Translated by P. Potter. LCL. Cambridge, MA: Harvard University Press, 1995.

——. *Hippocrates.* Vol. 10, *Generation; Nature of the Child; Diseases IV; Nature of Women; Barrenness.* Translated by P. Potter. LCL. Cambridge, MA: Harvard University Press, 2012.

——. *Hippocrates.* Vol. 11, *Diseases of Women I; Diseases of Women II.* Translated by P. Potter. LCL. Cambridge, MA: Harvard University Press, 2018.

——. *Hippocratic Writings.* Edited by G. E. R. Lloyd. Harmondsworth, UK: Pelican Books, 1978.

Hippolytus. *Hippolyte de Rome: La tradition apostolique d'après les anciennes versions. Texts Latin, introduction, traduction et notes.* Edited by B. Botte. 2nd ed. SC 11. Paris: Cerf, 1984. English: *Apostolic Tradition.*

——. *On the Apostolic Tradition: An English Version with Introduction and Commentary.* Translated, with commentary and introduction, by Alistair Stewart-Sykes. Popular Patristics Series 22. Crestwood, NY: St. Vladimir's Seminary Press, 2001.

John Chrysostom. *Six Books on the Priesthood.* Translated by Graham Neville. Crestwood, NY: St. Vladimir's Seminary Press, 1984.

Julian. *Letters. Epigrams. Against the Galilaeans. Fragments.* Translated by Wilmer C. Wright. LCL. Cambridge, MA: Harvard University Press, 1923.

Lactantius. *De Moribus Persecutorum.* Edited and translated by J. L. Creed. Oxford: Oxford University Press, 1984.

Layton, Bentley. *The Canons of Our Fathers: Monastic Rules of Shenoute.* OECS. Oxford: Oxford University Press, 2014.

LiDonnici, Lynn R. *The Epidaurian Miracle Inscriptions: Text, Translation, and Commentary.* Atlanta: Scholars Press, 1995.

The Lives of the Desert Fathers: The Historia Monachorum in Aegypto. Translated by N. Russell. Introduction by B. Ward. Kalamazoo, MI: Cistercian Publications, 1981.

Marcellus of Bordeaux. *De medicamentis.* Translated by Georgius Hermreich. Leipzig: Teubner, 1889.

Marcus Aurelius. *The Communings with Himself of Marcus Aurelius Antonius Emperor of Rome.* Translated and introduced by C. R. Haines. LCL. Cambridge, MA: Harvard University Press, 1970.

McGuckin, John A. *St. Cyril of Alexandria: The Christological Controversy; Its History, Theology, and Texts.* Leiden: Brill, 1994.

Morani, M. *Nemesius. De natura hominis.* Leipzig: Teubner, 1987.

Musurillo, Herbert. *Acts of the Christian Martyrs.* Oxford: Clarendon, 1972.

The Nag-Hammadi Library in English. Translated and introduced by J. M. Robinson. 3rd ed. San Francisco: HarperSanFrancisco, 1988.

Nemesius. *On the Nature of Man.* Translated with introductions and notes by R. W. Sharples and P. J. van der Eijk. Translated Texts for Historians 49. Liverpool: Liverpool University Press, 2008.

The Nicene and Post-Nicene Fathers. Edited by Philip Schaff. 1st ser. Vols. 1–14. Reprint, Peabody, MA: Hendrickson, 1994.

The Nicene and Post-Nicene Fathers. Edited by Philip Schaff, and Henry Wace. 2nd ser. Vols. 1–11. Reprint, Peabody, MA: Hendrickson, 1994.

Origen. *Contra Celsum.* Edited by M. Marcovich. Boston: Brill, 2001.

Pachomius. *Die Briefe Pachoms: Griechischer Text der Handschrift W. 145 der Ches-*

ter Beatty Library. Edited by Hans Quecke. Textus Patristici et Liturgici 11. Regensburg, Germany: Pustet, 1975. English: *Letters*.

———. *Pachomian Koinonia*. Vol. 1, *The Lives, Rules, and Other Writings of Saint Pachomius and His Disciples; The Life of Saint Pachomius and His Disciples*. Translated and introduced by Armand Veilleux. Kalamazoo, MI: Cistercian Publishing, 1980.

———. *Pachomian Koinonia*. Vol. 3, *Instructions, Letters, & Other Writings*. Translated and introduced by Armand Veilleux. Kalamazoo, MI: Cistercian Publishing, 1982.

The Passion of Saints Perpetua and Felicity. Translated by J. Farrell and C. Williams. In *Perpetua's Passions: Multidisciplinary Approaches to the* Passio Perpetuae et Felicitas, edited by Jan N. Bremmer and Marco Formisano, 14–23. Oxford: Oxford University Press, 2012.

Passio Sanctarum Perpetuae et Felicitatis. Edited by J. Farrell and C. Williams. In *Perpetua's Passions: Multidisciplinary Approaches to the* Passio Perpetuae et Felicitas, edited by Jan N. Bremmer and Marco Formisano, 24–32. Oxford: Oxford University Press, 2012.

Patrologia Graeca. Edited by J.-P. Migne. Vols. 5–83. Paris, 1857–1864.

Patrologia Latina. Edited by J.-P. Migne. Vols. 1–47. Paris, 1844–1849.

Philostratus. *The Life of Apollonius of Tyana*. Edited and translated by Christopher P. Jones. 2 Vols. LCL. Cambridge, MA: Harvard University Press, 2005.

Plato. *Lysis. Symposium. Gorgias*. Translated by W. R. M. Lamb. LCL. Cambridge, MA: Harvard University Press, 1925.

———. *Republic Books 1–5*. Translated by Christopher Emlyn-Jones and William Preddy. LCL. Cambridge, MA: Harvard University Press, 2013.

———. *Timaeus. Critias. Cleitophon. Menexenus. Epistles*. Translated by R. G. Bury. LCL. Cambridge, MA: Harvard University Press, 1929.

Plutarch. *Plutarch's Moralia*. Vol. 2. Translated by F. C. Babbitt. LCL. Cambridge, MA: Harvard University Press, 1971.

———. *Plutarch's Moralia*. Vol. 6. Translated by W. C. Helmbold. LCL. Cambridge, MA: Harvard University Press, 1970.

Preisendanz, Karl, trans. and ed. *Papyri Graecae Magicae: Die griechischen Zauberpapyri*. 3 vols. Leipzig: Teubner, 1928–1941.

Rebillard, Éric, ed. *Greek and Latin Narratives about the Ancient Martyrs*. Oxford: Oxford University Press, 2017.

Schiefsky, Mark John. *On Ancient Medicine*. Studies in Ancient Medicine 28. Leiden: Brill, 2005.

Seneca. *Ad Lucilium Epistulae Morales*. Translated by R. M. Gummere. 3 vols. LCL. Cambridge, MA: Harvard University Press, 1925 (reprint 1953).

———. *Moral Essays*. Translated by J. W. Basore. Vol. 1. LCL. Cambridge, MA: Harvard University Press, 1928 (reprint 1994).

Silvas, Anna M. *The Asketikon of St. Basil the Great*. OECS. Oxford: Oxford University Press, 2005.

Sophocles. *Sophocles*. Translated by H. Lloyd-Jones. Vol. 2. Cambridge, MA: Harvard University Press, 1994.

Stewards of the Poor: The Man of God, Rabbula, and Hiba in Fifth-Century Edessa. Translated and introduced by Robert Doran. Kalamazoo, MI: Cistercian Publications, 2006.

Theodoret of Cyrrhus. *A Cure for Pagan Maladies*. Translated and introduced by Thomas Halton. Ancient Christian Writers 67. New York and Mahwah, NJ, 2013.

———. *A History of the Monks of Syria*. Translated and introduced by R. M. Price. Trappist, KY: Cistercian Publications, 1985.

———. *On Divine Providence*. Translated and annotated by Thomas Halton. Ancient Christian Writers 49. New York and Mahwah, NJ, 1988.

Theophilus of Antioch. *Ad Autolycum*. Translated by R. M. Grant. Oxford: Clarendon, 1980.

Three Byzantine Saints: Contemporary Biographies of St. Daniel the Stylite, St. Theodore of Sykeon, and St. John the Almsgiver. Translated by Elizabeth Dawes and Norman H. Baynes. Crestwood, NY: St. Vladimir's Seminary Press, 1977.

Tieleman, Teun. *Chrysippus' On Affections: Reconstruction and Interpretation*. Leiden: Brill, 2003.

Wortley, John. *The Anonymous Sayings of the Desert Fathers: A Select Edition and Complete English Translation*. Cambridge: Cambridge University Press, 2013.

———. *Give Me a Word: The Alphabetical Sayings of the Desert Fathers*. Yonkers, NY: St. Vladimir's Seminary Press, 2014.

Wright, William. *The Chronicle of Joshua the Stylite*. Composed in Syriac AD 507 with English translation and notes. Reprint, Amsterdam: Philo, 1968.

Secondary Sources

Albl, Martin C. "'Are Any among You Sick?': The Health Care System in the Letter of James." *JBL* 121, no. 1 (2002): 123–43.

Allan, Nigel. "The Physician in Ancient Israel: His Status and Function." *Medical History* 45 (2001): 377–94.

Amundsen, D. W. "The Discourses of Early Christian Medical Ethics." In *The Cambridge World History of Medical Ethics*, edited by R. Baker and L. McCullough, 202–10. New York: Cambridge University Press, 2009.

———. "Medicine and Faith in Early Christianity." *BHM* 56, no. 3 (1982): 326–50.

———. *Medicine, Society, and Faith in the Ancient and Medieval Worlds*. Baltimore: Johns Hopkins University Press, 1996.

Amundsen, D. W., and G. B. Ferngren. "The Early Christian Tradition." In *Caring and Curing: Health and Medicine in the Western Religious Tradition*, edited by R. L. Numbers and D. W. Amundsen, 40–64. Baltimore: Johns Hopkins University Press, 1986.

———. "Medicine and Religion: Pre-Christian Antiquity." In *Health/Medicine and the Faith Traditions: An Inquiry into Religion and Medicine*, edited by M. E. Marty and K. L. Vaux, 53–92. Philadelphia: Fortress, 1982.

———. "The Perception of Disease and Disease Causality in the New Testament." *ANRW* II.37.3 (1996): 2934–56.

Andorlini, I. "Crossing the Borders between Egyptian and Greek Medical Practice." In *Popular Medicine in Graeco-Roman Antiquity: Explorations*, edited by William V. Harris, 161–72. Columbia Studies in the Classical Tradition. Leiden: Brill, 2016.

Arbresmann, Rudolph. "Christ the *Medicus Humilis* in St. Augustine." In *Augustinus Magister, Congrès International Augustinien, Paris, 21–24 September 1954*, 2:623–29. Paris, 1954.

———. "The Concept of 'Christus Medicus' in St. Augustine." *Traditio* 10 (1954): 1–28.

Asad, Talal. *Formations of the Secular: Christianity, Islam, Modernity*. Stanford, CA: Stanford University Press, 2003.

Astyrakaki, E., A. Papaioannou, and H. Askitopoulou. "References to Anesthesia, Pain, and Analgesia in the Hippocratic Collection." *International Anesthesia Research Society* 110, no. 1 (2010): 188–94.

Avalos, Hector. *Health Care and the Rise of Christianity*. Peabody, MA: Hendrikson, 1999.

———. *Illness and Health Care in the Ancient Near East: The Role of the Temple in Greece, Mesopotamia, and Israel*. HSM 54. Atlanta: Scholars Press, 1995.

Bae, Junghun. "John Chrysostom on Almsgiving and the Therapy of the Soul." PhD diss., Australian Catholic University, 2018.

Baker, Kimberly. "Augustine's Doctrine of the *Totus Christus*: Reflecting on the Church as Sacrament of Unity." *Hor* 37, no. 1 (2010): 7–24.

Ballester, Luis Garcia. "Soul and Body, Disease of the Soul and Disease of the Body in Galen's Medical Thought." In *Le Opere Psicologiche di Galeno: Atti del Terzo Colloquio Galenico Internazionale Pavia, 10–12 Settembre 1986*, edited by Paola Manuli and Mario Vegetti, 117–52. Naples: Bibliopolis, 1988.

Barasch, Mosche. *Blindness: The History of a Mental Image in Western Thought.* New York: Routledge, 2001.

Barrett-Lennard, R. J. S. "The Canons of Hippolytus and Christian Concern with Illness, Health, and Healing." *JECS* 13, no. 2 (2005): 137–64.

———. *Christian Healing after the New Testament: Some Approaches to Illness in the Second, Third, and Fourth Centuries.* Lanham, MD: University Press of America, 1994.

———. "Request for Prayer for Healing," In *New Documents Illustrating Early Christianity: A Review of Greek Inscriptions and Papyri Published in 1979*, edited by G. H. R. Horsley, 245–50. Ancient History Documentary Research Centre. Sydney: Macquarie University, 1987.

Beckwith, Carl L. "Suffering without Pain: The Scandal of Hilary of Poitiers' Christology." In *In the Shadow of the Incarnation: Essays on Jesus Christ in the Early Church in Honor of Brian E. Daley, S. J.*, edited by Peter W. Martens, 71–96. Notre Dame: University of Notre Dame Press, 2008.

Beeley, Christopher. *Gregory of Nazianzus on the Trinity and the Knowledge of God: In Your Light We Shall See the Light.* OSHT. Oxford: Oxford University Press, 2008.

———. "Gregory of Nazianzus on the Unity of Christ." In *In the Shadow of the Incarnation: Essays on Jesus Christ in the Early Church in Honor of Brian E. Daley, S. J.*, edited by Peter W. Martens, 97–120. Notre Dame: University of Notre Dame Press, 2008.

Behr, C. A. *Aelius Aristides and* The Sacred Tales. Amsterdam: Adolf M. Hakkert, 1968.

Bell, R. H. "Demon, Devil, Satan." In *Dictionary of Jesus and the Gospels*, edited by Joel B. Green, Jeannine K. Brown, and Nicholas Perrin, 193–202. Downers Grove, IL: IVP Academic, 2013.

Bendemann, Reinhard von. "Die Heilungen Jesu und die antike Medizin." *Early Christianity* 5, no. 3 (2014): 273–312.

Berlejung, Angelika. "Written on the Body: Body and Illness in the Physiognomic Tradition of the Ancient Near East and the Old Testament." In *Religion and Illness*, edited by Annette Weissenrieder and Gregor Etzelmüller, 137–72. Eugene, OR: Cascade Books, 2016.

Bishop, Jeffrey P. "Mind-Body Unity: Gregory of Nyssa and a Surprising Fourth-Century CE Perspective." *Perspectives in Biology and Medicine* 43, no. 4 (2000): 519–29.

Blowers, P. "Pity, Empathy, and the Tragic Spectacle of Human Suffering: Exploring the Emotional Culture of Compassion in Late Ancient Christianity." *JECS* 18, no. 1 (2010): 1–27.

Bochet, Isabelle. "Maladie de L'Âme et Thérapeutique Scripturaire selon Augustin." In *Les Pères de L'Église Face à la Science Médicale de Leur Temps*, edited by Véronique Boudon-Millot and Bernard Pouderon, 379–400. Paris: Beauchesne, 2005.

Böhme, Hartmut. "The Conquest of the Real by the Imaginary: On the *Passio Perpetuae*." In *Perpetua's Passions: Multidisciplinary Approaches to the* Passio Perpetuae et Felicitas, edited by Jan N. Bremmer and Marco Formisano, 220–44. Oxford: Oxford University Press, 2012.

Bonati, Isabella. "The (Un)Healthy Poor: Wealth, Poverty, Medicine and Health-Care in the Greco-Roman World." *Akroterion* 64 (2019): 15–43.

Borgen, Peder. "Miracles of Healing in the New Testament: Some Observations." *ST* 35 (1981): 91–106.

Bostock, David G. "Medical Theory and Theology in Origen." In *Origeniana Tertia: The Third International Colloquium for Origen Studies*, edited by Richard Hanson and Henri Crouzel, 191–99. Manchester: University of Manchester Press, 1981.

Boudon-Millot, Véronique. "Must We Suffer in Order to Stay Healthy? Pleasure and Pain in Ancient Medical Literature." In *Pain and Pleasure in Classical Times*, edited by William V. Harris, 36–54. Leiden: Brill, 2018.

Boudon-Millot, Véronique, and Bernard Pouderon, eds. *Les Pères de l'Église face à la science médicale de leur temps*. Paris: Beauchesne, 2005.

Bowersock, Glen W. *The Greek Sophists in the Roman Empire*. Oxford: Clarendon, 1969.

———. *Martyrdom and Rome*. Cambridge: Cambridge University Press, 1995.

———. "*Parabalani*: A Terrorist Charity in Late Antiquity." *Anabases* 12 (2010): 45–54.

Boyarin, Daniel. *Dying for God: Martyrdoms and the Making of Christianity and Judaism*. Stanford, CA: Stanford University Press, 1999.

Boyd, J. H. "A Biblical Theology of Chronic Illness." *TJ* 24, no. 2 (2003): 189–206.

Bradley, Mark, ed. *Rome, Pollution, and Propriety: Dirt, Disease, and Hygiene in the Eternal City from Antiquity to Modernity*. Cambridge: Cambridge University Press, 2016.

Brown, Michael L. *Israel's Divine Healer*. Studies in Old Testament Biblical Theology. Grand Rapids, MI: Zondervan, 1995.

Brown, Peter R. L. *Authority and the Sacred: Aspects of the Christianisation of the Roman World*. Cambridge: Cambridge University Press, 1995.

————. *The Cult of the Saints: Its Rise and Function in Latin Christianity.* Chicago: University of Chicago Press, 1982.

————. *The Making of Late Antiquity.* Cambridge, MA: Harvard University Press, 1978.

————. *Power and Persuasion in Late Antiquity: Towards a Christian Empire.* Madison: University of Wisconsin Press, 1992.

————. "The Rise and Function of the Holy Man in Late Antiquity." *JRS* (1971): 80–101. Reprint in Peter R. L. Brown, *Society and the Holy in Late Antiquity*, 166–95. Berkeley: University of California Press, 1982.

Browne, Stanley G. "Leprosy in the Bible." In *Medicine and the Bible*, edited by Bernard Palmer, 101–25. Exeter, UK: Paternoster, 1986.

Bruning, Bernard. "Die Einheit des Totus Christus bei Augustinus." In *Scientia Augustiniana: Studien über Augustinus, den Augustinismus und den Augustinorden*, edited by Cornelius Petrus Mayer and Willigis Eckermann, 43–75. Würzburg: Augustinus-Verlag, 1975.

Burrus, Virginia. *Saving Shame: Martyrs, Saints, and Other Abject Subjects.* Philadelphia: University of Pennsylvania Press, 2008.

————. "Torture and Travail: Producing the Christian Martyr." In *A Feminist Companion to Patristic Literature*, edited by Amy-Jill Levine and Maria M. Robbins, 56–71. London: T&T Clark, 2008.

Burt, Donald X. "Health, Sickness." In *Augustine through the Centuries*, edited by Allan D. Fitzgerald, 416–19. Grand Rapids: Eerdmans, 1999.

Byers, Sarah. "The Psychology of Compassion: Stoicism in *City of God* 9.5." In *Augustine's* City of God: *A Critical Guide*, edited by James Wetzel, 130–48. Cambridge: Cambridge University Press, 2012.

Cain, A. *The Letters of Jerome: Asceticism, Biblical Exegesis, and the Construction of Christian Authority in Late Antiquity.* OECS. Oxford: Oxford University Press, 2009.

Cain, Carole. "Personal Stories: Identity Acquisition and Self-Understanding in Alcoholics Anonymous." *Ethos* 19, no. 2 (1991): 210–53.

Caner, Daniel. "Not a Hospital but a Leprosarium: Basil's Basilias and an Early Byzantine Concept of the Deserving Poor." *DOP* 72 (2018): 25–48.

Carmichael, Ann G. "Public Health and Sanitation in the West before 1700." In *The Cambridge World History of Human Disease*, edited by Kenneth F. Kiple et al., 192–200. New York: Cambridge University Press, 1993.

Carroll, John T. "Sickness and Healing in the New Testament Gospels." *Int* 49, no. 2 (1995): 130–42.

Castelli, Elizabeth A. "Gender, Theory, and *The Rise of Christianity*: A Response to Rodney Stark." *JECS* 6, no. 2 (1998): 227–57.

————. "Mortifying the Body, Curing the Soul: Beyond Ascetic Dualism in *The*

Life of Saint Syncletica." *Differences: A Journal of Feminist Cultural Studies* 4, no. 2 (1992): 134–53.

Cavarnos, J. P. "Relation of the Body and Soul in the Thought of Gregory of Nyssa." In *Gregor von Nyssa und die Philosophie: Zweites Internationales Kolloquium über Gregor von Nyssa*, edited by H. Dörrie et al., 61–78. Leiden: Brill, 1976.

Chaniotis, Angelos. "Illness and Cures in the Greek Propitiatory Inscriptions and Dedications of Lydia and Phrygia." In *Ancient Medicine in Its Socio-Cultural Context*, edited by Ph. J. van der Ejik, H. F. J. Horstmanshoff, and P. H. Schrijvers, 2:323–44. Clio Medica 27. Amsterdam and Atlanta: Edition Rodopi, 1995.

Clark, Gillian. "Bodies and Blood: Late Antique Debate on Martyrdom, Virginity, and Resurrection." In *Changing Bodies, Changing Meanings: Studies on the Human Body in Antiquity*, edited by Dominic Montserrat, 99–115. London: Routledge, 1998.

———. "The Health of the Spiritual Athlete." In *Health in Antiquity*, edited by Helen King, 216–29. London: Routledge, 2005.

Clark, Patricia A., and M. Lynn Rose. "Psychiatric Disability and the Galenic Medieval Matrix." In *Disabilities in Roman Antiquity: Disparate Bodies A Capite ad Calcem*, edited by Christian Laes, C. F. Goodey, and M. Lynn Rose, 45–72. Mnemosyne Supplements, History and Archaeology of Classical Antiquity. Leiden: Brill, 2013.

Coakley, Sarah. Introduction to *Pain and Its Transformations: The Interface of Biology and Culture*, edited by Sarah Coakley and Kay K. Shelemay, 1–16. Cambridge, MA: Harvard University Press, 2007.

Cobb, L. Stephanie. *Divine Deliverance: Pain and Painlessness in Early Christian Martyr Texts*. Oakland: University of California Press, 2017.

———. *Dying to Be Men: Gender and Language in Early Christian Martyr Texts*. New York: Columbia University Press, 2008.

Cook, James. "'Hear and Shudder!': John Chrysostom's Therapy of the Soul." In *Revisioning John Chrysostom: New Perspective, Theories, and Approaches*, edited by Chris L. de Wet and Wendy Mayer, 247–75. Leiden: Brill, 2019.

Corrigan, Kevin. "The Soul-Body Relation in and before Augustine." *Studia Patristica* 63 (2006): 59–80.

Craffert, Pieter F. "Medical Anthropology as an Antidote for Ethnocentrism in Jesus Research? Putting the Illness-Disease Distinction into Perspective." *HTS Teologiese Studies/Theological Studies* 67, no. 1 (2011): art. #970, 14 pages. DOI: 10.4102/hts.v67i1.970.

Craik, E. M. *Hippocrates: Places in Man*. Oxford: Clarendon, 1998.

Crislip, Andrew T. *From Monastery to Hospital: Christian Monasticism and the Transformation of Health Care in Late Antiquity.* Ann Arbor: University of Michigan Press, 2005.

———. "'I Have Chosen Sickness': The Controversial Function of Sickness in Early Christian Ascetic Practice." In *Asceticism and Its Critics: Historical Accounts and Comparative Perspective,* edited by Oliver Freiberger, 179–209. AARCC. New York: Oxford University Press, 2006.

———. *Thorns in the Flesh: Illness and Sanctity in Late Ancient Christianity.* Philadelphia: University of Pennsylvania Press, 2012.

Crossan, John Dominic. *The Birth of Christianity: Discovering What Happened in the Years Immediately after the Execution of Jesus.* San Francisco: HarperSanFrancisco, 1998.

———. *Jesus: A Revolutionary Biography.* San Francisco: HarperSanFrancisco, 1994.

Czachesz, I. "The Grotesque Body in the Apocalypse of Peter." In *The Apocalypse of Peter,* edited by J. N. Bremmer and I. Czachesz, 108–26. Leuven: Peeters, 2003.

Daley, Brian E. "The Word and His Flesh: Human Weakness and the Identity of Jesus in Greek Patristic Christology." In *Suffering and Evil in Early Christian Thought,* edited by Nonna Verna Harrison and David G. Hunter, 171–90. Grand Rapids: Baker Academic, 2016.

D'Angelo, Tiziana. "Medicine, Religion and Magic in Two Inscribed Bronze Tablets from *Ticinum* (CIL V 6414–6415)." *ZPE* 202 (2017): 189–207.

Darling, Averell S. "The Levitical Code: Hygiene or Holiness." In *Medicine and the Bible,* edited by Bernard Palmer, 85–99. Exeter, UK: Paternoster, 1986.

Daunton-Fear, Andrew. *Healing in the Early Church: The Church's Ministry of Healing and Exorcism from the First to the Fifth Century.* Studies in Christian History and Thought. Eugene, OR: Wipf & Stock, 2009.

Davis, Stephen J. *The Cult of Saint Thecla: A Tradition of Women's Piety in Late Antiquity.* Oxford: Oxford University Press, 2001.

Debru, Armelle. "Physiology." In *The Cambridge Companion to Galen,* edited by R. J. Hankinson, 263–82. Cambridge: Cambridge University Press, 2008.

De Bruyn, Theodore. "Appeals to Jesus as the One 'Who Heals Every Illness and Every Infirmity' (MATT 4:23, 9:35) in Amulets in Late Antiquity." In *The Reception and Interpretation of the Bible in Late Antiquity: Proceedings of the Montréal Colloquium in Honour of Charles Kannengiesser, 11–13 October 2006,* edited by Lorenzo DiTommaso and Lucian Turcescu, 65–81. Leiden: Brill, 2008.

———. "Archaeological Views: Christian Amulets—a Bit of Old, a Bit of New." *BAR* 44, no. 5 (2018).

———. *Making Amulets Christian: Artefacts, Scribes, and Contexts*. Oxford: Oxford University Press, 2017.

Den Boeft, Jan, and Jan Bremmer. "Notiunculae Martyrologicae." *VC* 35 (1981): 43–56.

Denzey, Nicola. "Facing the Beast: Justin, Christian Martyrdom, and Freedom of the Will." In *Stoicism in Early Christianity*, edited by Tuomas Rasimus, Troels Engberg-Pedersen, and Ismo Dunderberg, 176–99. Grand Rapids: Baker Academic, 2010.

Devinant, Julien. "Mental Disorders and Psychological Suffering in Galen's Cases." In *Mental Illness in Ancient Medicine: From Celsus to Paul of Aegina*, edited by Chiara Thumigur and Peter Singer, 198–221. Studies on Ancient Medicine 50. Leiden: Brill, 2018.

D'Irsay, S. "Christian Medicine and Science in the Third Century." *JR* 10 (1930): 515–44.

———. "Patristic Medicine." *Annals of Medical History* 9 (1927): 264–378.

Donini, Pierluigi. "Psychology." In *The Cambridge Companion to Galen*, edited by R. J. Hankinson, 184–209. Cambridge: Cambridge University Press, 2008.

Dörnemann, Michael. "Einer ist Arzt, Christus: Medizinales Verständnis von Erlösung in der Theologie der griechischen Kirchenväter des zweiten Jahrhunderts." *ZAC* 17, no. 1 (2013): 102–24.

———. *Krankheit und Heilung in der Theologie der frühen Kirchenväter*. Tübingen: Mohr Siebeck, 2003.

Duncan-Jones, R. P. "The Impact of the Antonine Plague." *JRA* 9 (1996): 108–36.

Dunn, J. D. G., and G. H. Twelftree. "Demon Possession and Exorcism in the New Testament." *Churchman* 94 (1980): 210–25.

Edelstein, L. "Greek Medicine in Its Relation to Religion and Magic." In *Ancient Medicine: Selected Papers of Ludwig Edelstein*, edited by O. Temkin and C. L. Temkin, 205–46. Baltimore: Johns Hopkins University Press, 1967.

———. "The Relation of Ancient Philosophy to Medicine." In *Ancient Medicine: Selected Papers of Ludwig Edelstein*, edited by O. Temkin and C. L. Temkin, 349–66. Baltimore: Johns Hopkins University Press, 1967.

Edwards, Catharine. "The Suffering Body: Philosophy and Pain in Seneca's Letters." In *Constructions of the Classical Body*, edited by J. I. Porter, 252–68. Ann Arbor: University of Michigan Press, 2007.

Eijk, Philip J. van der. "Aristotle's Psycho-Physiological Account of the Soul-Body Relationship." In *Psyche and Soma: Physicians and Metaphysicians on the Mind-Body Problem,* edited by J. P. Wright and P. Potter, 57–77. Oxford: Oxford University Press, 2000.

———. "Galen and Early Christians on the Role of the Divine in the Causation and Treatment of Health and Disease." *Early Christianity* 5 (2014): 337–70.

———. "Galen on the Nature of Human Beings." In *Philosophical Themes in Galen,* edited by P. Adamson, R. Hansberger, and J. Wilberding, 89–134. Bulletin of the Institute of Classical Studies, Supp. 114. London: Institute of Classical Studies, University of London, 2014.

———. Introduction to *Medicine and Philosophy in Classical Antiquity: Doctors and Philosophers on Nature, Soul, Health, and Disease,* edited by Philip J. van der Eijk, 1–42. Cambridge: Cambridge University Press, 2005.

———. "Medicine and Health in the Graeco-Roman World." In *The Oxford Handbook of the History of Medicine,* edited by Mark Jackson, 21–39. Oxford: Oxford University Press, 2011.

———. *Medicine and Philosophy in Classical Antiquity: Doctors and Philosophers and Nature, Soul, Health, and Disease.* Cambridge: Cambridge University Press, 2005.

———. "The 'Theology' of the Hippocratic Treatise on the Sacred Disease." In *Medicine and Philosophy in Classical Antiquity: Doctors and Philosophers on Nature, Soul, Health, and Disease,* edited by Philip J. van der Eijk, 45–73. Cambridge: Cambridge University Press, 2005.

———. "Therapeutics." In *The Cambridge Companion to Galen,* edited by R. J. Hankinson, 283–303. Cambridge: Cambridge University Press, 2008.

Eijk, P. J. van der, H. F. J. Horstmanshoff, and P. H. Schrijvers, eds. *Ancient Medicine in Its Socio-Cultural Context.* 2 vols. Clio Medica 27. Amsterdam: Rodopi, 1995.

Elm, Susanna. "Roman Pain and the Rise of Christianity." In *Quo Vadis Medical Healing: Past Concepts and New Approaches,* edited by Susanna Elm and Stefan N. Willich, 41–54. New York: Springer, 2009.

———. *Sons of Hellenism, Fathers of the Church: Emperor Julian, Gregory of Nazianzus, and the Vision of Rome.* Transformation of the Classical Heritage 49. Berkeley: University of California Press, 2012.

Evans, M. "Plato and the Meaning of Pain." *Apeiron: A Journal for Ancient Philosophy and Science* 40 (2007): 71–93.

Eve, Eric. *The Healer from Nazareth: Jesus' Miracles in Historical Context.* London: SPCK, 2009.

Faraone, Christopher A. "Magic and Medicine in the Roman Imperial Period: Two Case Studies." In *Continuity and Innovation in the Magical Tradition,* edited by Gideon Bohak, Yuval Harari, and Shaul Shaked, 135–57. Leiden: Brill, 2011.

Farksafalvy, Denis. "Christological Content and Its Biblical Basis in the Letter of the Martyrs of Gaul." *SecCent* 9 (1992): 5–25.

Felix, B. "The Reception of Sophocles' Representation of Physical Pain." *AJP* 128, no. 4 (2007): 443–67.

Ferguson, Everett. *Demonology of the Early Christian World*. New York: Mellen, 1984.

———. *Early Christians Speak: Faith and Life in the First Three Centuries*. Vol. 2. Abilene, TX: Abilene Christian University Press, 2002.

Ferngren, Gary B. "Christianity and Healing in the Second Century." In *Actes/ Proceedings of the XXXIInd International Congress on the History of Medicine*, edited by E. Fierens et al., 131–37. Brussels: Societas Belgica Historiae Medicinae, 1991.

———. "Early Christianity as a Religion of Healing." *BHM* 66 (1992): 1–15.

———. "Early Christian Views of the Demonic Etiology of Disease." In *From Athens to Jerusalem: Medicine in Hellenized Jewish Lore and Early Christian Literature*, edited by S. S. Kottek, M. Horstmanschoff, et al., 183–201. Rotterdam: Erasmus, 2000.

———. *Medicine and Health Care in Early Christianity*. Baltimore: Johns Hopkins University Press, 2009.

———. "The Organisation of the Care of the Sick in Early Christianity." In *Actes/ Proceedings of the XXX International Congress of the History of Medicine*, edited by H. Schadewaldt and K.-H. Leven, 192–98. Düsseldorf: Vicom KG, 1988.

———. "Virtue and Health/Medicine in Pre-Christian Antiquity." In *Virtue and Medicine: Explorations in the Character of Medicine*, edited by E. E. Shelp, 3–22. Dordrecht: Reidel, 1985.

Ferngren, G. B., and D. W. Amundsen. "Medicine and Christianity in the Roman Empire: Compatibilities and Tensions." *ANRW* II.37.3 (1996): 2957–80.

Figura, Michael. "The Suffering of God in Patristic Theology." *Comm* 30, no. 3 (2003): 366–85.

Finn, Richard. *Almsgiving in the Later Roman Empire: Christian Promotion and Practice, 313–450*. Oxford: Oxford University Press, 2006.

Fitzgerald, John T. "Galen's *De indolentia* in the Context of Greco-Roman Medicine, Moral Philosophy, and Physiognomy." In *Galen's* De indolentia: *Essays on a Newly Discovered Letter*, edited by Clare K. Rothschild and Trevor W. Thompson, 203–20. Tübingen: Mohr Siebeck, 2014.

Flemming, Rebecca. "Demiurge and Emperor in Galen's World of Knowledge."

In *Galen and the World of Knowledge*, edited by Christopher Gill, Tim Whitmarch, and John Wilkins, 59–84. Cambridge: Cambridge University Press, 2009.

———. *Medicine and the Making of Roman Women: Gender, Nature, and Authority from Celsus to Galen.* Oxford: Oxford University Press, 2001.

Flower, Richard. "Medicalizing Heresy: Doctors and Patients in Epiphanius of Salamis." *Journal of Late Antiquity* 11, no. 2 (2018): 251–73.

Foucault, M. *The Birth of the Clinic: An Archaeology of Medical Perception.* Translated by A. M. Sheridan Smith. New York: Pantheon, 1973.

———. *The Care of the Self.* Vol. 3 of *The History of Sexuality.* Translated by Robert Hurley. New York: Pantheon, 1986.

———. "Technologies of the Self." In *Technologies of the Self: A Seminar with Michel Foucault*, edited by L. H. Martin, H. Gutman, and P. H. Hutton, 16–49. Amherst: University of Massachusetts Press, 1988.

———. *The Use of Pleasure.* Vol. 2 of *The History of Sexuality.* Translated by Robert Hurley. New York: Pantheon, 1986.

Fox, Bethany McKinney. *Disability and the Way of Jesus: Holistic Healing in the Gospels and the Church.* Downers Grove, IL: IVP Academic, 2019.

Francis, James A. *Subversive Virtue: Asceticism and Authority in the Second-Century Pagan World.* University Park: Pennsylvania State University Press, 1995.

Frankfurter, David. "Amuletic Invocations of Christ for Health and Fortune." In *Religions of Late Antiquity in Practice*, edited by R. Valantasis, 340–43. Princeton: Princeton University Press, 2000.

———, ed. *Guide to the Study of Ancient Magic.* Leiden: Brill, 2019.

———. "Syncretism and the Holy Man in Late Antique Egypt." *JECS* 11, no. 3 (2003): 339–85.

Frede, Michael. "Galen's Theology." In *Galien et la Philosophie*, edited by Jonathan Barnes and Jacques Jouanna, 73–126. Entretiens sur L'antiquité Classique 39. Vandoeuvres and Geneva: Fondation Hardt, 2003.

Freisenbruch, Annelise. "Back to Fronto: Doctor and Patient in His Correspondence with an Emperor." In *Ancient Letters: Classical and Late Antique Epistolography*, edited by Ruth Morello and A. D. Morrison, 235–56. Oxford: Oxford University Press, 2007.

Frevel, C. "Krankheit/Heilung." In *Handbuch theologischer Grundbegriffe zum Alten und Neuen Testament*, edited by Angelika Berlejung and C. Frevel, 284–88. Darmstadt: Wissenschaftliche Buchgesellschaft, 2006.

Frings, H. J. "Medizin und Arzt bei den griechischen Kirchenvätern bis Chrysostomos." PhD diss., University of Bonn, 1959.

Fulford, Ben. "Gregory of Nazianzus and Biblical Interpretation." In *Re-Reading Gregory of Nazianzus: Essays on History, Theology, and Culture*, edited by Christopher A. Beeley, 31–48. Washington, DC: Catholic University of America Press, 2012.

Gaiser, Frederick J. *Healing in the Bible: Theological Insight for Christian Ministry*. Grand Rapids: Baker Academic, 2010.

Garland, Robert. *The Eye of the Beholder: Deformity and Disability in the Graeco-Roman World*. Ithaca, NY: Cornell University Press, 1995.

Garrett, Jan Edward. "Is the Sage Free from Pain?" *Volga Journal of Philosophy and Social Sciences*, no. 6 (1999). http://people.wku.edu/jan.garrett/painst.htm.

Garrett, Susan R. "Paul's Thorn and Cultural Models of Affliction." In *The Social World of the First Christians: Essays in Honor of Wayne A. Meeks*, edited by O. L. Yarbrough and L. M. White, 82–99. Minneapolis: Fortress, 1995.

Garro, Linda C. "Chronic Illness and the Construction of Narratives." In *Pain as Human Experience: An Anthropological Perspective*, edited by Mary-Jo D. Good, Paul E. Brodwin, Byron J. Good, and Arthur Klenman, 100–137. Berkeley and Los Angeles: University of California Press, 1992.

———. "Cognitive Medical Anthropology." In *Encyclopedia of Medical Anthropology: Health and Illness in the World's Cultures*, edited by C. R. Ember and M. Ember, 12–23. New York: Kluwer, 2004.

Gask, George, and John Todd. "The Origin of Hospitals." In *Science, Medicine, and History: Essays on the Evolution of Scientific Thought and Medical Practice Written in Honour of Charles Singe*, edited by E. Ashworth Underwood, 1:122–30. Oxford: Oxford University Press, 1953.

Gautherie, Aurélien. "Physical Pain in Celsus' *On Medicine*." In *"Greek" and "Roman" in Latin Medical Texts: Studies in Cultural Change and Exchange in Ancient Medicine*, edited by Brigitte Maire, 137–54. Leiden: Brill, 2014.

Gavrilyuk, Paul L. *The Suffering of the Impassible God: The Dialectics of Patristic Thought*. Oxford: Oxford University Press, 2004.

Gill, Christopher. "Galen's Περὶ Ἀλυπίας as Philosophical Therapy: How Coherent Is It?" In *A Tale of Resilience: Galen's De Indolentia in Context*, edited by C. Petit, 135–54. Leiden: Brill, 2018.

———. *Naturalistic Psychology in Galen and Stoicism*. Oxford: Oxford University Press, 2010.

———. "Philosophical Therapy as Preventive Psychological Medicine." In *Mental Disorders in the Classical World*, edited by William V. Harris, 339–60. Columbia Studies in the Classical Tradition 38. Leiden: Brill, 2013.

Good, Byron J. "A Body in Pain—the Making of a World of Chronic Pain." In *Pain as Human Experience: An Anthropological Perspective*, edited by

Mary-Jo D. Good, Paul E. Brodwin, Byron J. Good, and Arthur Klenman, 29–48. Berkeley and Los Angeles: University of California Press, 1992.

———. *Medicine, Rationality, and Experience: An Anthropological Perspective.* Lewis Henry Morgan Lectures, 1990. Cambridge: Cambridge University Press, 1994.

Goodine, Elizabeth A., and Matthew W. Mitchell. "The Persuasiveness of a Woman: The Mistranslation and Misinterpretation of Eusebius; *HE* 5.1–41." *JECS* 13 (2005): 1–19.

Gordon, Richard. "The Healing Event in Graeco-Roman Folk-Medicine." In *Ancient Medicine in Its Socio-Cultural Context*, edited by Philip J. van der Eijk, H. F. J. Horstmanshoff, and P. H. Schrijvers, 2:363–76. Amsterdam: Rodopi, 1995.

Gorrini, Maria Elena. "The Hippocratic Impact on Healing Cults: The Archaeological Evidence in Attica." In *Hippocrates in Context: Papers Read at the XIth International Hippocrates Colloquium*, edited by Philip van der Eijk, 135–56. Leiden: Brill, 2005.

Gourevitch, Danielle. "Popular Medicines and Practices in Galen." In *Popular Medicine in Graeco-Roman Antiquity: Explorations*, edited by William V. Harris, 251–71. Columbia Studies in the Classical Tradition. Leiden: Brill, 2016.

Grams, L. "Medical Theory in Plato's *Timaeus*." *Rhizai* 6 (2009): 161–92.

Grant, Mark. *Galen on Food and Diet*. London: Routledge, 2000.

Green, Joel B. "Healing and Healthcare." In *The World of the New Testament*, edited by Joel B. Green and Lee Martin McDonald, 330–41. Grand Rapids: Baker Academic, 2013.

Griffith, Susan B. "Medical Imagery in the 'New' Sermons of Augustine." *Studia Patristica* 63 (2006): 107–12.

Griffith, Sydney H. "Ephraem, the Deacon of Edessa, and the Church of the Empire." In *Diakonia: Studies in Honor of Robert T. Meyer*, edited by T. Halton, 22–52. Washington, DC: Catholic University Press, 1986.

Grig, Lucy. *Marking Martyrs in Late Antiquity*. London: Duckworth, 2004.

Grmek, M. D. *Diseases in the Ancient Greek World*. Baltimore: Johns Hopkins University Press, 1989.

Grohmann, Marianne. *Fruchtbarkeit und Geburt in den Psalmen*. FAT 53. Tübingen: Mohr Siebeck, 2007.

Gundert, Beate. "Soma and Psyche in Hippocratic Medicine." In *Psyche and Soma: Physicians and Metaphysicians on the Mind-Body Problem from Antiquity to Enlightenment*, edited by John P. Wright and Paul Potter, 13–35. Oxford: Clarendon, 2000.

Haas, Christopher. *Alexandria in Late Antiquity: Topography and Social Conflict.* Baltimore: Johns Hopkins University Press, 2006.

Hadot, P. *Philosophy as a Way of Life: Spiritual Exercises from Socrates to Foucault.* Edited and introduced by A. I. Davidson. Translated by M. Chase. Oxford: Blackwell, 1995.

Hager, Kathryn. "John Cassian: The Devil in the Details." *Studia Patristica* 64 (2013): 59–64.

Hallman, Joseph M. "The Seed of Fire: Divine Suffering in the Christology of Cyril of Alexandria and Nestorius of Constantinople." *JECS* 5, no. 3 (1997): 369–91.

Hankinson, R. J. "Philosophy of Nature." In *The Cambridge Companion to Galen,* edited by R. J. Hankinson, 210–41. Cambridge: Cambridge University Press, 2008.

Hanson, A. E. "Greek Medical Papyri from the Fayum Village of Tebtunis: Patient Involvement in a Local Health-Care System?" In *Hippocrates in Context,* edited by Philip J. van der Eijk, 387–402. Leiden: Brill, 2005.

Hanson, R. P. C. *The Search for the Christian Doctrine of God: The Arian Controversy, 318–381.* Edinburgh: T&T Clark, 1988.

Harakas, S. S. "The Eastern Orthodox Tradition." In *Caring and Curing: Health and Medicine in the Western Religious Tradition,* edited by R. L. Numbers and D. W. Amundsen, 146–72. Baltimore: Johns Hopkins University Press, 1986.

———. *Health and Medicine in the Eastern Orthodox Tradition.* New York: Crossroad, 1990.

Harnack, A. von. *Medicinisches aus der ältesten Kirchengeschichte.* TUGAL. Leipzig: J. C. Hinrichs, 1892.

Harris, William V. "Pain and Medicine in the Classical World." In *Pain and Pleasure in Classical Times,* edited by William V. Harris, 55–82. Leiden: Brill, 2018.

———, ed. *Popular Medicine in Graeco-Roman Antiquity: Explorations.* Columbia Studies in the Classical Tradition. Leiden: Brill, 2016.

———. "Popular Medicine in the Classical World." In *Popular Medicine in Graeco-Roman Antiquity: Explorations,* edited by William V. Harris, 1–64. Columbia Studies in the Classical Tradition. Leiden: Brill, 2016.

———. *Restraining Rage: The Ideology of Anger Control in Classical Antiquity.* Cambridge, MA: Harvard University Press, 2001.

Harrison, Nonna Verna, and David G. Hunter, eds. *Suffering and Evil in Early Christian Thought.* Grand Rapids: Baker Academic, 2016.

Harrison, R. K. "Disease." In *The Interpreter's Dictionary of the Bible: An Illustrated Encyclopedia,* 1:847–54. New York and Nashville: Abingdon, 1962.

———. "Healing, Health." In *The Interpreter's Dictionary of the Bible: An Illustrated Encyclopedia*, 2:541–48. New York and Nashville: Abingdon, 1962.

Harvey, Susan A. "On Holy Stench: When the Odor of Sanctity Sickens." *Studia Patristica* 35 (2001): 90–101.

———. "Physicians and Ascetics in John of Ephesus: An Expedient Alliance." In *Symposium on Byzantine Medicine*, edited by J. Scarborough, 87–93. Washington, DC: Dumbarton Oaks, 1984.

———. *Scenting Salvation: Ancient Christianity and the Olfactory Imagination*. Berkeley and Los Angeles: University of California Press, 2006.

Hasel, Gerhard F. "Health and Healing in the Old Testament." *AUSS* 21, no. 3 (1983): 191–202.

Head, P. M. "On the Christology of the Gospel of Peter." *VC* 46, no. 3 (1992): 209–24.

Heffernan, Thomas J. *The Passion of Perpetua and Felicity*. New York: Oxford University Press, 2012.

———. *Sacred Biography: Saints and Their Biographers in the Middle Ages*. New York: Oxford University Press, 1988.

Helm, J. "Sickness in Early Christian Healing Narratives: Medical, Religious and Social Aspects." In *From Athens to Jerusalem: Medicine in Hellenized Jewish Lore and Early Christian Literature*, edited by S. S. Kottek, M. Horstmanschoff, et al., 241–58. Rotterdam: Erasmus, 2000.

Hemer, Colin J. "Medicine in the New Testament World." In *Medicine and the Bible*, edited by Bernard Palmer, 43–83. Exeter, UK: Paternoster, 1986.

Henning, Meghan. "Metaphorical, Punitive, and Pedagogical Blindness in Hell." *Studia Patristica* 81 (2017): 139–51.

Herrin, Judith. "Ideals of Charity, Realities of Welfare: The Philanthropic Activity of the Byzantine Church." In *Church and People in Byzantium: Society for the Promotion of Byzantine Studies; Twelfth Spring Symposium of Byzantine Studies, Manchester, 1986*, edited by Rosemary Morris, 151–64. Manchester: University of Manchester Press, 1986.

Hess, Richard. S. "The COVID-19 Virus, Illness, and Biblical Interpretation in Its Ancient Context." *Canon and Culture: A Journal of Biblical Interpretation in Context* 14, no. 2 (2020): 51–83.

Heyne, Thomas. "Tertullian and Medicine." *Studia Patristica* 50 (2011): 137–74.

———. "Were Second-Century Christians 'Preoccupied' with Physical Healing and the Asclepian Cult?" *Studia Patristica* 44 (2010): 63–69.

Hill, Judith L. "Health, Sickness and Healing in the New Testament: A Brief Theology." *Africa Journal of Evangelical Theology* 26, no. 2 (2007): 151–95.

Hogan, L. P. *Healing in the Second Temple Period*. Freiburg, Switzerland: Universitätsverlag; Göttingen: Vandenhoeck & Ruprecht, 1992.

Holman, Susan R. *The Hungry Are Dying: Beggars and Bishops in Roman Cappadocia*. OSHT. Oxford: Oxford University Press, 2001.

———. "The Hungry Body: Famine, Poverty, and Identity in Basil's *Hom.* 8." *JECS* 7, no. 3 (1999): 337–63.

———. "Rich and Poor in Sophronius of Jerusalem's *Miracles of Saints Cyrus and John*." In *Wealth and Poverty in Early Church and Society*, edited by Susan R. Holman, 103–24. Grand Rapids: Baker Academic, 2008.

———. "The Social Leper in Gregory of Nyssa's and Gregory of Nazianzus's 'περὶ Φιλοπτωχίας.'" *HTR* 92, no. 3 (1999): 283–309.

Holmes, Brooke A. "Aelius Aristides' Illegible Body." In *Aelius Aristides between Greece, Rome, and the Gods*, edited by William V. Harris and B. Holmes, 81–113. Leiden: Brill, 2008.

———. "Disturbing Connections: Sympathetic Affections, Mental Disorder, and the Elusive Soul in Galen." In *Mental Disorders in the Classical World*, edited by William V. Harris, 147–76. Columbia Studies in the Classical Tradition 38. Leiden: Brill, 2013.

Honecker, M. "Christus Medicus." *KD* 31 (1985): 307–23.

Hope, V., and E. Marshall. *Death and Disease in the Ancient City*. London: Routledge, 2000.

Horden, Peregrine. "The Christian Hospital in Late Antiquity: Break or Bridge?" In *Gesundheit—Krankheit: Kulturtransfer medizinischen Wissens von der Spätantike bis in die Frühe Neuzeit*, edited by F. Steger and K. P. Jankrift, 77–99. Cologne: Böhlau, 2004.

———. "The Death of Ascetics: Sickness and Monasticism in the Early Byzantine Middle East." In *Monks, Hermits, and the Ascetic Tradition*, edited by W. J. Sheils, 41–52. SCH 22. Oxford: Blackwell, 1985.

———. "The Earliest Hospitals in Byzantium, Western Europe, and Islam." *Journal of Interdisciplinary History* 35, no. 3 (2005): 361–89.

———. *Hospitals and Healing from Antiquity to the Later Middle Ages*. Aldershot, UK: Ashgate, 2008.

———. "Pain in Hippocratic Medicine." In *Religion, Health, and Suffering*, edited by J. R. Hinnells and R. Porter, 295–315. London and New York: Kegan Paul International, 1999.

———. "Poverty, Charity, and the Invention of the Hospital." In *The Oxford Handbook of Late Antiquity*, edited by Scott F. Johnson, 715–43. Oxford: Oxford University Press, 2012.

———. "Saints and Doctors in the Early Byzantine Empire: The Case of Theodore of Sykeon." In *The Church and Healing*, edited by W. J. Sheils, 1–13. SCH 19. Oxford: Blackwell, 1982.

Horn, Cornelia B. "A Nexus of Disability in Ancient Greek Miracle Stories: A Comparison of Accounts of Blindness from the Asklepieion in Epidauros and the Shrine of Thecla in Seleucia." In *Disabilities in Roman Antiquity: Disparate Bodies A Capite ad Calcem*, edited by Christian Laes, C. F. Goodey, and M. Lynn Rose, 115–44. Leiden: Brill, 2013.

Horstmanshoff, H. F. J. "Did the God Learn Medicine? Asclepius and Temple Medicine in Aelius Aristides' *Sacred Tales*." In *Magic and Rationality in Ancient Near Eastern and Graeco-Roman Medicine*, edited by H. F. J. Horstmanshoff and M. Stol, 325–41. Leiden: Brill, 2004.

Hout, M. P. J. van den. *A Commentary of the Letters of M. Cornelius Fronto*. Leiden: Brill, 1999.

Howard, J. Keir. *Disease and Healing in the New Testament: An Analysis and Interpretation*. Lanham, MD: University Press of America, 2001.

Israelowich, Ido. "Medical Care in the Roman Army during the High Empire." In *Perspectives on Popular Medicine in Classical Antiquity*, edited by William V. Harris, 215–30. Leiden: Brill, 2016.

———. *Patients and Healers in the High Roman Empire*. Baltimore: Johns Hopkins University Press, 2015.

———. *Society, Medicine, and Religion in the* Sacred Tales *of Aelius Aristides*. Leiden: Brill, 2012.

Jackson, R. *Doctors and Diseases in the Roman Empire*. Norman: University of Oklahoma Press, 1988.

Jacobs, Andrew S. *Epiphanius of Cyprus: A Cultural Biography of Late Antiquity*. Oakland: University of California Press, 2016.

Janowski, Bernd. "'Heal Me, for I Have Sinned against You!' (Psalm 41:5 MT): On the Concept of Illness and Healing in the Old Testament." In *Religion and Illness*, edited by Annette Weissenrieder and Gregor Etzelmüller, 173–97. Eugene, OR: Cascade Books, 2016.

Johnston, P. S. "Life, Disease and Death." In *Dictionary of the Old Testament: Pentateuch*, edited by T. Desmond Alexander and David W. Baker, 532–36. Downers Grove, IL: InterVarsity Press, 2003.

Jones, Christopher P. "Apollonius of Tyana in Late Antiquity." In *Greek Literature in Late Antiquity: Dynamism, Didacticism, Classicism*, edited by Scott Fitzgerald Johnson, 49–64. London: Routledge, 2006.

Jouanna, Jacques. "Dietetics in Hippocratic Medicine: Definition, Main Problems, Discussion." In *Greek Medicine from Hippocrates to Galen*, edited by Philip van der Eijk, 137–53. Leiden: Brill, 2012.

———. "Galen's Concept of Nature." In *Greek Medicine from Hippocrates to Galen*, edited by Philip van der Eijk, 287–311. Leiden: Brill, 2012.

———. *Hippocrates*. Translated by M. B. DeBevoise. Baltimore: Johns Hopkins University Press, 1999.

———. "The Legacy of the Hippocratic Treatise *The Nature of Man*: The Theory of the Four Humours." In *Greek Medicine from Hippocrates to Galen*, edited by Philip van der Eijk, 335–59. Leiden: Brill, 2012.

———. "Water, Health and Disease in the Hippocratic Treatise *Airs, Waters, Places*." In *Greek Medicine from Hippocrates to Galen*, edited by Philip van der Eijk, 155–72. Leiden: Brill, 2012.

Kalospyros, Nicholas A. E. "The Threshold of Pain: The Literary Embodiment of Pain and Its Cognates in the Hippocratic Corpus." In *Medicine and Healing in the Ancient Mediterranean World*, edited by Demetrios Michaelides, 84–91. Oxford: Oxbow Books, 2014.

Karayannopoulos, I. "St. Basil's Social Activity: Principles and Praxis." In *Basil of Caesarea: Christian, Humanist, Ascetic*, edited by P. J. Fedwick, 375–91. Toronto: Pontifical Institute of Mediaeval Studies, 1981.

Kaufman, David H. "Galen on the Therapy of Distress and the Limits of Emotional Therapy." *Oxford Studies on Ancient Philosophy* 47 (2014): 275–96.

Kearsley, R. A. "Magic, Medicine and Cults." In *New Documents Illustrating Early Christianity*, vol. 6, *A Review of the Greek Inscriptions and Papyri Published in 1980–81*, edited by S. R. Llewelyn, 190–96. Grand Rapids: Eerdmans, 2007.

Kee, Howard Clark. "Medicine and Healing." In *The Anchor Bible Dictionary*, edited by David Noel Freedman, 4:659–64. New York: Doubleday, 1992.

———. *Medicine, Miracle, and Magic in New Testament Times*. SNTSMS 55. Cambridge: Cambridge University Press, 1986.

———. *Miracle in the Early Christian World*. New Haven: Yale University Press, 1983.

Keenan, Mary E. "St. Gregory of Nyssa and the Medical Profession." *BHM* 15, no. 2 (1944): 150–61.

Kelley, Nicole. "Philosophy as Training for Death: Reading the Ancient Christian Martyr Acts as Spiritual Exercises." *CH* 75, no. 4 (2006): 723–47.

Kelsey, Morton. *Healing and Christianity: A Classic Study*. Minneapolis: Augsburg, 1995.

Kim, Young Richard. *Epiphanius of Cyprus: Imagining an Orthodox World*. Ann Arbor: University of Michigan Press, 2015.

King, Daniel. *Experiencing Pain in Imperial Greek Culture*. OCM. Oxford: Oxford University Press, 2018.

King, Helen. "Chronic Pain and the Creation of Narrative." In *Construction of the Classical Body*, edited by J. Porter, 269–86. Ann Arbor: University of Michigan Press, 1999.

———. "The Early Anodynes: Pain in the Ancient World." In *The Management of Pain: The Historical Perspective*, edited by R. Mann, 51–62. Carnforth, UK: Parthenon, 1988.

———, ed. *Health in Antiquity*. London: Routledge, 2005.

———. *Hippocrates' Woman: Reading the Female Body in Ancient Greece*. London: Routledge, 1998.

———. "Introduction: What Is Health?" In *Health in Antiquity*, edited by Helen King, 1–11. London: Routledge, 2005.

———. "Medical Texts as a Source for Women's History." In *The Greek World*, edited by Anton Powell, 199–218. London: Routledge, 1995.

———. "Producing Woman: Hippocratic Gynaecology." In *Women in Ancient Societies: An Illusion of the Night*, edited by Léonie Archer, Susan Fischler, and Maria Wyke, 102–14. New York: Macmillan, 1994.

———. "Women's Health and Recovery in the Hippocratic Corpus." In *Health in Antiquity*, edited by Helen King, 150–61. London: Routledge, 2005.

King, Karen L. "Christianity and Torture." In *The Oxford Handbook of Religion and Violence*, edited by Mark Juergensmeyer, Margo Kitts, and Michael Jerryson, 293–305. New York: Oxford University Press, 2013.

———. "Rethinking the Diversity of Ancient Christianity: Responding to Suffering and Persecution." In *Beyond the Gnostic Gospels: Studies Building on the Work of Elaine Pagels*, edited by Eduard Incinschil et al., 60–78. STAC 82. Tübingen: Mohr Siebeck, 2013.

Kinnard, I. "*Imitatio Christi* in Christian Martyrdom and Asceticism: A Critical Dialogue." In *Asceticism and Its Critics: Historical Accounts and Comparative Perspectives*, edited by Oliver Freiberger, 131–50. New York: Oxford University Press, 2006.

Kleinman, Arthur. "Concepts and a Model for the Comparison of Medical Systems as Cultural Systems." In *Concepts of Health, Illness, and Disease: A Comparative Perspective*, edited by C. Currer and M. Stacey, 29–47. New York: Berg, 1986.

———. *The Illness Narratives: Suffering, Healing, and the Human Condition*. New York: Basic Books, 1988.

———. *Patients and Healers in the Context of Culture: An Exploration of the Borderland between Anthropology, Medicine, and Psychiatry*. Berkeley and Los Angeles: University of California Press, 1980.

———. "Some Uses and Misuses of the Social Sciences in Medicine." In *Meta-*

theory in Social Science: Pluralisms and Subjectivities, edited by D. W. Fiske and R. A. Shweder, 222–45. Chicago: University of Chicago Press, 1986.

Kolbet, Paul R. *Augustine and the Cure of Souls: Revising a Classical Ideal*. Notre Dame: University of Notre Dame Press, 2010.

Kosmenniemi, Erkki. "Apollonius of Tyana: A Typical ΘΕΙΟΣ ΑΝΗΡ?" *JBL* 117, no. 13 (1998): 455–67.

Kotsopoulos, Sotiris. "Aretaeus the Cappadocian on Mental Illness." *Comprehensive Psychiatry* 27, no. 2 (1986): 171–79.

Kottek, S. S. "Hygiene and Healing among the Jews in the Post Biblical Period: A Partial Reconstruction." *ANRW* II.37.3 (1996): 2843–65.

Kouloumentas, Stavros. "The Body and Polis: Alcmaeon on Health and Disease." *British Journal for the History of Philosophy* 22, no. 5 (2014): 867–87.

Kudlien, Fridolf. "Beichte und Heilung." *Medizin Historisches Journal* 13 (1978): 1–14.

———. "Galen's Religious Belief." In *Galen: Problems and Prospects*, edited by Vivian Nutton, 117–30. London: Wellcome Institute for the History of Medicine, 1981.

LaCapra, Dominick. *History and Its Limits: Human, Animal, Violence*. Ithaca, NY: Cornell University Press, 2009.

Lagrée, Jacqueline. "Wisdom, Health, Salvation: The Medical Model in the Works of Clement of Alexandria." In *From Athens to Jerusalem: Medicine in Hellenized Jewish Lore and Early Christian Literature*, edited by S. S. Kottek, M. Horstmanschoff, et al., 227–40. Rotterdam: Erasmus, 2000.

Laird, Raymond. *Mindset, Moral Choice, and Sin in the Anthropology of John Chrysostom*. Early Christian Studies 15. Strathfield, Australia: St. Paul's Publications, 2012.

Langslow, D. R. *Medical Latin in the Roman Empire*. Oxford: Oxford University Press, 2006.

Leyerle, Blake. "The Etiology of Sorrow and Its Therapeutic Benefits in the Preaching of John Chrysostom." *Journal of Late Antiquity* 8, no. 2 (2015): 368–85.

Lieber, Elinor. "Old Testament 'Leprosy,' Contagion and Sin." In *Contagion: Perspectives from Pre-Modern Societies*, edited by Lawrence I. Conrad and Dominik Wujastyk, 99–136. Aldershot, UK: Ashgate, 2000.

Littman, R. J., and M. L. Littman. "Galen and the Antonine Plague." *AJP* 94 (1973): 243–55.

Lloyd, G. E. R. *In the Grip of Disease: Studies in the Greek Imagination*. Oxford: Oxford University Press, 2003.

Loewe, Ron. "Illness Narratives." In *Encyclopedia of Medical Anthropology: Health and Illness in the World's Cultures*, edited by C. R. Ember and M. Ember, 42–49. New York: Kluwer, 2004.

Lössl, Josef. "Dolor, dolere." In *Augustinus-Lexikon* 2.3/4, edited by C. Mayer, 581–91. Basel: Schwabe, 1999.

———. "Julian of Aeclanum on Pain." *JECS* 10, no. 2 (2000): 203–43.

Luijendijk, A. *"Greetings in the Lord": Early Christians and the Oxyrhynchus Papyri.* HTS 60. Cambridge, MA: Harvard University Press, 2008.

Lyman, Rebecca J. "The Making of a Heretic: The Life of Origen in Epiphanius' *Panarion* 64." *Studia Patristica* 31 (1997): 445–51.

MacMullen, Ramsay. *Christianizing the Roman Empire (AD 100–400).* New Haven: Yale University Press, 1984.

———. *Paganism in the Roman Empire.* New Haven: Yale University Press, 1981.

———. "The Place of the Holy Man in the Later Roman Empire." *HTR* 112, no. 1 (2019): 1–32.

Maraval, Pierre. *Actes et Passions des martyrs chrétiens des premiers siècles.* Sagesses Chrétiennes. Paris: Cerf, 2010.

Marganne, M.-H. "The Role of Papyri in the History of Medicine." *Histoire des sciences médicales* 38 (2004): 157–64.

Markschies, Christoph. "Demons and Disease." *Studia Patristica* 81 (2017): 11–35.

Martin, Thomas F. "Paul the Patient: *Christus Medicus* and the '*Stimulus Carnis*' (2 Cor. 12:7): A Consideration of Augustine's Medicinal Christology." *AugStud* 32, no. 2 (2001): 219–56.

Mattern, Susan P. *Galen and the Rhetoric of Healing.* Baltimore: Johns Hopkins University Press, 2008.

———. *The Prince of Medicine: Galen in the Roman Empire.* Oxford: Oxford University Press, 2013.

Mattingly, C., and L. C. Carro, eds. *Narrative and the Cultural Construction of Illness and Healing.* Berkeley: University of California Press, 2000.

Mayer, Wendy. "John Chrysostom: Moral Philosopher and Physician of the Soul." In *John Chrysostom: Past, Present, Future*, edited by Doru Costache and Mario Baghos, 193–216. Sydney: AIOCS Press, 2017.

———. "Madness in the Works of John Chrysostom: A Snapshot from Late Antiquity." In *Concept of Madness from Homer to Byzantium: Manifestations and Aspects of Mental Illness and Disorder*, edited by Hélène Perdicoyianni-Paléologou, 349–73. Byzantinische Forschungen 32. Amsterdam: Adolf M. Hakkert, 2016.

———. "Medicine in Transition: Christian Adaptation in the Later Fourth-Century East." In *Shifting Genres in Late Antiquity*, edited by Geoffrey Greatrex and Hugh Elton, 11–26. Farnham, UK: Ashgate, 2015.

————. "Persistence in Late Antiquity of Medico-Philosophical Psychic Therapy." *Journal of Late Antiquity* 8, no. 2 (2015): 337–51.

————. "Shaping the Sick Soul: Reshaping the Identity of John Chrysostom." In *Christians Shaping Identity from the Roman Empire to Byzantium: Studies Inspired by Pauline Allen*, edited by Geoffrey Dunn and Wendy Mayer, 140–64. Supplements to Vigiliae Christianae 132. Leiden: Brill, 2015.

McDonald, Glenda Camille. "The 'Locus Affectus' in Ancient Medical Theories of Disease." In *Medicine and Space: Body, Surroundings, and Borders in Antiquity and the Middle Ages*, edited by Patricia A. Baker, Han Nijdam, and Karine van 't Land, 63–83. Leiden: Brill, 2012.

Meconi, David Vincent. *The One Christ: St. Augustine's Theology of Deification*. Washington, DC: Catholic University of America Press, 2013.

Meggitt, Justin. "The Historical Jesus and Healing: Jesus' Miracles in Psychosocial Context." In *Spiritual Healing: Scientific and Religious Perspectives*, edited by Fraser Watts, 17–43. Cambridge: Cambridge University Press, 2011.

Mena, Peter A. "Scenting Saintliness: The Ailing Body, Chicana Feminism, and Communal Identity in Ancient Christianity." *JFSR* 33, no. 2 (2017): 5–20.

Merideth, Anne Elizabeth. "Illness and Healing in the Early Christian East." PhD diss., Princeton University, 1999.

Metzger, Nadine. "Not a *Daimōn*, but a Severe Illness: Oribasius, Posidonius and Later Ancient Perspective on Superhuman Agents Causing Disease." In *Mental Illness in Ancient Medicine: From Celsus to Aegina*, edited by C. Thumiger and P. N. Singer, 79–106. Leiden: Brill, 2018.

Miller, Patricia Cox. *Dreams in Late Antiquity: Studies in the Imagination of a Culture*. Princeton: Princeton University Press, 1994.

Miller, Samantha L. *Chrysostom's Devil: Demons, the Will, and Virtue in Patristic Soteriology*. Downers Grove, IL: IVP Academic, 2020.

Miller, Timothy S. *The Birth of the Hospital in the Byzantine Empire*. Baltimore: Johns Hopkins University Press, 1985. Reprinted with new introduction, 1997.

Minnen, P. van. "Medical Care in Late Antiquity." *Clio Medica: Acta Academia Internationalis Historiae Medicinae* 27 (1995): 153–69.

Moberg, Sean. "The Use of Illness in the *Apophthegmata Patrum*." *JECS* 26, no. 4 (2018): 571–600.

Montserrat, Dominic. "'Carrying on the Work of the Earlier Firm': Doctors, Medicine and Christianity in the Thaumata of Sophronius of Jerusalem." In *Health and Antiquity*, edited by Helen King, 230–42. London: Routledge, 2005.

Moore, Peter. "Chrysostom's Concept of γνώμη: How 'Chosen Life's Orientation' Undergirds Chrysostom's Strategy in Preaching." *Studia Patristica* 67 (2013): 351–58.

Morley, N. "The Salubriousness of the Roman City." In *Health in Antiquity,* edited by Helen King, 192–204. London: Routledge, 2005.

Morris, David B. *The Culture of Pain.* Berkeley and Los Angeles: University of California Press, 1991.

Moscoso, Javier. *Pain: A Cultural History.* New York: Palgrave Macmillan, 2012.

Moss, Candida R. *Ancient Christian Martyrdom: Diverse Practices, Theologies, and Traditions.* New Haven: Yale University Press, 2012.

———. "Miraculous Events in Early Christian Stories about Martyrs." In *Credible, Incredible: The Miraculous in the Ancient Mediterranean,* edited by T. Nicklas and J. E. Spittler, 283–301. Tübingen: Mohr Siebeck, 2013.

Mowvley, H. "Health and Salvation." *Baptist Quarterly* 22 (1967): 100–113.

Newmyer, S. T. "Talmudic Medicine and Greco-Roman Science: Cross-Currents and Resistance." *ANRW* II.37.3 (1996): 2895–911.

Nicolae, Jan. "'Christus Praedicator/Medicator': Homiletical, Patristic and Modern Elements of *Theologia Medicinalis.*" *European Journal of Science and Theology* 8, no. 2 (2012): 15–27.

Nijhuis, Karin. "Greek Doctors and Roman Patient: A Medical Anthropological Approach." In *Ancient Medicine in Its Socio-Cultural Context,* edited by Ph. J. Van der Ejik, H. F. J. Horstmanshoff, and P. H. Schrijvers, 1:49–67. Clio Medica 27. Amsterdam and Atlanta: Edition Rodopi, 1995.

Noorda, S. "Illness and Sin, Forgiving and Healing: The Connection of Medical Treatment and Religious Beliefs in Ben Sira 38:1–15." In *Studies in Hellenistic Religions,* edited by M. J. Vermaseren, 215–24. Leiden: Brill, 1979.

Nussbaum, Martha C. *The Therapy of Desire: Theory and Practice in Hellenistic Ethics.* Princeton: Princeton University Press, 1994.

Nutton, Vivian. *Ancient Medicine.* 2nd ed. London: Routledge, 2013.

———. "From Galen to Alexander: Aspects of Medicine and Medical Practice in Late Antiquity." *DOP* 38 (1984): 1–14.

———. "Galenic Madness." In *Mental Disorders in the Classical World,* edited by William V. Harris, 119–27. Columbia Studies in the Classical Tradition 38. Leiden: Brill, 2013.

———. "God, Galen and the Depaganization of Ancient Medicine." In *Religion and Medicine in the Middle Ages,* edited by P. Biller and J. Ziegler, 17–32. York, UK: York Medieval Press, 2001.

———. "Healers in the Medical Marketplace: Towards a Social History of Graeco-Roman Medicine." In *Medicine in Society: Historical Essays,* edited by Andrew Wear, 15–58. Cambridge: Cambridge University Press, 1992.

———. "The Medical Meeting Place." In *Ancient Medicine in Its Socio-Cultural Context: Papers Read at the Congress Held at Leiden University, 13–15 April 1992*, edited by Philip van der Eijk, H. F. J. Horstmanshoff, and P. H. Schrijvers, 1:3–25. Amsterdam and Atlanta: Rodopi, 1995.

———. "Murders and Miracles: Lay Attitudes towards Medicine in Classical Antiquity." In *Patients and Practitioners*, edited by R. Porter, 24–53. Cambridge: Cambridge University Press, 1985.

———. "Roman Medicine: Tradition, Confrontation, Assimilation." *ANRW* II.37.1 (1993): 49–78.

Oates, J. F., R. S. Bagnall, S. J. Clackson, A. A. O'Brien, J. D. Sosin, T. G. Wilfong, and K. A. Worp, eds. *Checklist of Editions of Greek, Latin, Demotic, and Coptic Papyri, Ostraca, and Tablets.* Last updated June 1, 2011. http://scriptorium.lib.duke.edu/papyrus/texts/clist.html.

Oberhelman, Steven M. "Anatomical Votive Reliefs as Evidence for Specialization at Healing Sanctuaries in the Ancient Mediterranean World." *Athens Journal of Health* 1, no. 1 (2014): 47–62.

———. "Dreams in Greco-Roman Medicine." *ANRW* II.37.1 (1993): 121–56.

———, ed. *Dreams, Healing, and Medicine in Greece: From Antiquity to the Present.* Surrey, UK, and Burlington, VT: Ashgate, 2013.

———. "Galen, on Diagnosis from Dreams." *Journal of the History of Medicine and Allied Sciences* 38, no. 1 (1983): 36–47.

O'Keefe, John J. "Impassible Suffering? Divine Passion and Fifth-Century Christology." *TS* 58 (1997): 39–60.

Olyan, S. M. *Disability in the Hebrew Bible: Interpreting Mental and Physical Differences.* Cambridge: Cambridge University Press, 2008.

Omiya, T. "Leprosy." In *Dictionary of Jesus and the Gospels*, edited by Joel B. Green, Jeannine K. Brown, and Nicholas Perrin, 517–18. Downers Grove, IL: IVP Academic, 2013.

Pagels, Elaine. "Gnostic and Christian Views of Christ's Passion." In *The Rediscovery of Gnosticism*, edited by B. Layton, 1:262–83. Leiden: Brill, 1980.

Panagiotidou, Olympia. "Asclepius: A Divine Doctor, a Popular Healer." In *Popular Medicine in Graeco-Roman Antiquity: Explorations*, edited by William V. Harris, 86–104. Columbia Studies in the Classical Tradition. Leiden: Brill, 2016.

Papadogiannakis, Yannis. *Christianity and Hellenism in the Fifth-Century Greek East: Theodoret's Apologetics against the Greeks in Context.* Cambridge and London: Center for Hellenic Studies, 2012.

Perkins, Judith. *Roman Imperial Identities in the Early Christian Era.* London: Routledge, 2009.

————. "The 'Self' as Sufferer." *HTR* 85, no. 3 (1992): 245–72.

————. *The Suffering Self: Pain and Narrative Representation in the Early Christian Era.* London: Routledge, 1995.

Perrin, Michel Jean-Louis. "Médecine, Maladie et Théologie chez Lactance (250–325)." In *Les Pères de L'Église Face à la Science Médicale de Leur Temps*, edited by Véronique Boudon-Millot and Bernard Pouderon, 335–50. Paris: Beauchesne, 2005.

Petridou, Georgia. "Asclepius the Divine Healer, Asclepius the Divine Physician: Epiphanies as Diagnostic and Therapeutic Tools." In *Medicine and Healing in the Ancient Mediterranean*, edited by Demetrios Michaelides, 291–301. Oxford: Oxbow Books, 2014.

————. "Becoming a Doctor, Becoming a God: Religion and Medicine in Aelius Aristides' *Hieroi Logoi.*" In *Religion and Illness*, edited by Annette Weissenrieder and Gregor Etzelmüller, 306–35. Eugene, OR: Cascade Books, 2016.

————. "The Curious Case of Aelius Aristides: The Author as Sufferer and Illness as 'Individualizing Motif.'" In *Autoren in religiösen literarischen Texten der späthellenistischen und der frühkaiserzeitlichen Welt*, edited by Eve-Marie Becker and Jörg Rüpke, 199–219. Tübingen: Mohr Siebeck, 2018.

————. "The Healing Shrines." In *A Companion to Greek Science, Technology, and Medicine in Ancient Greece and Rome*, edited by Georgia L. Irby, 1:434–49. Malden, MA, and Oxford: Wiley Blackwell, 2016.

Petridou, Georgia, and Chiara Thumger, eds. *Homo Patiens: Approaches to Patients in the Ancient World.* Leiden: Brill, 2015.

Pilch, John J. *Healing in the New Testament: Insights from Medical and Mediterranean Anthropology.* Minneapolis: Fortress, 2000.

————. "Understanding Healing in the Social World of Early Christianity." *BTB* 22 (1992): 26–33.

Podlecki, Anthony J. "The Power of the Word in Sophocles' *Philoctetes.*" *GRBS* 7 (1966): 233–50.

Porterfield, Amanda. *Healing in the History of Christianity.* Oxford: Oxford University Press, 2005.

Price, R. M. "Illness Theodicies in the New Testament." *Journal of Religion and Health* 25 (1986): 309–15.

Prince, Brian D. "The Metaphysics of Bodily Health and Disease in Plato's *Timaeus.*" *British Journal for the History of Philosophy* 22, no. 5 (2014): 908–28.

Quinn, Dennis P. "In the Names of God and His Christ: Evil Daemons, Exorcism, and Conversion in Firmicus Maternus." *Studia Patristica* 69 (2013): 3–14.

Reff, Daniel T. *Plagues, Priests, and Demons: Sacred Narratives and the Rise of*

Christianity in the Old World and the New. Cambridge: Cambridge University Press, 2005.

Remus, Harold. *Jesus as Healer*. Cambridge: Cambridge University Press, 1997.

———. *Pagan-Christian Conflict over Miracle in the Second Century CE*. Cambridge: Philadelphia Patristic Foundation, 1983.

Renberg, Gil H. "Public and Private Places of Worship in the Cult of Asclepius at Rome." *Memoirs of the American Academy in Rome* 51/52 (2006/2007): 87–172.

Rey, Roselyne. *The History of Pain*. Translated by Louise E. Wallace, J. A. Cadden, and S. W. Cadden. Cambridge, MA: Harvard University Press, 1995.

Rhee, Helen. *Early Christian Literature: Christ and Culture in the Second and Third Centuries*. London: Routledge, 2005.

———. *Loving the Poor, Saving the Rich: Wealth, Poverty, and Early Christian Formation*. Grand Rapids: Baker Academic, 2012.

———. "Portrayal of Patients in Early Christian Writings." *Studia Patristica* 81 (2017): 127–38.

Riddle, J. M. "High Medicine and Low Medicine in the Roman Empire." *ANRW* II.37.1 (1993): 102–20.

Riethmüller, Jürgen W. *Asklepios: Heiligtümer und Kulte*. 2 vols. Heidelberg: Verlag Archäologie und Geschichte, 2005.

Risse, Guenter B. "Asclepius at Epidaurus: The Divine Power of Healing Dreams." Pages 1–22 in lecture delivered May 13, 2008. Updated March 10, 2015. Accessed on May 19, 2021. https://ucsf.academia.edu/Guenter Risse/papers.

———. *Mending Bodies, Saving Souls: A History of Hospitals*. Oxford: Oxford University Press, 1999.

Rist, John M. *Augustine: Ancient Thought Baptized*. Cambridge: Cambridge University Press, 1994.

———. *Stoic Philosophy*. Cambridge: Cambridge University Press, 1969.

Roby, Courtney. "Galen on the Patient's Role in Pain Diagnosis: Sensation, Consensus, and Metaphor." In *Homo Patiens: Approaches to Patients in the Ancient World*, edited by Georgia Petridou and Chiara Thumger, 304–22. Leiden: Brill, 2015.

Rocca, J. *Galen on the Brain*. Leiden: Brill, 2003.

Roots, Peter A. "The *De Opificio Dei*: The Workmanship of God and Lactantius." *ClQ* 37, no. 2 (1987): 466–86.

Rosen, Ralph M. "Spaces of Sickness in Greco-Roman Medicine." In *Medicine and Space: Body, Surroundings, and Borders in Antiquity and the Middle Ages*, edited by Patricia A. Baker, Han Nijdam, and Karine van 't Land, 227–43. Leiden: Brill, 2012.

———. "Towards a Hippocratic Anthropology: *On Ancient Medicine* and the

Origins of Humans." In *Ancient Concepts of the Hippocratic: Papers Presented at the XIIIth International Hippocrates Colloquium, Austin, Texas, 11–13 August, 2008*, edited by Lesley Dean-Jones and Ralph M. Rosen, 242–57. Leiden: Brill, 2016.

Rosner, F. "Jewish Medicine in the Talmudic Period." *ANRW* II.37.3 (1996): 2866–94.

Rousselle, A. "From Sanctuary to Miracle-Worker: Healing in Fourth-Century Gaul." In *Ritual, Religion, and the Sacred: Selections from the Annales, Économies, Sociétés, Civilisations*, edited by R. Forster and O. Ranum, translated by E. Forster and P. M. Ranum, 7:95–127. Baltimore: Johns Hopkins University Press, 1982.

Rubenson, Samuel. "Argument and Authority in Early Monastic Correspondence." In *Foundations of Power and Conflicts of Authority in Late-Antique Monasticism*, edited by A. Camplani and G. Filoramo, 75–87. OLA 157. Leuven: Peeters, 2004.

Rylaarsdam, David. *John Chrysostom on Divine Pedagogy: The Coherence of His Theology and Preaching*. Oxford: Oxford University Press, 2014.

Sage, Michael M. *Cyprian*. Patristic Monograph Series. Cambridge, MA: Philadelphia Patristic Foundation, 1975.

Samellas, Antigone. "Public Aspects of Pain in Late Antiquity: The Testimony of Chrysostom and the Cappadocians in Their Graeco-Roman Context." *ZAC* 19, no. 2 (2015): 260–96.

Sanzo, Joseph E. "Early Christianity." In *Guide to the Study of Ancient Magic*, edited by David Frankfurter, 198–239. Leiden: Brill, 2019.

Scarborough, J. *Roman Medicine*. Ithaca, NY: Cornell University Press, 1969.

Scarry, Elaine. *The Body in Pain*. Oxford: Oxford University Press, 1987.

Scheidel, Walter. "Disease and Death in the Ancient City of Rome." Version 2.0. Princeton/Stanford Working Papers in Classics, April 2009.

———. "A Model of Demographic and Economic Change in Roman Egypt after the Antonine Plague." *JRA* 15 (2002): 97–114.

———. "Physical Well-Being." In *The Cambridge Companion to the Roman Economy*, edited by Walter Scheidel, 321–33. Cambridge: Cambridge University Press, 2012.

———. "Roman Wellbeing and the Economic Consequences of the 'Antonine Plague.'" Version 3.0. Princeton/Stanford Working Papers in Classics, January 2010.

Schleifer, Ronald. *Pain and Suffering*. Routledge Series Integrating Science and Culture. London: Routledge, 2014.

Scobie, Alex. "Slums, Sanitation, and Mortality in the Roman World." *Klio* 68 (1986): 399–433.

Seeliger, Hans Reinhard, and Wolfgang Wischmeyer, eds. *Märtyrerliteratur.* TU-GAL 172. Berlin: de Gruyter, 2015.

Segal, Charles. *Tragedy and Civilization: An Interpretation of Sophocles.* Cambridge, MA: Harvard University Press, 1981.

Serfass, Adam. "Wine for Widows: Papyrological Evidence for Christian Charity in Late Antique Egypt." In *Wealth and Poverty in Early Church and Society,* edited by Susan R. Holman, 88–102. Grand Rapids: Baker Academic, 2008.

Seybold, Klaus, and Ulrich B. Mueller. *Sickness and Healing.* Translated by Douglas W. Scott. Nashville: Abingdon, 1981.

Shaw, Brent. "Body/Power/Identity." *JECS* 4 (1996): 269–312.

Shaw, Teresa. "*Askesis* and Appearance of Holiness." *JECS* 6 (1998): 485–99.

Siegel, Rudoph E. *Galen on Sense Perception: His Doctrines, Observations, and Experiments on Vision, Hearing, Smell, Taste, Touch, and Pain, and Their Historical Sources.* Basel: S. Karger, 1970.

Sigerist, Henry E. "The Special Position of the Sick." In *Culture, Disease, and Healing,* edited by D. Landy, 388–94. New York: Macmillan, 1977.

Singer, Peter N. "The Fight for Health: Tradition, Competition, Subdivision and Philosophy in Galen's Hygienic Writings." *British Journal for the History of Philosophy* 22 (2014): 974–95.

———. "Galen's Pathological Soul: Diagnosis and Therapy in Ethical and Medical Texts and Contexts." In *Mental Illness in Ancient Medicine: From Celsus to Paul of Aegina,* edited by Chiara Thumigur and Peter Singer, 381–420. Studies on Ancient Medicine 50. Leiden: Brill, 2018.

Slaveva-Griffin, Svetla, and Ilaria L. E. Ramelli, eds. *Lovers of the Soul, Lovers of the Body: Philosophical and Religious Perspectives in Late Antiquity.* Cambridge, MA: Harvard University Press, 2020.

Smith, J. Warren. "Suffering Impassibly: Christ's Passion in Cyril of Alexandria's Soteriology." In *Suffering and Evil in Early Christian Thought,* edited by Nonna Verna Harrison and David G. Hunter, 191–202. Grand Rapids: Baker Academic, 2016.

Sorabji, R. *Emotion and Peace of Mind: From Stoic Agitation to Christian Temptation.* Gifford Lectures. Oxford: Oxford University Press, 2000.

Staden, H. von. "Body, Soul, and Nerves: Epicurus, Herophilus, Erasistratus, the Stoics, and Galen." In *Psyche and Soma: Physicians and Metaphysicians on the Mind-Body Problem from Antiquity to Enlightenment,* edited by John P. Wright and Paul Potter, 79–116. Oxford: Clarendon, 2000.

————. "Incurability and Hopelessness: The Hippocratic Corpus." In *La Maladie et les Maladies dans la Collection hippocratique, Actes du VI colloque internationale hippocratique*, edited by P. Porter, G. Maloney, and J. Desautels, 75–112. Quebec: Editions du Sphinx, 1990.

Stambaugh, John E. *Serapis under the Early Ptolemies*. Leiden: Brill, 1972.

Stannard, Jerry. "Marcellus of Bordeaux and the Beginnings of Medieval Materia Medica." *Pharmacy in History* 15, no. 2 (1973): 47–53.

Stark, Rodney. *The Rise of Christianity: A Sociologist Reconsiders History*. Princeton: Princeton University Press, 1996.

Steger, Florian. "Aristides, Patient of Asclepius in Pergamum." In *In Praise of Asclepius: Aelius Aristide, Selected Prose Hymns*, edited by Donald A. Russell et al., 130–42. Tübingen: Mohr Siebeck, 2016.

————. *Asclepius: Medicine and Cult*. Translated by Margot M. Saar. Stuttgart: Franz Steiner Verlag, 2018.

Straw, Carole. "Martyrdom and Christian Identity: Gregory the Great, Augustine, and Tradition." In *The Limits of Ancient Christianity: Essays on Late Antique Culture and Thought in Honor of Robert Markus*, edited by W. Kliggsheim and M. Vessey, 250–66. Ann Arbor: University of Michigan Press, 1999.

————. "'A Very Special Death': Christian Martyrdom in Its Classical Context." In *Sacrificing the Self: Perspectives on Martyrdom and Religion*, edited by M. Cormack, 39–57. Oxford: Oxford University Press, 2002.

Sussman, M. "Sickness and Disease." In *The Anchor Bible Dictionary*, edited by David Noel Freedman, 6:6–15. New York: Doubleday, 1992.

Sweeney, Christopher. "The Function of the Holy Men in Context: Inscriptions, *Eulogiai*, and *Vitae*." Paper presented at the North American Patristic Society Annual Meeting, Chicago, May 2014.

Temkin, Owsei. *Hippocrates in a World of Pagans and Christians*. Baltimore: Johns Hopkins University Press, 1991.

Tenney, Merrill C. "Diseases." In *The New International Dictionary of the Bible*, 272–78. 1987.

Thaden, Robert H. von, Jr. "Glorify God in Your Body: The Redemptive Role of the Body in Early Christian Ascetic Literature." *Cistercian Studies Quarterly* 38, no. 2 (2003): 191–209.

Tieleman, Teun. "Galen's Psychology." In *Galien et la Philosophie: Huit Exposés Suivis de Discussions*, edited by Jonathan Barnes, 131–61. Entretiens sur L'Antiquité Classique 49. Vandoeuvres and Geneva: Fondation Hardt, 2003.

————. "Religion and Therapy in Galen." In *Religion and Illness*, edited by Gregor Etzemüller and Annette Weissenrieder, 15–31. Eugene, OR: Wipf & Stock, 2016.

———. "Wisdom and Emotion: Galen's Philosophical Position in *Avoiding Distress.*" In *A Tale of Resilience: Galen's* De Indolentia *in Context*, edited by C. Petit, 199–216. Leiden: Brill, 2018.

Tilley, Maureen. "The Ascetic Body and the (Un)Making of the World of the Martyr." *JAAR* 59 (1991): 467–79.

Trembovler, L. "A Sound Mind in a Diseased Body: A 'Medical' Aspect of the Soul-Body Relationship in Later Greek and Early Christian Philosophy." In *From Athens to Jerusalem: Medicine in Hellenized Jewish Lore and Early Christian Literature*, edited by S. S. Kottek, M. Horstmanschoff, et al., 171–79. Rotterdam: Erasmus, 2000.

Twelftree, Graham H. "Healing, Illness." In *Dictionary of Paul and His Letters*, edited by Gerald F. Hawthorne and Ralph P. Martin, 378–81. Downers Grove, IL: IVP Academic, 1993.

———. *In the Name of Jesus: Exorcism among Early Christians.* Grand Rapids: Baker Academic, 2007.

———. *Jesus the Miracle Worker: A Historical and Theological Study.* Downers Grove, IL: InterVarsity Press, 1999.

Uitti, Roger W. "Health and Wholeness in the Old Testament." *Consensus* 17, no. 2 (1991): 47–62.

Underwood, Norman. "Medicine, Money, and Christian Rhetoric: The Socio-Economic Dimensions of Healthcare in Late Antiquity." *Studies in Late Antiquity* 2, no. 3 (2018): 342–84.

Upson-Saia, Kristi. "Wounded by Divine Love." In *Melania: Early Christianity through the Life of One Family*, edited by Catherine M. Chin and Caroline T. Schroeder, 86–106. Oakland: University of California Press, 2017.

Vakaloudi, Anastasia D. "Illnesses, Curative Methods and Supernatural Forces in the Early Byzantine Empire (4th–7th C. A. D.)." *Byzantion* 73, no. 1 (2003): 172–200.

Van Babel, Tarsicius. "The 'Christus Totus' Idea: A Forgotten Aspect of Augustine's Spirituality." In *Studies in Patristic Christology: Proceedings of the Third Maynooth Patristic Conference*, edited by Thomas Finan and Vincent Twomey, 84–94. Portland, OR: Four Courts, 1998.

Van Dam, R. *Saints and Their Miracles in Late Antique Gaul.* Princeton: Princeton University Press, 1993.

Van Hoof, Lieve. *Plutarch's Practical Ethics: The Social Dynamics of Philosophy.* Oxford: Oxford University Press, 2010.

Verhey, Allen. "Health and Healing in Memory of Jesus." *ExAud* 21 (2005): 24–48.

Vikan, Gary. "Icons and Icon Piety in Early Byzantium." In *Sacred Images and Sacred Power in Byzantium*, 1–16. New York: Aldershot, 2003.

Vlahogiannis, N. "'Curing' Disability." In *Health in Antiquity*, edited by H. King, 180–91. London: Routledge, 2005.

Vogt, Sabine. "Drug and Pharmacology." In *The Cambridge Companion to Galen*, edited by R. J. Hankinson, 304–22. Cambridge: Cambridge University Press, 2008.

Vööbus, Arthur. *History of Asceticism in the Syrian Orient*. Vol. 2, *A Contribution to the History of Culture in the Near East*. Leuven: Secrétariat du CorpusSCO, 1960.

Wahlen, C. "Healing." In *Dictionary of Jesus and the Gospels*, edited by Joel B. Green, Jeannine K. Brown, and Nicholas Perrin, 362–70. Downers Grove, IL: IVP Academic, 2013.

Wailoo, Keith. *Pain: A Political History*. Baltimore: Johns Hopkins University Press, 2014.

Wainwright, Elaine M. *Women Healing/Healing Women: The Genderization of Healing in Early Christianity*. London: Equinox, 2006.

Wall, Patrick. *Pain: The Science of Suffering*. New York: Columbia University Press, 2000.

Wallace-Hadrill, D. S. *The Greek Patristic View of Nature*. Manchester: Manchester University Press; New York: Barnes and Noble, 1968.

Ware, Kallistos. "The Impassible Suffers." In *Suffering and Evil in Early Christian Thought*, edited by Nonna Verna Harrison and David G. Hunter, 213–33. Grand Rapids: Baker Academic, 2016.

———. "The Meaning of 'Pathos' in Abba Isaias and Theodoret of Cyrus." *Studia Patristica* 20 (1989): 315–22.

Weissenrieder, Annette. *Images of Illness in the Gospel of Luke: Insights from Ancient Medical Texts*. Tübingen: Mohr Siebeck, 2003.

———. "Interior Views of a Patient: Illness and Rhetoric in 'Autobiographical' Texts (L. Annaeus Seneca, Marcus Cornelius Fronto and the Apostle Paul)." In *Religion and Illness*, edited by Annette Weissenrieder and Gregor Etzelmüller, 336–57. Eugene, OR: Cascade Books, 2016.

Wells, Louise. *The Greek Language of Healing from Homer to New Testament Times*. 1998. Reprint, Berlin: de Gruyter, 2014.

Wessel, Susan. "The Reception of Greek Science in Gregory of Nyssa's '*De hominis opificio*.'" *VC* 63, no. 1 (2009): 24–46.

Wet, Chris L. de. "Medical Discourse, Identity Formation, and Otherness in Late Ancient Christianity." Paper presented at the International Patristic Conference at Oxford. August 2019.

———. "The Preacher's Diet: Gluttony, Regimen, and Psycho-Somatic Health in the Thought of John Chrysostom." In *Revisioning John Chrysostom: New Approaches, New Perspectives*, edited by Chris L. de Wet and Wendy Mayer, 410–63. Leiden: Brill, 2019.

White, L. Michael. "The Pathology and Cure of Grief (λύπη): Galen's *De indolentia* in Context." In *Galen's* De indolentia: *Essays on a Newly Discovered Letter*, edited by Clare K. Rothschild and Trevor W. Thompson, 221–50. Tübingen: Mohr Siebeck, 2014.

Whitehorne, J. E. G. "Was Marcus Aurelius a Hypochondriac?" *Latomus* 36 (1977): 413–21.

Wickkiser, B. *Asklepios, Medicine, and the Politics of Healing in Fifth-Century Greece*. Baltimore: Johns Hopkins University Press, 2008.

Wilkinson, John R. *The Bible and Healing: A Medical and Theological Commentary*. Grand Rapids: Eerdmans, 1998.

Williams, Craig. "Perpetua's Gender: A Latinist Reads the *Passio Perpetuae et Felicitatis*." In *Perpetua's Passions: Multidisciplinary Approaches to the* Passio Perpetuae et Felicitas, edited by Jan N. Bremmer and Marco Formisano, 54–77. Oxford: Oxford University Press, 2012.

Wilson, Nicole. "The Semantics of Pain in Greco-Roman Antiquity." *Journal of History of the Neurosciences* 22 (2013): 129–43.

Wiseman, Donald J. "Medicine in the Old Testament World." In *Medicine and the Bible*, edited by Bernard Palmer, 13–42. Exeter, UK: Paternoster, 1986.

Wolterstorff, Nicholas. "Augustine's Rejection of Eudaimonism." In *Augustine's* City of God: *A Critical Guide*, edited by James Wetzel, 149–66. Cambridge: Cambridge University Press, 2012.

———. "The Place of Pain in the Space of Good and Evil." In *Pain and Its Transformations: The Interface of Biology and Culture*, edited by Sarah Coakley and Kay K. Shelemay, 406–19. Cambridge, MA: Harvard University Press, 2007.

Woolf, Clifford. "Deconstructing Pain: A Deterministic Dissection of the Molecular Basis of Pain." In *Pain and Its Transformations: The Interface of Biology and Culture*, edited by Sarah Coakley and Kay K. Shelemay, 27–35. Cambridge, MA: Harvard University Press, 2007.

Worman, Nancy. "Infection in the Sentence: The Discourse of Disease in Sophocles' *Philoctetes*." *Arethusa* 33 (2000): 1–36.

Wright, Jessica. "Between Despondency and the Demon: Diagnosing and Treating Spiritual Disorder in John Chrysostom's Letter to Stageirios." *Journal of Late Antiquity* 8, no. 2 (2015): 352–67.

———. "Brain and Soul in Late Antiquity." PhD diss., Princeton University, 2016.

———. "John Chrysostom and the Rhetoric of Cerebral Vulnerability." *Studia Patristica* 81 (2017): 109–24.

———. "Preaching Phrenitis: Augustine's Medicalization of Religious Difference." *JECS* 28, no. 4 (2020): 525–53.

Young, Allan. "The Anthropologies of Illness and Sickness." *Annual Review of Anthropology* 11 (1982): 257–85.

Young, Robin Darling. *In Procession before the World: Martyrdom as Public Liturgy in Early Christianity*. Milwaukee: Marquette University Press, 2001.

Zacher, Jonathan L. "Medical Art in Spiritual Direction: Basil, Barsanuphios, and John on Diagnosis and Meaning in Illness." *JECS* 28, no. 4 (2020): 591–623.

Zias, Joseph. "Death and Disease in Ancient Israel." *BA* 54 (1991): 147–59.

Index of Authors

Index of Subjects

INDEX OF SCRIPTURE
AND OTHER ANCIENT SOURCES